Physical Chemistry of Functionalised Biomedical Nanoparticles

University of Bristol, UK

17–19 September 2014

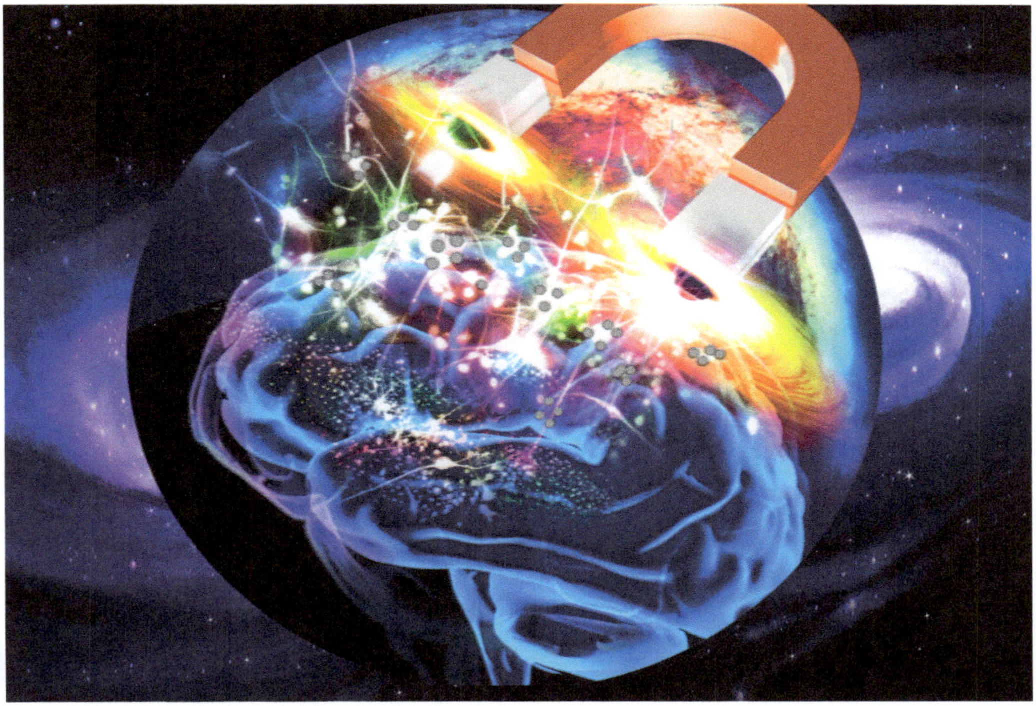

FARADAY DISCUSSIONS

Volume 175, 2014

ROYAL SOCIETY OF CHEMISTRY

The Faraday Division of the Royal Society of Chemistry, previously the Faraday Society, founded in 1903 to promote the study of sciences lying between Chemistry, Physics and Biology.

Faraday Discussions documents a long-established series of Faraday Discussion meetings which provide a unique international forum for the exchange of views and newly acquired results in developing areas of physical chemistry, biophysical chemistry and chemical physics.

Editorial staff

Faraday Discussions (Print ISSN 1359-6640, Electronic ISSN 1364-5498) is published 8 times a year by the Royal Society of Chemistry, Thomas Graham House, Science Park, Milton Road, Cambridge, CB4 0WF, UK.

Volume 175 ISBN: 978-1-78262-101-0

2014 annual subscription price: print+electronic £827, US $1542; electronic only £785, US $1465. Customers in Canada will be subject to a surcharge to cover GST. Customers in the EU subscribing to the electronic version only will be charged VAT. All orders, with cheques made payable to the Royal Society of Chemistry, should be sent to RSC Order Department, Royal Society of Chemistry, Thomas Graham House, Science Park, Milton Road, Cambridge, CB4 0WF, UK.
Tel +44 (0) 1223 432398; E-mail www.orders@rsc.org

If you take an institutional subscription to any RSC journal you are entitled to free, site-wide web access to that journal. You can arrange access via Internet Protocol (IP) address at www.rsc.org/ip. Customers should make payments by cheque in sterling payable on a UK clearing bank or in US dollars payable on a US clearing bank.

Printed in the UK

Scientific committee volume 175

Faraday standing committee on conferences

Physical Chemistry of Functionalised Biomedical Nanoparticles

Faraday Discussions

www.rsc.org/faraday_d

A General Discussion on the Physical Chemistry of Functionalised Biomedical Nanoparticles was held in Bristol, UK on the 17th, 18th and 19th of September 2014.

CONTENTS

ISSN 1359-6640; ISBN 978-1-78262-101-0

Cover

See S. C. Edman Tsang *et al.*, *Faraday Discuss.*, 2014, **175**, 13.

Multifunctional silica-coated iron oxide nanoparticles: a facile four-in-one system for *in situ* study of neural stem cell harvesting

Image courtesy of Yung-Kang Peng, Cathy N. P. Lui, Tsen-Hsuan Lin, Chen Chang, Pi-Tai Chou, Ken K. L. Yung and S. C. Edman Tsang, University of Oxford, UK.

INTRODUCTORY LECTURE

PAPERS AND DISCUSSIONS

SUPPORTING ORGANISATION

SCI

where science meets business

Faraday Discussions

PAPER

Introductory Lecture: The physical chemistry of functionalised biomedical nanoparticles

Peter J. Dobson*

Received 9th December 2014, Accepted 9th December 2014
DOI: 10.1039/c4fd90076f

It cannot have escaped the attention of many chemists that there is a strong convergence of the chemistry and materials science of nanoparticles and the needs of medicine. This is leading to a new interdisciplinary subject called "nanomedicine" and it is poised to change the future of medicine in many ways and provide outlets and opportunities for many chemists, chemical engineers and materials scientists. Ironically, much of this activity might have been driven by some of the hype surrounding the imagined dangers of nanotechnology, with lobby groups warning of dire consequences if new man-made, engineered nanoparticles 'escaped' into the environment. Soon, however, scientists realised that designing nanoparticles to actually enter, modify and destroy cells is difficult, but, if successful, it could open up many new and beneficial diagnoses and treatments for disease. This is now developing into a very exciting time for scientists, engineers and the whole healthcare community. In the past decade we have witnessed the synthesis of new nanoparticles for enhancing images and hence diagnosis *via* MRI, PET, luminescence and ultrasound imaging. There were examples presented in this Faraday Discussion meeting. Some of these new entities are now undergoing trials for approval by regulatory bodies. In parallel with this has been the realisation that if we can induce particles to enter parts of the body to detect disease we can also engineer them to have therapeutic properties, either by carrying a small drug payload or by inducing hyperthermia, creating free radicals to destroy cancerous cells or to modify the genetic make-up of cells. Examples of many new methods for nanoparticle synthesis specifically in these areas were presented at this Discussion meeting.

All of these ideas depend on the targeting of nanoparticles to cells, in order attached them to specific cell types, and this is where the subject of this Faraday Discussion comes into play. There is a growing understanding about what is needed to functionalise the surface of a nanoparticle in order for it to attach to a cell or to enter a cell. The tools of the physical and materials scientists are now highly developed to quantify the size, shape and composition of the nanoparticle

The Queen's College, High Street, Oxford, OX1 4AW. E-mail: peter.dobson@queens.ox.ac.uk

and to give a fairly accurate picture of its surface in terms of the chemical ligands and electronic charge. We are still struggling to fully characterise and understand the exact details of the outer membranes of cells and to understand how to precisely target them, but this gap in knowledge is closing.

Having now started to develop the means of making nanoparticles with different functionalities, we have reached a stage where we have to ask "what happens next?" Concepts such as the blood–brain barrier have to be questioned and quantified. If we can design particles to cross this barrier or enter any other cells in the body, we have to ask what will be the fate of the particles we have introduced after they have delivered diagnosis or drugs or some other form of therapy. This is compelling us to learn more about the physiology of the body and to find ways of tracking the fate of nanoparticles or parts of them as they pass through the gut, blood or lymph system. We also have to work out if there are critical sizes involved in crossing the blood–brain barrier or designing particles to be purposely trapped in the vasculature around tumours. Targeting modalities can vary from antibody recognition to the use of smaller aptamers or peptides and we have to recognise the possible issues with non-specific binding and its implications for normal or healthy cells.

Although the chemists and materials scientists/engineers have now devised clever protocols for making nanoparticles that can combine several functionalities, they have yet to translate these into a form that lends itself to scale-up under "Good Manufacturing Practice" (GMP) conditions. This will be a key role for many of the new start-up companies or the larger and well-established pharmaceutical companies who have not yet embraced nanotechnology. It is not clear how this translation or innovation in medicine will occur. It is clearly a role for governments to take *via* funding that is beyond the usual sources for basic research. Furthermore, there needs to be significant funding support to give confidence to investors entering what is perceived to be a fairly long and drawn-out regulatory-controlled stage.

One specific area where nanotechnology can play a powerful role is in antimicrobial control and there were good examples of this in the Discussion, particularly from the group at UCL led by Professor Ivan Parkin, who demonstrated coatings based on very small gold nanoparticles and titania (*Faraday Discuss.*, 2014, DOI: 10.1039/C4FD00113C). This field is assuming increasing importance as microbes develop some immunity to the antibiotics that have been in use in the past 50 years. It is relatively straightforward to design and make several types of nanoparticles to attack microbes *in vitro*, and these methods include particles that produce either oxidising free radicals or solvated electrons that can act as powerful reducing agents. There are also options to stabilise many natural antimicrobial molecules or use nanosilver. The ultimate challenge will be to see if we can help develop new antibiotics for delivery inside the body. One should not underestimate the potential of some of the new forms of carbon, such as graphene and nanodiamond, to have antimicrobial properties, for reasons that are not fully explained. Nanodiamonds with nitrogen-vacancy centres have also been shown to be a very powerful new image enhancing agent, because their luminescent properties can be modulated with quite weak magnetic fields. This permits the detection of bio-labelled nanodiamond in any near-surface imaging technique using either image subtraction or modulated lock-in camera techniques, and is a good early example of "quantum imaging".[1]

The sobering fact in much of nanomedicine is the time it will take to see a lot of the new exciting paradigms reach the market or be applied regularly in healthcare. The regulatory pathway is long and potentially expensive. It is likely that the promising ideas of today will take a further 5–8 years to reach clinical practice. There is much to be done to establish the fate of all the components that are being used in nanoparticle-based imaging systems and drug delivery systems. This concern also has to go beyond the human body to ensure that any fragments from such treatment are environmentally safe if they enter the waste-water stream.

References

1 S. Sarkar, A. Bumb, X. Wu, K. Sochacki, P. Kellman, M. Brechbiel and K. Neuman, *Biomed. Opt. Express*, 2014, **5**, 1190–1202, DOI: 10.1364/BOE.5.001190.

Faraday Discussions

PAPER

Multifunctional silica-coated iron oxide nanoparticles: a facile four-in-one system for *in situ* study of neural stem cell harvesting†

Yung-Kang Peng,‡[a] Cathy N. P. Lui,‡[b] Tsen-Hsuan Lin,[c] Chen Chang,[c] Pi-Tai Chou,[d] Ken K. L. Yung[b] and S. C. Edman Tsang*[a]

Received 13th June 2014, Accepted 28th July 2014

DOI: 10.1039/c4fd00132j

Neural stem cells (NSCs), which generate the main phenotypes of the nervous system, are multipotent cells and are able to differentiate into multiple cell types *via* external stimuli from the environment. The extraction, modification and re-application of NSCs have thus attracted much attention and raised hopes for novel neural stem cell therapies and regenerative medicine. However, few studies have successfully identified the distribution of NSCs in a live brain and monitored the corresponding extraction processes both *in vitro* and *in vivo*. To address those difficulties, in this study multi-functional uniform nanoparticles comprising an iron oxide core and a functionalized silica shell (Fe_3O_4@SiO_2(FITC)-CD133, FITC: a green emissive dye, CD133: anti-CD133 antibody) have been strategically designed and synthesized for use as probe nanocomposites that provide four-in-one functionality, *i.e.*, magnetic agitation, dual imaging (both magnetic resonance and optical) and specific targeting. It is shown that these newly synthesized Fe_3O_4@SiO_2(FITC)-CD133 particles have clearly demonstrated their versatility in various applications. (1) The magnetic core enables magnetic cell collection and T_2 magnetic resonance imaging. (2) The fluorescent FITC embedded in the silica framework enables optical imaging. (3) CD133 anchored on the outermost surface is demonstrated to be capable of targeting neural stem cells for cell collection and bimodal imaging.

1. Introduction

Neural stem cells (NSCs) are multipotent cells that are capable of self-replicating and generating various phenotypes of the nervous system.[1–3] Numerous studies

[a]Department of Chemistry, University of Oxford, Oxford, OX1 3QR, UK. E-mail: edman.tsang@chem.ox.ac.uk
[b]Department of Biology, Hong Kong Baptist University, Kowloon Tong, Hong Kong, China
[c]Institute of Biomedical Sciences, Academia Sinica, Taipei 115, Taiwan
[d]Department of Chemistry, National Taiwan University, Taipei 106, Taiwan

† Electronic supplementary information (ESI) available. See DOI: 10.1039/c4fd00132j

‡ These authors contributed equally to this work.

have suggested that a subpopulation of cells surrounding the lateral ventricle (LV) and sub-ventricular zone (SVZ) is capable of producing neurons for the olfactory bulb in adults.[4–6] Furthermore, it has been proposed that long-term potentiation is induced by the proliferation of NSCs in the dentate gyrus of the hippocampus during memory formation. These findings potentiate the use of adult NSCs as a therapeutic tool in neurological disorders.[5,6] However, isolation of neural stem cells from a live brain without slaughtering the donor has been a great challenge.

To circumvent these difficulties, we have recently introduced a novel, precise and safe method to detach and collect NSCs from the ependymal lining in the adult rodent brain *in situ*.[7] We used magnetic nanoparticles conjugated with anti-CD133 antibodies to selectively target the CD133 positive NSCs. After incubation for 6 hours, the particles were shown to specifically bind to the target cells. The animals were then placed under a weak external rotational magnetic field generated from the magnetic stirrer plate for the detachment of the stem cells localized along the SVZ/LV. The extracted stem cells were also demonstrated to be differentiated into different neuronal phenotypes after culture. Unfortunately, at that time we were unable to monitor the real time distribution of the nanoparticles in the rat brain and the fate of these particles, since the selective binding and removal of labels/cells from the ependymal linings of SVZ/LV by the magnetic particles could be dynamic in the rat body. An *in situ* observation technique is therefore necessary to understand the time course and distribution of targeted stem cells in the rat brain.

Magnetic resonance imaging (MRI) has received much attention over the past two decades as a technique for clinical diagnosis, owing to its high resolution, unlimited tissue penetration, and absence of ionizing radiation.[8–10] Contrast agents, such as superparamagnetic iron oxide (Fe_3O_4) nanoparticles (NPs) with high r_2 relaxivity, can generate (dark) signal in a corresponding area compared to the background and have been widely used in clinical imaging. Since the r_2 relaxivity is proportional to the net magnetization (M_z) of the Fe_3O_4 NPs, a larger M_z of Fe_3O_4 nanoparticle can give a stronger T_2-weighted enhancement. To expand its applications in biomedicine, the design of multifunctional Fe_3O_4 NPs has been intensively pursued in the last ten years.[11–14] For instance, multifunctional magnetic NPs, which can render functions such as targeting, therapy, optical and MR imaging, have been used in *in vivo* tumor theranostic applications.[15,16]

In this study, first of all we synthesized hydrophobic Fe_3O_4 NPs *via* thermal decomposition (high temperature) of iron oleate precursor, instead of the hydrophilic co-precipitation method we used previously (room temperature).[7] Since the magnetization of Fe_3O_4 NPs depends strongly on its crystallinity, the higher the temperature at which the Fe_3O_4 nanoparticle is synthesized, the higher the magnetization it possesses.[17] The harvesting of targeted cells and their corresponding T_2-weighted MR signal can thus both be improved. Secondly, in order to combine both MR and optical imaging in a single nanoparticle (important to follow the fate of the labeled particle), we also incorporated amine-functionalized fluorescein isothiocyanate (FITC) in the silica shell during the sol–gel reaction of tetraethyl orthosilicate (TEOS). The outermost silica surface of the resulting product, Fe_3O_4@SiO_2(FITC), was further functionalised with immobilized anti-CD133 antibodies for specific targeting. The final product, Fe_3O_4@SiO_2(FITC)-anti-CD133 antibody (abbreviated as Fe_3O_4@SiO_2(FITC)-CD133 hereafter), combines four functions in a single unit, *i.e.* (1) The magnetic core serves not only as a magnet for magnetic cell collection, but also provides the capability for T_2 magnetic resonance imaging. (2) The fluorescent FITC embedded in the silica framework enables optical imaging. (3) CD133 anchored on the outermost surface is shown to be capable of targeting neural stem cells. Finally, Fe_3O_4@SiO_2(FITC)-CD133 is demonstrated to be versatile in monitoring the time course distribution and the collection process of targeted stem cells in a living rat brain.

2. Experimental details

Chemicals

1-Octadecene (technical grade, 90%, Acros), oleic acid (90%, Acros), iron chloride ($FeCl_3 \cdot 6H_2O$, 40 mmol, Aldrich, 98%), hexanol (98%, Acros), Triton X-100 (Acros), tetraethyl orthosilicate (98%, Acros), ammonium hydroxide (28–30 wt%, Fluka), triethoxysilane ($HSi(OEt)_3$, Alfa Aesar), 3-aminopropyltrimethoxysilane (APTMS, 95%, Acros), 3-aminopropyltrimethoxysilane (APTES, 97%, Aldrich), fluorescein isothiocyanate isomer I (FITC, 90%, Aldrich), *N*-(3-dimethylaminopropyl)-*N*′-ethylcarbodiimide hydrochloride (EDC, Aldrich), *N*-hydroxysuccinimide (NHS, 98%+, Acros), hexane and isopropyl alcohol were used without further purification.

Fe_3O_4 NPs

Iron oxide NPs were prepared by the method developed by Park *et al.* with minor modifications.[18] Iron oleate precursor was prepared by the reaction of iron chloride and oleic acid in methanol under basic conditions. Typically, 1.24 g of the iron oleate was used and dissolved in 10 g of 1-octadecene. The solution mixture was degassed at 100 °C in order to remove any moisture and oxygen. The reaction mixture was subsequently treated with a standard temperature program. First of all, the solution was rapidly heated to 200 °C at a rate of 5 °C min^{-1}. The solution was then further heated to 300 °C at a heating rate of 1.2 °C min^{-1} with vigorous stirring and maintained at the temperature for 1 h before cooling to room temperature. Solid samples were collected *via* centrifuging at 9000 rpm for 3 min, washed, and then the precipitate was redispersed with hexane and isopropyl alcohol several times. Finally, the purified Fe_3O_4 nanoparticles were dried in the oven.

Fe_3O_4@SiO_2(FITC) nanocomposites

The preconjugated *N*-1-(3-triethoxy-silylpropyl)-*N*-fluoresceyl thiourea (FITC–APTES) was prepared by combining 5 μL of APTES and 2 mL of 0.023 M FITC ethanolic solution under continuous stirring and dark conditions.[19] Fe_3O_4@SiO_2(FITC) nanocomposites were prepared from reverse micelles using a previously reported procedure.[20] The as-prepared Fe_3O_4 nanoparticles (see above) were used as seeds for growth of the SiO_2 shell. Briefly, Fe_3O_4 nanoparticles (2 mg), 100 μL of TEOS, and 100 μL of FITC–APTES ethanolic solution were added to a heterogeneous solution containing cyclohexane (24 mL), hexanol (4.8 mL), Triton X-100 (6 mL), and water (1 mL). After 6 h of stirring, NH_4OH (100 mL) was added to initiate the hydrolysis of both TEOS and FITC–APTES. Due to the $(EtO)_3Si$ functional groups on FITC–APTES, it is reasonable to expect the successful encapsulation of the small green-emitting FITC molecules by the SiO_2 shell. The reaction was allowed to continue for another 24 h with stirring at room temperature. The product, Fe_3O_4@SiO_2(FITC), was well dispersed in ethanol and was further purified by centrifugation.

Fe_3O_4@SiO_2(FITC)-CD133 nanocomposites

To externally coat the amino group on the surface of silica particles, 3-aminopropyltrimethoxysilane (10 μL) was added to a mixture of ethanol (30 mL), DI water (6 mL), NH_4OH (300 μL), and Fe_3O_4@SiO_2(FITC) (20 mg).[21] After 24 h, the sample was centrifuged several times to remove the unreacted chemicals. The precipitate of particles was then collected and redispersed in DI water. The anti-CD133 antibodies were covalently conjugated onto the silica coated magnetic nanoparticles by EDC/NHS chemistry.[7] Typically, Fe_3O_4@SiO_2(FITC)-NH_2 (16 mg) were incubated with 10 mg EDC and 10 mg NHS for 30 min. Then, anti-CD133 antibodies (1 mL) were added to the mixture and incubated for 1 h at room temperature. The antibody-conjugated Fe_3O_4@SiO_2(FITC) nanocomposites were purified by centrifugation at 10 000 rpm for 15 min and washed 3–4 times with PBS (pH 7.4).

Particle characterization

The shape and size distributions of the nanoparticles were measured with a JEOL JEM 1230 transmission electron microscope (TEM). X-Ray energy dispersive spectroscopy (EDS) was performed using a GENESIS 2000 EDS detector connected to the TEM instrument. Conventional Formvar coated 200 mesh Cu grids were used for the above-mentioned electron microscope (EM) experiments. Powder X-ray diffraction data was collected on a Bruker D8 Advance diffractometer. The workup procedure was carried out with Cu Kr radiation (λ) 1.54178 Å. Magnetic measurements were performed by superconducting quantum interference device (SQUID) magnetometer (MPMS, Quantum Design). The measurements were recorded between −15 000 and 15 000 Oe at 300 K. The ζ-potential was measured in a Malvern Zetasizer 3000 HS. Suspensions (200 μg mL^{-1}) of each material in DI water were prepared. The ζ-potential was measured immediately after ultrasonication for 5 min, and 10 scans were taken per sample. Measurements of r_1 and r_2 relaxation times were made at 40 °C using a 0.47 T Minispec spectrometer (Bruker Minispec mq series relaxometer); r_2 relaxation times were determined using a CarrPurcellMeiboomGill (CPMG) sequence, recycle time 10 s, eight averages with phase cycling, and 180° pulse separation of 1 ms. Monoexponential fitting was performed to even echoes over 250 ms acquisition window. The r_1 relaxation was estimated using inversion recovery techniques, recycle time 10 s, four averages with phase cycling, and eight inversion times logarithmically spaced over the interval 0–2000 ms. Linear regression between r_1 and r_2 and iron concentration was performed using standard techniques.

In vivo experiments

Detail of the *in vivo* experimental procedures can be found in the ESI† and our previous report.[7] For *in vivo* MRI experiments, all animals were initially anesthetized with 5% isoflurane at 1 L min^{-1} air flow. When fully anesthetized, the animal was placed in a prone position and fitted with a custom designed head holder inside the magnet. Isoflurane was then maintained with 1–1.5% for rat at

1 L min^{-1} air flow throughout the experiments. Sprague–Dawley rat brain experiments were performed in a Biospec 4.7 T spectrometer before and at different times after the injection of Fe_3O_4@SiO_2(FITC)-CD133 with (T1/T2) TR = 500/5000 ms, (T1/T2) TE_{eff} = 8/70 ms, FOV = 7 cm × 7 cm, slice thickness = 1.3 mm, and matrix size = 256 × 128 (zero-padded to 256 × 256) with six repetitions. Typically, nanoparticles suspended in PBS buffer (5 μL) were administrated stereotaxically into the SVZ/LV of the Institute of Cancer Research rats (8 weeks of age, male, body mass ~300 g) at a dose of 0.19 mg kg^{-1}.

3. Results and discussion

Synthesis and characterization

Scheme 1 illustrates the overall synthetic protocol (detail of the synthesis is described in the Experimental section). First of all, iron oxide NPs were synthesized by the thermal decomposition of precursor Fe(oleate)$_3$ complex in octadecene. As shown in Fig. 1a, the oleic acid-capped Fe_3O_4 NPs are nearly monodisperse (σ < 7.5%) and well-dispersed. The average diameter is about 9.28 ± 0.71 nm (Fig. S1†; 50 particles are used for statistics). The silica shell was then coated onto the as-prepared Fe_3O_4 NP by a sol–gel reaction of tetraethyl orthosilicate (TEOS) *via* a reverse micro-emulsion method. During the condensation reaction of TEOS, the amine functionalized fluorescence molecules *N*-1-(3-trimethyloxy-silylpropyl)-*N*′-fluoresceyl thiourea (APTES–FITC) can be simultaneously encapsulated into the silica framework. Due to the protection of the silica, the embedded fluorophores are expected to be stable against enzymatic digestion and thus allow long term cell tracking. Fig. 1b shows the TEM image of the resulting product, Fe_3O_4@SiO_2(FITC), which has a homogeneous 10 nm silica

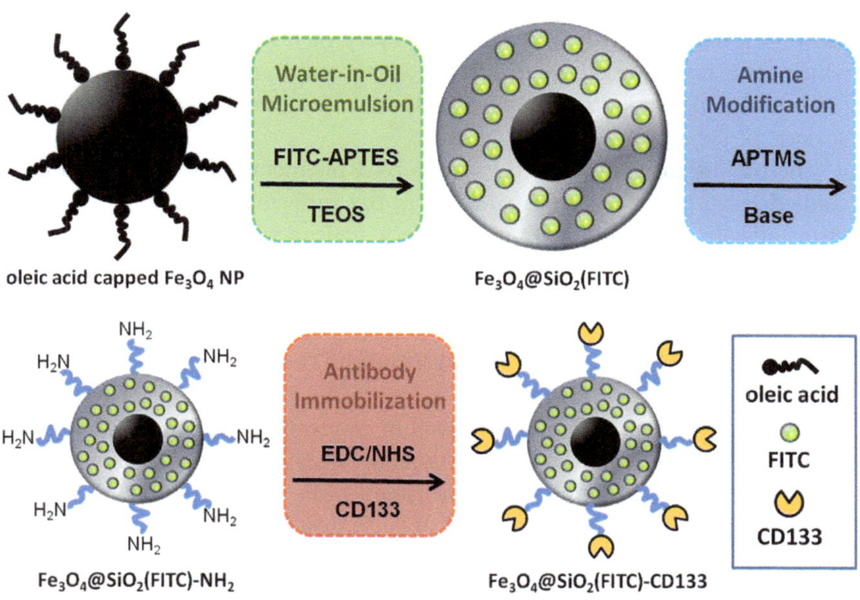

Scheme 1 Schematic illustration of the synthesis of Fe_3O_4@SiO_2(FITC)-CD133 nanocomposites.

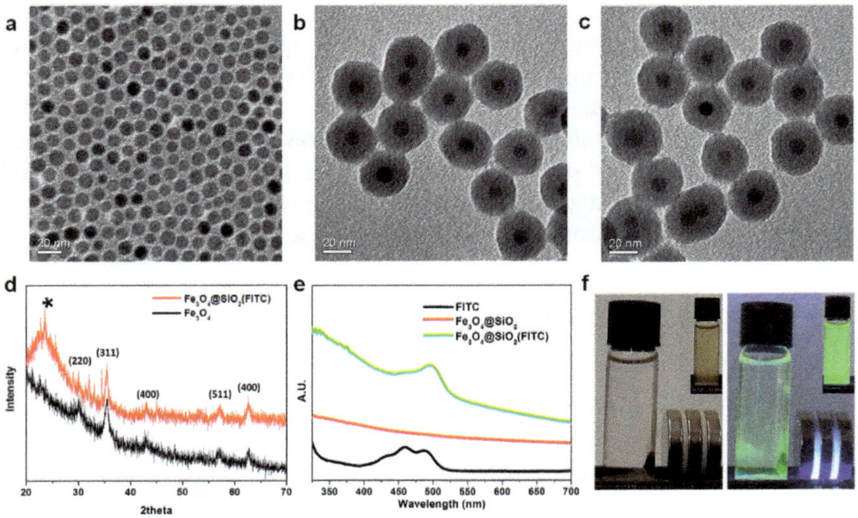

Fig. 1 TEM of (a) oleic-acid-capped Fe_3O_4 NPs, (b) Fe_3O_4@SiO_2(FITC) and (c) Fe_3O_4@SiO_2(FITC)-CD133. (d) X-ray diffraction spectra of Fe_3O_4 NPs and Fe_3O_4@SiO_2(FITC). (e) Absorption spectra of FITC (black), Fe_3O_4@SiO_2 (red) and Fe_3O_4@SiO_2(FITC) (green). (f) The appearance of Fe_3O_4@SiO_2(FITC)-CD133 NPs upon placement of a magnet at one side of the vial (left) and the corresponding green emission produced under excitation by a 366 nm UV lamp (right). Inset: the appearance of the Fe_3O_4@SiO_2(FITC)-CD133 suspension in aqueous solution (left) and under UV excitation (right).

shell around each Fe_3O_4 core. The average diameter of the nanocomposite is 31.71 $\pm$ 1.54 nm (Fig. S2†; 50 particles are used for statistics). The corresponding XRD pattern (Fig. 1d) confirms the existence of Fe_3O_4, and a broad peak around 23° is attributed to the amorphous silica shell. The composition of Fe_3O_4@SiO_2(FITC) nanocomposite was further analyzed by EDX spectroscopy (Fig. S3†), which confirms the existence of silicon, oxygen and iron elements. Both results (XRD and EDX) support that silica is successfully coated onto the Fe_3O_4 NP surface.

The coating mechanism of this water-in-oil microemulsion system has been studied recently.[22] It has been shown that the surfactant TritonX-100 not only forms micelles in the cyclohexane solution, but also replaces the surfactant oleic acid once the Fe_3O_4 NPs are added. Later, the TEOS that is added will hydrolyze at the oil–water interface and perform the ligand exchange with TritonX-100, which is chemically absorbed on the Fe_3O_4 NP surface, and then transfer the NPs to the water phase. Finally, the hydrolyzed TEOS on the Fe_3O_4 NP surface undergoes a condensation process and forms silica shells. In Fig. 1e, it is clear that the characteristic absorption bands at ~490 nm corresponding to the absorption of FITC in significant quantities also appear in the Fe_3O_4@SiO_2(FITC) sample, while Fe_3O_4@SiO_2 shows no prominent absorption peak at this wavelength. Fig. 1f demonstrates the magnetic separation of Fe_3O_4@SiO_2(FITC), in which the green emission appears on the same side as the magnetically-separated powder (Fig. 1f right), while there is no conspicuous emission remaining in the solution, thus eliminating the possibility of unencapsulated FITC. These results verify the successful encapsulation of FITC in the silica framework.

The Fe_3O_4@SiO_2(FITC) NPs were then further modified with (3-aminopropyl)-triethoxysilane (APTMS) under basic condition to yield Fe_3O_4@SiO_2(FITC)-NH_2 (Scheme 1). It is believed that the triethoxysilane groups of APTMS can undergo condensation with hydroxyl groups on the outermost surface of Fe_3O_4@SiO_2 (FITC) under basic conditions, in which some hydroxyl groups would therefore be replaced by amine groups. Since the hydroxyl groups are negatively charged at neutral pH (Si–O^-), the modification of an amine group (positive charged in neutral pH, NH_3^+) can be identified with zeta potential (ζ-potential) measurement.[23] As shown in Fig. S4,† the value of ζ-potential for bare Fe_3O_4@SiO_2(FITC) increases positively from −44.5 mV (Fig. S4a†) to −19.8 mV (Fig. S4b†) after APTMS modification, suggesting the successful replacement of surface hydroxyl groups with amine groups. To supply the core–shell particles with bio-targeting capability, antibodies CD133 were then conjugated with the primary amine of Fe_3O_4@SiO_2(FITC)-NH_2 *via* an amide bond formation between the amine and the carboxylic acid of CD133. As shown in Fig. 1c, the final nanoparticles, Fe_3O_4@SiO_2(FITC)-CD133, remain well-shaped and no degradation of the silica shell is observed after the amine and CD133 modification.

The magnetic properties of Fe_3O_4@SiO_2(FITC)-CD133 were assessed by SQUID (Fig. S5†). Compared to the magnetization value of naked Fe_3O_4 (~100 emu g^{-1} at 300 K),[24] the magnetization of Fe_3O_4@SiO_2(FITC)-CD133 drops to ~70 emu g^{-1}, which is plausibly due to the introduction of the silica host and antibody CD133. However, the magnetization of Fe_3O_4@SiO_2(FITC)-CD133 is much higher than the previous Fe_3O_4 NP we used (~30 emu g^{-1}).[7] It should be noted that the magnetic moment of the Fe_3O_4 NP is strongly dependent on the size, composition, and magneto-crystalline phase of the Fe_3O_4 NP.[17] Since both particle size and composition are comparable to our previous Fe_3O_4 NP (prepared by co-precipitation method), the high magnetization value achieved in this study should be due to the high temperature preparation of Fe_3O_4 NP in oil phase, which provides its high crystallinity. The response of Fe_3O_4@SiO_2(FITC)-CD133 to an external magnet was also demonstrated in Fig. 1f. As the transverse relaxivity (r_2) is strongly related to the net magnetization ($M_z \propto r_2$), the larger magnetization of the Fe_3O_4 nanoparticle provides a stronger T_2-weighted enhancement. Fig. S6† shows the corresponding longitudinal (r_1) and transverse relaxivity (r_2) of Fe_3O_4@SiO_2(FITC)-CD133 relaxation properties. The suppressed r_1 value (1.41 mM^{-1} S^{-1}), the high r_2 value (254.86 mM^{-1} S^{-1}) and the high r_2/r_1 ratio (as high as 180) clearly reveal that the nano-composite is a superior T_2 contrast agent.[17]

Since the Fe_3O_4@SiO_2(FITC)-CD133 should be able to selectively target the CD133 positive neural stem cells, we therefore assessed the specificity of the Fe_3O_4@SiO_2(FITC)-CD133 particles by employing primary astrocytes ($CD133^+$) and SH-SY5Y cells ($CD133^-$) in an *in vitro* test. The distribution of Fe_3O_4@SiO_2(FITC)-CD133 in each cell solution was monitored by merging the differential interference contrast (DIC) (left column of Fig. 2a) and confocal images (right column of Fig. 2a). As shown in Fig. 2, the Fe_3O_4@SiO_2(FITC)-CD133 can favorably bind to the cell surface of the $CD133^+$ astrocytes. With a magnetic field applied to the cell culture, the $CD133^+$ cells tagged with Fe_3O_4@SiO_2(FITC)-CD133 could also be effectively driven in solution. In contrast, the CD133 negative SH-SY5Y cells were evenly distributed in the solution and the cells were immobile on the application of a magnetic field. Furthermore, lactate dehydrogenase (LDH) cytotoxicity assay was applied to assess the cytotoxicity of the NPs after 24 h-exposure

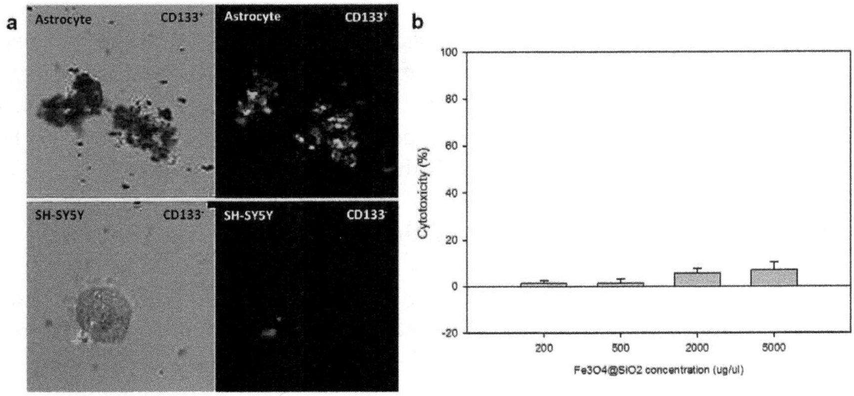

Fig. 2 (a) Selective labelling of CD133 positive cells with Fe_3O_4@SiO_2(FITC)-CD133. The cells were incubated with particles for 6 hours *in vitro*. The Fe_3O_4@SiO_2(FITC)-CD133 can specifically bind onto (a) CD133 positive primary astrocytes (upper row) from PND (postnatal day) 1 rats but not onto CD 133 negative SH-SY5Y cells (bottom row). (b) The LDH cytotoxicity assay illustrates the percentage of cell death in CD133 positive primary astrocytes after 24 hours treatment with different concentrations of NPs ranging from 200 to 5000 μg mL^{-1}.

of the particles in the concentration range of 200–5000 μg mL^{-1} to CD133$^+$ astrocytes (Fig. 2b). The percentages of cell death were comparable among different groups and controls after the treatment. The as-prepared Fe_3O_4@SiO_2(FITC)-CD133 showed no sign of toxicity.

To study the time course distribution of injected Fe_3O_4@SiO_2(FITC)-CD133 *in situ*, we established *in vivo* MR imaging of animal with a 4.7 T MR system. We have unilaterally injected the Fe_3O_4@SiO_2(FITC)-CD133 (2000 μg mL^{-1} in 5 mL of PBS) into the left ventricle of a live rat brain. It was envisaged that the Fe_3O_4@SiO_2(FITC)-CD133 would specifically bind to the NSCs. As shown in Fig. 3, the time course spatial distribution of the Fe_3O_4@SiO_2(FITC)-CD133 tagged NSCs could be monitored in real time without sacrificing the subjects. Coronal views of the T_2-weighted MRI images were acquired from posterior (row 1) to anterior (row 4) of the ventricles at different time points, *i.e.* 1 hour, 6 hours, 24 hours and 48 hours. The shape of the left ventricle in T_2-weighted imaging became clear as we prolonged the induction time from 1 hour to 48 hours. Fig. S7† shows a time course of T_2 signal intensity of Fig. 3 row 3 (where we collected NSCs). Accordingly, the contrast to noise ratio (CNR) of the left ventricle reached ~70% at the first hour after administration with reference to the CNR before the administration. It is interesting to note that the highest contrast was indeed found on the SVZ/LV, indicative of the preferential binding of the NSCs on this surface (no significant contrast was found in the fluid and other brain regions). In addition, the control experiment using the same particles, but without the anti-CD133 antibodies, did not seem to bind on the SVZ/LV region (but evenly distributed). This confirms our previous observation from using confocal microscopy that the labeled magnetic particles are mainly deposited on the choroid plexus lining along the SVZ/LV region.

It is thus believed that the contrast value of the MRI reflected the number of particles that can be rapidly attached on the surface containing the NSCs after the first hour of administration (excess particles in the fluid were quickly removed).

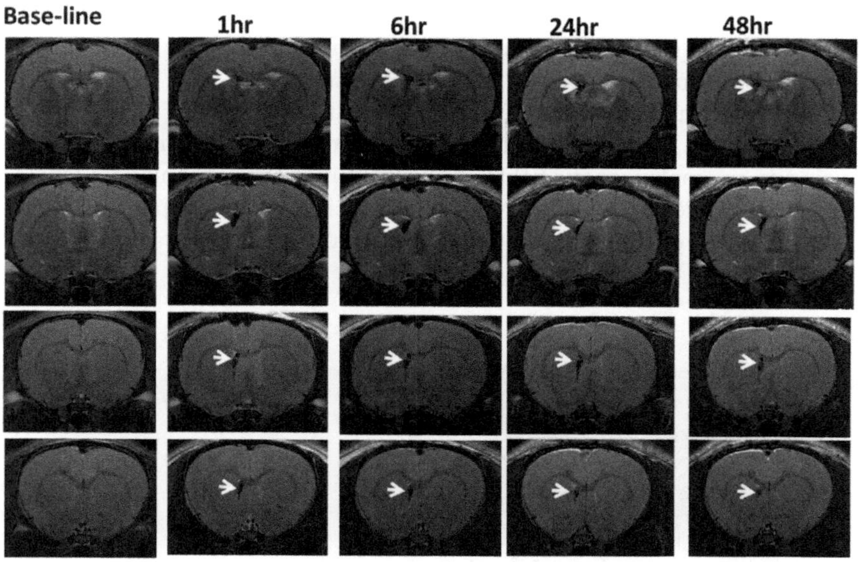

Fig. 3 A spatial distribution (from posterior (row 1) to anterior (row 4), slice thickness = 1.3 mm) of T_2-weighted MR imaging of a rat brain (dark region) before and at different incubation times (1 h, 6 h, 24 h and 48 h) after injection of Fe_3O_4@SiO_2(FITC)-CD133.

However, this value declined to 60% at 6 hour and further dropped to ~40% after 24 h incubation. The data thus clearly suggest some surface bound particles are unstable and can be continuously removed from the surface to the moving cerebrospinal fluid (CSF). In contrast, if we further prolonged the incubation time for another 24 h, the CNR did not show further decline (~48%), as presumably the particles became strongly attached as time progressed (or became engulfed by the cells). This result indicates that the labels/cells on the ependymal lining of the ventricle are rather dynamic, and can be flushed by fluid or taken up by inner cells. Several processes should therefore be considered: (1) the diffusion and adsorption of particles on NSCs, (2) the removal of particles by CSF, (3) dynamically binding/(4) unbinding to NSCs and (5) the internalization of bound particles by NSC. Accordingly, the initial decrease of T_2 CNR signal can be attributed to the higher rate of particle removal (processes 2 and 4) compared to the counter processes (1), (3) and (5). However, the latter rates cannot be ignored as time progresses further (Fig. S7†). Based on the above discussion, the best operational time to monitor or extract the labeled surface region NSCs should not be longer than 24 h after administration, otherwise nearly 50% of the particles will have been removed. In addition, the internalization of bound particles is rather significant over longer time periods, and such particles may not be detached to the fluid easily with or without applying external magnetic field. It should be noted here that no change of T_2 signal (no labeled particles in this region) was observed in the right ventricle throughout the whole study period (Fig. S7†). These results clearly indicate that (1) the particles could specifically bind to the CD133 positive cells on the left ventricle in a live subject and (2) cell collection should be carried out within 24 h of administration of Fe_3O_4@SiO_2(FITC)-CD133. It should be noted that the above conclusions are supported by our previous observation that a higher NSC collection efficiency was actually found when using a shorter induction time (6 h) rather than a longer induction time (24 h) in the post-mortem analysis of the brain samples by confocal microscopy.[7]

Besides evaluating the best time period for the incubation of these particles using MRI, the magnetic agitation time required to extract the bound NSCs using an external magnetic field was also examined. The animals with labeled magnetic particles in the left ventricles of their brains were placed under a weak external rotational magnetic field, generated from a laboratory magnetic

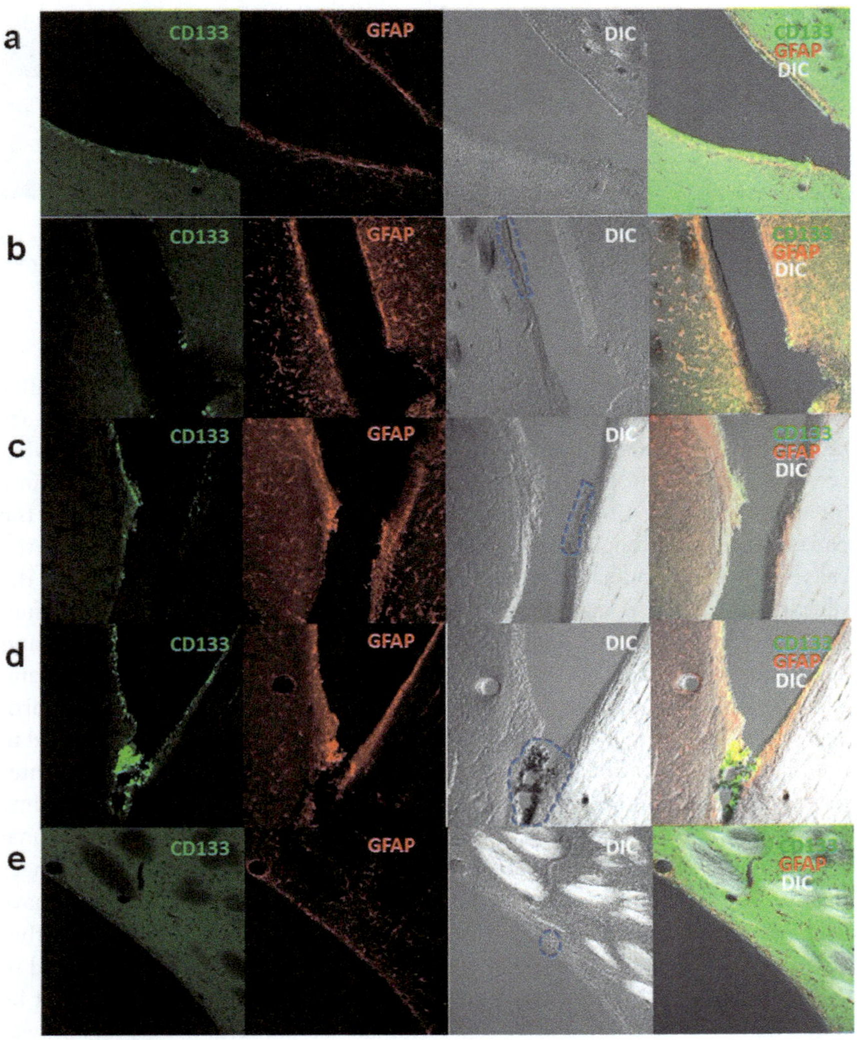

Fig. 4 The efficiency of neural stem cell isolation with different magnetic agitation time: (a) 0 min; (b) 5 min; (c) 10 min; (d) 15 min with $Fe_3O_4@SiO_2$(FITC)-CD133 and (e) 15 min with $Fe_3O_4@SiO_2$(FITC). An effective detachment of the CD133 positive neural stem cells from the choroid plexus lining along the SVZ/LV can be achieved after 15 min agitation under a weak external rotational magnetic field. The dashed circles illustrate the presence of the nanoparticles.

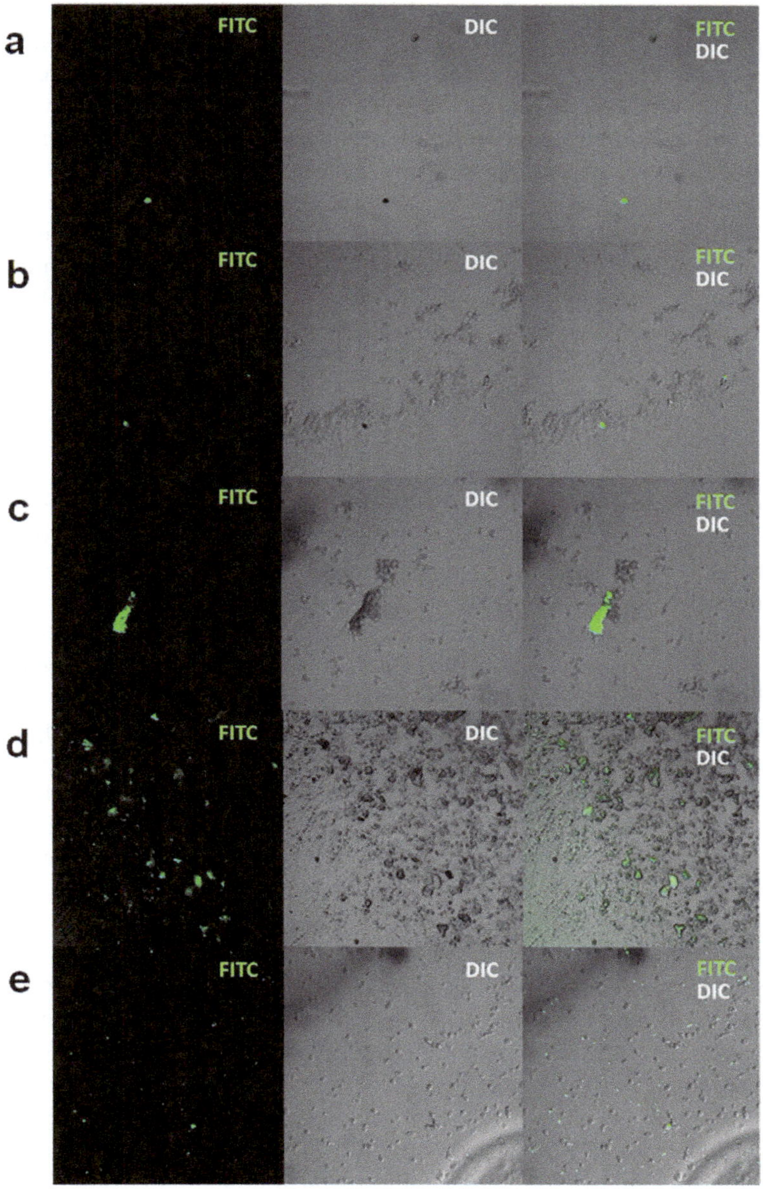

Fig. 5 An *in vivo* extraction of CD133 positive neural stem cell is performed after magnetic agitation: (a) 0 min; (b) 5 min; (c) 10 min; (d) 15 min with $Fe_3O_4@SiO_2$(FITC)-CD133 and (e) 15 min with $Fe_3O_4@SiO_2$(FITC). Around 1 μL of extract can be collected from SVZ/LV by micro-syringe. The longer the agitation, the more CD133 positive cells can be isolated and collected for further cell culture. Without conjugation of the antibody, $Fe_3O_4@SiO_2$(FITC) can be found freely in the extract.

stirrer plate, to allow the detachment of stem cells from the choroid plexus lining along the SVZ/LV. Notably, the duration of this "magnetic agitation" is thought to be critical for an efficient stem cell extraction. We have compared several agitation times (0 min, 5 min, 10 min and 15 min) and 15 min magnetic agitation was shown to effectively isolate the neural stem cells from SVZ/LV (Fig. 4 and 5) without inducing great damage to the cells and linings. Our previous study also demonstrated that the extracted cells could be collected for cell culture using a micro-syringe. The extracted stem cells were able to be cultured and further differentiated into different neuronal phenotypes *in vitro*.[7] With these promising findings, we propose that patients could supply their own neural stem cells at some stages of their lives for later cell replacement treatments. The risk of immune rejection could be greatly reduced, as the cells all originate from the same individual. An innovative tailor-made neurological disorder therapy can thus be envisaged by using stem cells originating from the subjects themselves.

4. Conclusions

In summary, we demonstrate a biocompatible magnetofluorescent T_2 MRI contrast agent for targeted MR and fluorescence imaging of neural stem cells and their extraction based on silica-coated Fe_3O_4 NPs. This multifunctional nanoparticle, Fe_3O_4@SiO_2(FITC)-CD133 is shown to facilitate *in situ* imaging and harvesting of the neural stem cells. (1) The magnetic core facilitates the capability for magnetic cell collection and T_2 MRI. (2) The fluorescent FITC embedded in the silica framework enables optical imaging. (3) Anti-CD133 antibodies anchored on the outermost surface are shown to be capable of targeting neural stem cells for cell collection and bimodal imaging. The spatial distribution of T_2-weighted MR imaging of a rat brain clearly indicates that the binding/removal of labels/cells from the ependymal linings of the ventricles is dynamic and an incubation time is required for the elementary surface binding processes to reach equilibrium. The best operation time (to monitor or extract the labeled NSCs) is not longer than 24 h, otherwise nearly half of the particles have been removed by the CSF. According to the MRI and optical data, our studies clearly show that the Fe_3O_4@SiO_2(FITC)-CD133-targeted neural stem cells can be efficiently isolated by external magnetic agitation. Finally, it is also worth noting that the current technology for stem cell collection in a live brain may also be extended to other cell extracts in particular areas of a subject by modifying the corresponding targeting ligand on the outermost silica surface. Notably, the cell distribution and the process of cell extraction could also be monitored by MRI and optical imaging using related particles prepared by the nano-chemistry described above.

Acknowledgements

We thank the EPSRC of UK for initial funding of this project. YKP acknowledges the University of Oxford Clarendon Fund Scholarship for his DPhil study. We are indebted to the support from Taiwan Mouse Clinic which is currently funded through the National Research Program for Biopharmaceuticals (NRPB) at the National Science Council of Taiwan, particularly for the MRI mouse experiments presented in this paper.

Notes and references

1 F. Doetsch, I. Caille, D. A. Lim, J. M. Garcia-Verdugo and A. Alvarez-Buylla, *Cell*, 1999, **97**, 703–716.

2 S. Pluchino, L. Zanotti, B. Rossi, E. Brambilla, L. Ottoboni, G. Salani, M. Martinello, A. Cattalini, A. Bergami, R. Furlan, G. Comi, G. Constantin and G. Martino, *Nature*, 2005, **436**, 266–271.

3 G. L. Ming and H. J. Song, *Neuron*, 2011, **70**, 687–702.

4 C. B. Johansson, S. Momma, D. L. Clarke, M. Risling, U. Lendahl and J. Frisen, *Cell*, 1999, **96**, 25–34.

5 M. Carlen, K. Meletis, C. Goritz, V. Darsalia, E. Evergren, K. Tanigaki, M. Amendola, F. Barnabe-Heider, M. S. Y. Yeung and L. Naldini, *Nat. Neurosci.*, 2009, **12**, 259–267.

6 L. C. Fuentealba, K. Obernier and A. Alvarez-Buylla, *Cell Stem Cell*, 2012, **10**, 698–708.

7 C. N. P. Lui, Y. P. Tsui, C. T. Wu, A. S. L. Ho, D. K. Y. Shum, Y. S. Chan, H. W. Li, S. C. E. Tsang and K. K. L. Yung, *Angew. Chem., Int. Ed.*, 2013, **52**, 12298–12302.

8 P. Caravan, J. J. Ellison, T. J. McMurry and R. B. Lauffer, *Chem. Rev.*, 1999, **99**, 2293–2352.

9 P. Caravan, *Chem. Soc. Rev.*, 2006, **35**, 512–523.

10 Y.-K. Peng, C.-L. Liu, H.-C. Chen, S.-W. Chou, W.-H. Tseng, Y.-J. Tseng, C.-C. Kang, J.-K. Hsiao and P.-T. Chou, *J. Am. Chem. Soc.*, 2013, **135**, 18621–18628.

11 M. F. Kircher, U. Mahmood, R. S. King, R. Weissleder and L. Josephson, *Cancer Res.*, 2003, **63**, 8122–8125.

12 M. S. Martina, J. P. Fortin, C. Menager, O. Clement, G. Barratt, C. Grabielle-Madelmont, F. Gazeau, V. Cabuil and S. Lesieur, *J. Am. Chem. Soc.*, 2005, **127**, 10676–10685.

13 C. W. Lai, Y. H. Wang, C. H. Lai, M. J. Yang, C. Y. Chen, P. T. Chou, C. S. Chan, Y. Chi, Y. C. Chen and J. K. Hsiao, *Small*, 2008, **4**, 218–224.

14 Y. Chen, H. R. Chen, D. P. Zeng, Y. B. Tian, F. Chen, J. W. Feng and J. L. Shi, *ACS Nano*, 2010, **4**, 529–539.

15 J. E. Lee, N. Lee, H. Kim, J. Kim, S. H. Choi, J. H. Kim, T. Kim, I. C. Song, S. P. Park, W. K. Moon and T. Hyeon, *J. Am. Chem. Soc.*, 2010, **132**, 552–557.

16 C. R. Thomas, D. P. Ferris, J.-H. Lee, E. Choi, M. H. Cho, E. S. Kim, J. F. Stoddart, J.-S. Shin, J. Cheon and J. I. Zink, *J. Am. Chem. Soc.*, 2010, **132**, 10623–10625.

17 Y.-W. Jun, J.-H. Lee and J. Cheon, *Angew. Chem., Int. Ed.*, 2008, **47**, 5122–5135.

18 J. Park, K. An, Y. Hwang, J.-G. Park, H.-J. Noh, J.-Y. Kim, J.-H. Park, N.-M. Hwang and T. Hyeon, *Nat. Mater.*, 2004, **3**, 891–895.

19 Y.-S. Lin and C. L. Haynes, *Chem. Mater.*, 2009, **21**, 3979–3986.

20 Y.-K. Peng, C.-W. Lai, C.-L. Liu, H.-C. Chen, Y.-H. Hsiao, W.-L. Liu, K.-C. Tang, Y. Chi, J.-K. Hsiao, K.-E. Lim, H.-E. Liao, J.-J. Shyue and P.-T. Chou, *ACS Nano*, 2011, **5**, 4177–4187.

21 M.-L. Ho, J.-M. Hsieh, C.-W. Lai, H.-C. Peng, C.-C. Kang, I.-C. Wu, C.-H. Lai, Y.-C. Chen and P.-T. Chou, *J. Phys. Chem. C*, 2009, **113**, 1686–1693.

22 H. L. Ding, Y. X. Zhang, S. Wang, J. M. Xu, S. C. Xu and G. H. Li, *Chem. Mater.*, 2012, **24**, 4572–4580.

23 Y.-K. Peng, C.-W. Lai, Y.-H. Hsiao, K.-C. Tang and P.-T. Chou, *Mater. Express*, 2011, **1**, 136–143.
24 J.-H. Lee, Y.-M. Huh, Y.-W. Jun, J.-W. Seo, J.-T. Jang, H.-T. Song, S. Kim, E.-J. Cho, H.-G. Yoon, J.-S. Suh and J. Cheon, *Nat. Med.*, 2007, **13**, 95–99.

Faraday Discussions

PAPER

Synthesis of functionalized, dispersible carbon-coated cobalt nanoparticles for potential biomedical applications†

Quirin M. Kainz,[a] Soraia Fernandes,[a] Corina M. Eichenseer,[a] Francesca Besostri,[a] Helmut Körner,[b] Rainer Müller[c] and Oliver Reiser*[a]

Received 13th May 2014, Accepted 25th June 2014
DOI: 10.1039/c4fd00108g

Ferromagnetic carbon-coated cobalt nanoparticles have been ligated by a covalent or non-covalent strategy with dendrons or polymers bearing amino-, hydroxy- or polyether-functionalities, aiming at synthesizing materials that are dispersible in aqueous solutions, being a prerequisite for applications in biomedicine. Most efficiently for this purpose proved to be a covalently anchored polyethyleneimine (PEI) with a loading of approximately 10 mmol of amino groups per gram of hybrid nanomaterial, resulting in dispersions that are stable in water for several days.

Introduction

Nanoparticles are increasingly recognized for applications in biomedicine, either as the principle itself or as smart platforms to carry biomolecules or synthetic drugs.[1–3] Owing to their size, which can range from a few to several hundred nanometers, and shape, nanoparticles offer unique properties when interacting with a chemical or biological environment.

The most widely used nanoparticles contain a metal or metal oxide core, which is often stabilized by an outer inorganic or organic shell, such as silica, graphene, thiols, alkyl phosphonates, or fatty acids. Further modification of such nanoparticles can then be achieved by anchoring molecules to the shell of the nanoparticles, opening the way to the design of materials that can exhibit specific functions, *e.g.* as diagnostics tools,[4,5] or act as carriers for drugs.[6] Nowadays, nanoparticles can be produced from a variety of metals, offering a wide diversity of platforms to choose from, taking into account the desired properties such as a

[a]*Institut für Organische Chemie, Universität Regensburg, Universitätsstr. 31, 93053 Regensburg, Germany. E-mail: oliver.reiser@chemie.uni-regensburg.de; Fax: +49 941 9434121; Tel: +49 941 9434631*
[b]*Institut für Physik, Universität Regensburg, 93040 Regensburg, Germany*
[c]*Institut für Physikalische und Theoretische Chemie, Universität Regensburg, 93040 Regensburg, Germany*

† Electronic supplementary information (ESI) available: Procedures for the preparation of **7c**, **11** and **15**. See DOI: 10.1039/c4fd00108g

specific and uniform size, stability in various environments (air, solvents, pH), and low production costs involved, especially on a large-scale.

One additional dimension that is increasingly catching the attention of researchers in both chemistry and biomedicine is the fact that nanoparticles can be magnetic.[7,8] This property allows inimitable possibilities for their manipulation, *e.g.* controlled agitation or movement in a chemical or biological system using an external magnet.[9] Moreover, by applying external fluctuating magnetic fields, magnetic nanoparticles can be specifically heated, which is promising for applications both in chemistry and biomedicine, *e.g.* for hyperthermia.[10] Finally, the magnetic properties of nanoparticles also bear great potential for magnetic resonance imaging,[11] which allows non-invasive tracing, for example of nanoparticle–drug hybrids. Taking advantage of their superparamagnetic properties and thus their diminished agglomeration as well as their low toxicity, superparamagnetic iron oxide nanoparticles (SPIONs) have been used extensively in biomedicine and biotechnology over the past years.[12] For applications in medicinal chemistry, *e.g.* for hyperthermia by inductive heating, ferromagnetic materials are nevertheless very attractive due to their considerably elevated magnetization, provided that their high tendency to agglomerate can be overcome in order to arrive at materials that are dispersible in water.

Over the past few years we have been investigating ferromagnetic cobalt nanoparticles that contain a graphene-like carbon layer.[13] Such nanoparticles, originally invented by Stark *et al.*[14] and now being commercially available, have a number of attractive features: The metal core of these particles (20–50 nm in size) renders them highly magnetic, having a saturation magnetization of 130–160 emu g^{-1}. This allows their easy recovery, even when a heavy load such as a dendrimer or polymer is attached to these particles, which generally decreases the magnetization to levels of unfunctionalized superparamagnetic iron oxide particles (30–50 emu g^{-1}).[15] The carbon shell, having a thickness of only 1–3 nm, provides a high stability against oxygen as well as acidic and basic pH, and makes them compatible with both aqueous and organic solvents. On the other hand, due to their high magnetization as well as their hydrophobic surface these particles are prone to agglomeration, especially in water or buffer, the latter being the relevant medium for biomedical applications.

We therefore examined if one can attenuate the properties of these nanoparticles by attaching to them polar groups, aiming at achieving good dispersibility in water. Stark *et al.* recently showed that carbon-coated nanoparticles that are covalently functionalized with highly charged polymers allow the formation of stable dispersions in aqueous media.[16] We report herein various complementary synthetic strategies for the functionalization of these nanoparticles with hydroxy, amine and ammonium terminal groups, and the evaluation of such materials with respect to their dispersibility in water.

The initial functionalization of carbon-coated nanoparticles has been achieved so far by two strategies (Fig. 1): (1) in a covalent way, where the carbon shell can be reacted with phenyl radicals generated by the thermal decomposition of arene diazonium salts, which can serve as a starting point for the introduction of azide moieties for their further functionalization by azide–alkyne-cycloadditions (click reactions);[17] and (2) in a non-covalent way, by ligating polyarenes to the carbon shell utilizing the attractive π-stacking interactions with the graphene-like layer of the nanoparticle surface. This approach is particularly effective in water due to

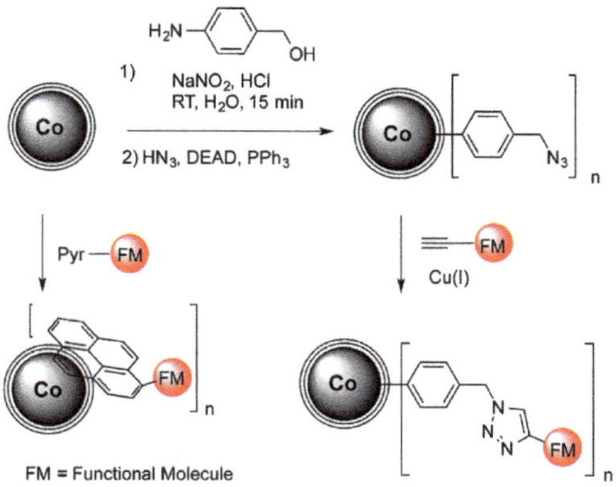

Fig. 1 Principal strategies for covalent and non-covalent attachment of functional molecules to carbon-coated iron and cobalt nanoparticles.

the hydrophobic effect, as demonstrated by the successful immobilization of metal complexes.[18,19]

We report here the application of these strategies for the synthesis of carbon-coated cobalt nanoparticles having multiple polyether, amino or hydroxy groups on their surface. In addition we disclose a new strategy for the functionalization of carbon-coated nanoparticles based on the [3 + 2]-cycloaddition of a carbonyl ylide generated from 3-phenyloxirane-2,2-dicarbonitrile (**12**). The as-prepared nanomaterials are then evaluated for their dispersibility in aqueous solutions.

Experimental

Carbon-coated cobalt nanoparticles were purchased from Turbobeads Llc (Co/C, 20.5 $m^2\ g^{-1}$, mean particle size ≈ 25 nm). Prior to use, they were washed in a concentrated HCl (Merck, puriss)/deionized water (Millipore) mixture (1 : 1) 5 times for 24 h. The acid residuals were removed by washing with Millipore water (5×) and the particles were dried at 50 °C in a vacuum oven. Pent-4-ynoic acid anhydride,[20] azide-functionalized Co/C nanoparticles (loading: 0.14 mmol g^{-1}),[17] propargyl-[G3]-$(OH)_8$ **2**,[21] propargyl-[G3]-$(NH_3{}^+TFA^-)_8$ **3**,[22] Co/C-[G3]-$(OH)_8$ **4**,[23] Co/C-[G3]-$(NH_3{}^+TFA^-)_8$ **5**,[23] 2-azidoethyl-β-D-glucopyranoside,[24] pyrene-[G3]-$(OH)_8$ **7a**,[25] pyrene-[G3]-$(alkyne)_8$ **7b**,[25] phenethylamine-functionalized carbon-coated cobalt nanoparticles **21**,[16] and aziridine **22**[26] were prepared according to literature procedures. Magnetic nanobeads were dispersed using an ultrasound bath (Sonorex RK 255 H-R, Bandelin) and recovered with the aid of a neodymium-based magnet (N48, W-12-N, Webcraft GmbH, side length 12 mm) unless indicated otherwise.

[G3]-glucose-functionalized carbon-coated cobalt nanoparticles 6

[G3]-$(OH)_8$-functionalized cobalt nanoparticles **4** (100 mg) were sonicated in 3 mL of CH_2Cl_2 for 10 min, then 1.5 mg (12 μmol) DMAP and 28 mg (160 μmol) pent-4-ynoic anhydride were added and the resulting slurry sonicated for one hour at ambient temperature. After shaking for 24 h on the laboratory shaker, the particles were washed with CH_2Cl_2 (5 × 5 mL) and dried under high vacuum, after

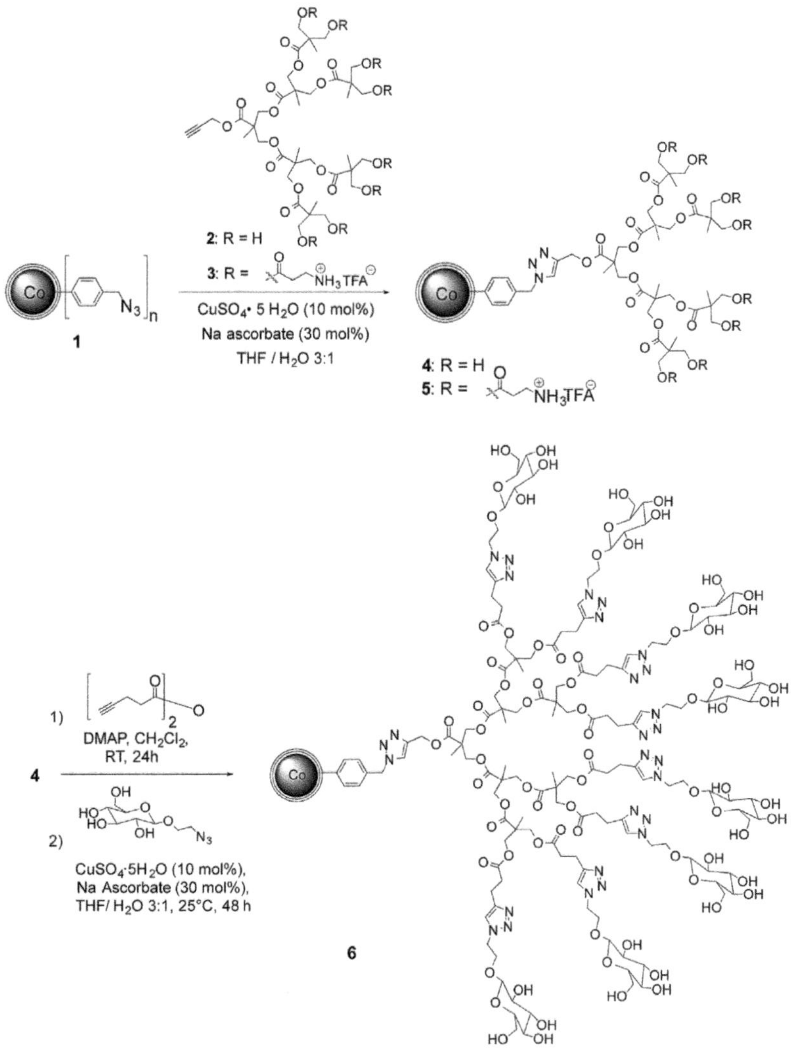

Scheme 1 Synthesis of C@Co nanoparticles **4–6** bearing covalently attached dendrimers with terminal hydroxy, ammonium and glucose groups.

which 99 mg of nanobeads could be isolated. IR (ν/cm^{-1}): 2928, 1749, 1467, 1376, 1246, 1161, 1003, 776, 652; elemental microanalysis (%): C, 9.41; H, 0.32; N, 0.38.

50 mg of these particles, 2-azidoethyl-β-D-glucopyranoside (10 mg, 40 μmol), $CuSO_4 \cdot 5H_2O$ (0.5 mg, 2 μmol) and sodium ascorbate (1.2 mg, 6 μmol) were dispersed in 3 mL of a degassed THF–H_2O 1.5 : 1 mixture by sonication for one hour. The reaction flask was put on the laboratory shaker overnight, followed by magnetic decantation and washing with acetone (5 × 5 mL) and H_2O (5 × 5 mL). Drying yielded 51 mg of functionalized nanoparticles **6**. IR (ν/cm^{-1}): 3379, 2918, 1730, 1560, 1463, 1370, 1234, 1129, 1037, 829, 756; elemental microanalysis (%): C, 10.46; H, 0.57; N, 1.06, corresponding to a loading of 0.2 mmol dendron per gram or 0.64 mmol g^{-1} of hydroxyl end groups.

General procedure (GPA) for the non-covalent functionalization of carbon-coated cobalt nanoparticles with pyrene-tagged molecules

50 mg of washed carbon-coated cobalt nanoparticles were dispersed in 5 mL H_2O by sonication (10 min) in an ultrasonic bath at ambient temperature, followed by addition of the pyrene-tagged compound and further sonication of the reaction mixture for 1 h. The nanoparticles were recovered from the reaction mixture with the aid of a magnet. For washing purposes, 5 mL of H_2O were added and the slurry was sonicated for 15 min using an ultrasonic bath. The magnetic particles were collected with the aid of a magnet and the washing procedure was repeated five times. Then the particles were dried thoroughly under high vacuum.

Pyrene-[G3]-$(OH)_8$-functionalized carbon-coated cobalt nanoparticles 8

Carbon-coated cobalt nanoparticles (50 mg) were treated according to the GPA with pyrene-[G3]-$(OH)_8$ **7a** (18.6 mg, 15 μmol) to yield 52 mg of nanoparticles **8.** DRIFTS (ν/cm^{-1}): 3478, 2981, 1751, 1526, 1425, 1235, 1063; elemental microanalysis (%): C, 14.55; H, 0.36; N, 0.52, corresponding to a loading of 0.16 mmol dendron per gram or 1.28 mmol of hydroxyl groups per gram of nanomaterial.

Pyrene-[G3]-$(NH_3^+TFA^-)_8$-functionalized carbon-coated cobalt nanoparticles 9

Carbon-coated cobalt nanoparticles (30 mg) were treated according to the GPA with pyrene-[G3]-$(NH_3^+TFA^-)_8$ **7c** (40.8 mg (15 μmol)) to yield 29.5 mg of nanoparticles **9**. IR (ν/cm^{-1}): 1765, 1687, 1537, 1197, 1085, 966, 863, 794; elemental microanalysis (%): C, 9.1; H, 0.29; N, 0.36, corresponding to a loading of 0.02 mmol dendron per gram or 0.16 mmol of ammonium groups per gram of nanomaterial.

Pyrene-[G3]-$(Glc)_8$-functionalized carbon-coated cobalt nanoparticles 10

Carbon-coated cobalt nanoparticles (15 mg) were treated according to the GPA with pyrene-[G3]-$(Glc)_8$ **7b** (30 mg, 8 μmol) to yield 16 mg of nanoparticles **10.** IR (ν/cm^{-1}): 1758, 1655, 1542, 1199, 1092, 936, 814, 652, 602; elemental microanalysis (%): C, 9.88; H, 0.38; N, 0.30, corresponding to a loading of 0.02 mmol dendron per gram or 0.64 mmol of hydroxy groups per gram of nanomaterial.

PEG_{2000}-OMe-functionalized carbon-coated nanoparticles 12

Carbon-coated cobalt nanoparticles **1** (100 mg) were pre-dispersed in 3 mL of degassed CH_2Cl_2 by sonication (10 min). Thereafter **11** (204 mg, 0.1 mmol), CuI (2 mg, 0.01 mmol) and Et_3N (14 μL, 0.1 mmol) were added. After sonication for 1 h at ambient temperature, the reaction mixture was placed on the laboratory shaker for 24 h. The magnetic nanoparticles were recovered using a magnet and washed with acetone (5 × 5 mL) and H_2O (5 × 5 mL) as described above. Then, the particles were separated magnetically and dried under high vacuum (yield: 105 mg). IR (ν/cm^{-1}): 2919, 2866, 1651, 1462, 1352, 1253, 1113, 844; elemental microanalysis (%): C, 8.1; H, 0.47; N, 0.18, corresponding to a loading of 0.02 mmol PEG_{2000}-OMe per gram or 0.90 mmol of ether groups per gram of nanomaterial.

PEG_{2000}-OMe-functionalized carbon-coated nanoparticles 16

Pyrene-PEG_{2000}-OMe **14** (58 mg, 25 μmol) was added to a dispersion of 50 mg carbon-coated cobalt nanoparticles in 5 mL H_2O and the resulting mixture was sonicated for 1 h at ambient temperature using an ultrasonic bath or alternatively for 5 min using a tip sonicator (Branson Sonic Power J-17A, level 4) at 0 °C. After washing with water (5 × 5 mL) as described above 51 mg of particles **16** were recovered. DRIFTS (ν/cm^{-1}): 2968, 2922, 1750, 1510, 1159, 853; elemental microanalysis (%): C, 15.14; H, 0.12; N, 0.58, corresponding to a loading of 0.02 mmol PEG_{2000}-OMe per gram or 0.90 mmol of ether groups per gram of nanomaterial.

Amine-functionalized carbon-coated cobalt nanoparticles 19

In a microwave vial 200 mg Co/C nanoparticles were mixed with 3-phenyloxirane-2,2-dicarbonitrile **17** (100 mg, 0.59 mmol). The sealed tube was heated in the microwave to 120 °C for 4 min. After cooling down the particles were recovered with the aid of a magnet and washed with ethyl acetate (5 × 3 mL) and diethyl ether (3 × 2 mL). After evaporation of the solvent 204 mg of the particles **18** were obtained. IR (ν/cm^{-1}): 2367, 2361, 2338. Elemental microanalysis (%): C, 10.72; H, 0.25; N, 1.15, corresponding to a loading of 0.41 mmol tetrahydrofuran per gram or 0.82 mmol nitrile groups per gram of nanomaterial.

Under N_2 a Schlenk flask was charged with 8.5 mL THF and 320 μL TMSCl (2.56 mmol, 18 equiv.). While stirring $NaBH_4$ (96.8 mg. 2.56 mmol, 18 equiv.) was added. The resulting solution was heated at 80 °C for 4 h. After cooling down to room temperature 160 mg of **18** (0.166 mmol nitrile groups, 1 equiv.) were added and the reaction mixture was heated at 80 °C for 24 h. The nanobeads were washed with THF (3 × 2 mL) and diethyl ether (5 × 3 mL), dried under vacuum to yield 157 mg of the desired Co/C-epoxy-NH_2 nanobeads **19**.

Polyethyleneimine (PEI)-functionalized carbon-coated cobalt nanoparticles 23

Amino-functionalized carbon-coated cobalt nanoparticles **21** (100 mg, 0.015 mmol, 1.0 equiv.) were pre-dispersed in 10 mL DCM using an ultrasonic bath for 15 min. Under stirring, aziridine **22** (778 μL, 15 mmol, 1000 equiv.) and HCl (15 μL, 15 M) were added to the reaction mixture, which then was heated to 80 °C for 24 h. The nanobeads were collected using an external magnet, washed with DCM (2 × 50 mL), H_2O (5 × 50 mL) and again DCM (3 × 50 mL) and dried under vacuum at 50 °C to yield 284 mg PEI-functionalized particles **23**. IR (ν/cm^{-1}): 3417, 2934, 2821, 2362, 1648, 1604, 1458, 1351, 1298, 1014. Elemental microanalysis [%]: C, 28.69; H, 5.30; N, 13.27, corresponding to a loading of 9.5 mmol amine per gram of nanomaterial. TGA (N_2): 59.7 wt% polymer.

Results and discussion

Azide-functionalized Co@C nanoparticles **1** (azide loading 0.14 mmol g^{-1}) were readily prepared from the corresponding unfunctionalized nanoparticles by covalently grafting benzyl alcohol onto the particles *via* diazonium chemistry followed by hydroxy azide exchange (*cf.* Fig. 1).[17] Dendrimers **2**[21] and **3**[22] having an acetylene linker at the focal point and hydroxy or ammonium groups as terminal groups were covalently ligated with **1a** *via* a copper-catalyzed alkyne–azide

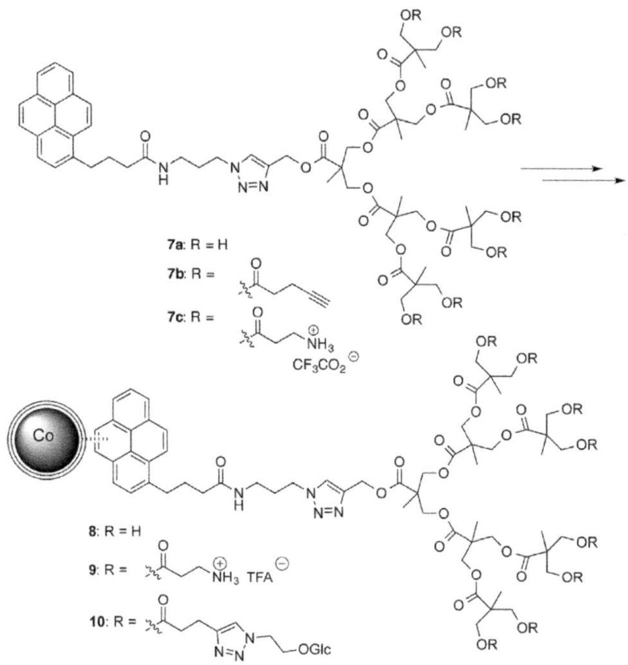

Scheme 2 Synthesis of C@Co-coated nanoparticles bearing non-covalently attached dendrimers with hydroxy, ammonium and glucose end groups.

cycloaddition to give rise to **4** and **5**, as previously reported by us.[23] Nanoparticles **4** were esterified with pent-4-ynoic acid anhydride,[20] which allowed the decoration of the nanoparticles with glucose end groups by performing a second copper-catalyzed alkyne–azide cycloaddition, this time with 2-azidoethyl-β-D-glucopyranoside[24] to give rise to **6** (Scheme 1).

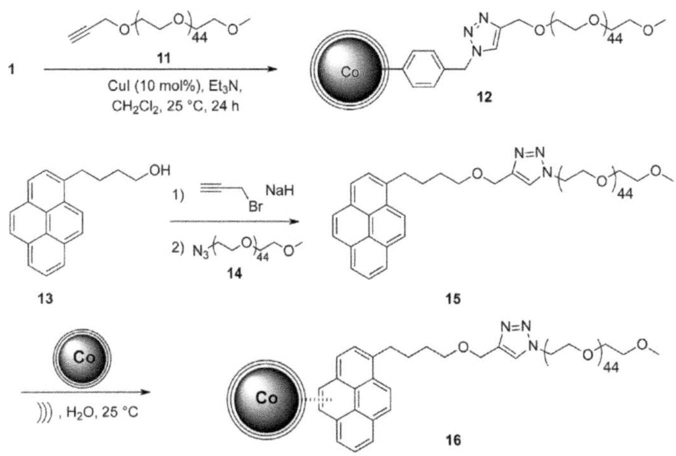

Scheme 3 Synthesis of PEG_{2000}-functionalized C/Co-nanoparticles **12** and **16**.

Alternatively, we synthesized analogues of **4–6** in which the dendrons were non-covalently attached to nanoparticles, utilizing π–π-interactions between their carbon shell and the pyrene moieties. Thus, starting from the pyrene-tagged dendron **7**, its functionalization with terminal ammonium groups was achieved by coupling with *N*-Boc-3-aminopropionic acid anhydride followed by Boc-deprotection. Likewise, terminal glucose groups could be attached to **7** following the same strategy as applied for the transformation of **4** to **6**. The as-prepared pyrene-tagged dendrons were immersed together with pristine carbon-coated iron or cobalt nanoparticles in water. For every reaction, 2.5 equiv. of pyrene-tagged molecules were used in respect to the estimated maximum loading (0.2 mmol g^{-1}) that is possible by the non-covalent ligation strategy.[27] After sonication for 1 h with the use of an ultrasonication bath at ambient temperature, the particles were collected uaing a magnet, yielding non-covalently functionalized Co/C **8–10** nanoparticles after washing (Scheme 2). The loading that was obtained for **8–10** in these experiments was quite different but seems to correlate well with steric and electronic factors reflecting the types of end groups attached. The neutral and sterically undemanding **7a** allowed the coverage of the nanoparticle surface with dendrons at 0.16 mmol g^{-1}, whereas **7b**, due to its considerably higher steric bulk, and **7c**, due to the repulsion of positive charges of the end groups, resulted in a substantially decreased loading of 0.02 mmol dendron per gram.

Additionally, we also synthesized poly(ethylene glycol) (PEG)-coated C/Co nanoparticles **12** and **16** by both strategies as described above (Scheme 3). Poly(ethylene glycol) (PEG) is known as a highly hydrophilic and biocompatible polymer. Surface modification using such highly hydrophilic polymers reduces

Table 1 Dispersibility of functionalized carbon-coated nanoparticles in water

Nanoparticle		Functionalization	Loading of dendron/polymer (mmol g^{-1})	Functional group loading (mmol g^{-1})	Dispersion stability in water (min)
Covalent dendrimer	C@Co	Carbon shell	N/A	N/A	0
	4	[G3]-$(OH)_8$	0.04	0.32	15
	5	[G3]-$(NH_3^+TFA^-)_8$	0.02	0.16	30
	6	[G3]-$(Glc)_8$	0.02	0.64	30
Non-covalent dendrimer	**8**	Pyrene-[G3]-$(OH)_8$	0.16	1.28	15
	9	Pyrene-[G3]-$(NH_3^+TFA^-)_8$	0.02	0.16	15
	10	Pyrene-[G3]-$(Glc)_8$	0.02	0.64	30
Covalent PEG	**12**	PEG_{2000}-OMe	0.02	0.90	60
Non-covalent PEG	**16**	PEG_{2000}-OMe	0.02	0.90	30
Covalent furan	**19**	Tetrahydrofuran-$(CH_2NH_2)_2$	0.41	0.82	<15
Covalent PEI	**23**	$[RN(CH_2CH_2)NH_2]_n$	0.02	10	>1200

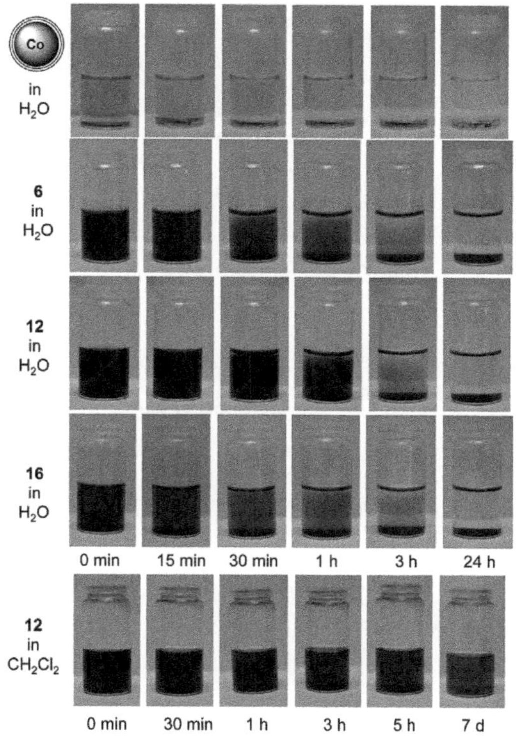

Fig. 2 Dispersibility of unfunctionalized carbon-coated nanoparticles and selected functionalized nanoparticles having the same loading of dendrimers or PEG_{2000}-OMe and overall similar number of functional groups. Covalently bound glucose dendrimers **6** proved less effective than covalently bound PEG_{2000}-OMe **12**, which was also superior to non-covalently bound PEG_{2000}-OMe *via* π-stacking through pyrene linkers **16**. The dispersibility of **12** was by far superior in dichloromethane than in water.

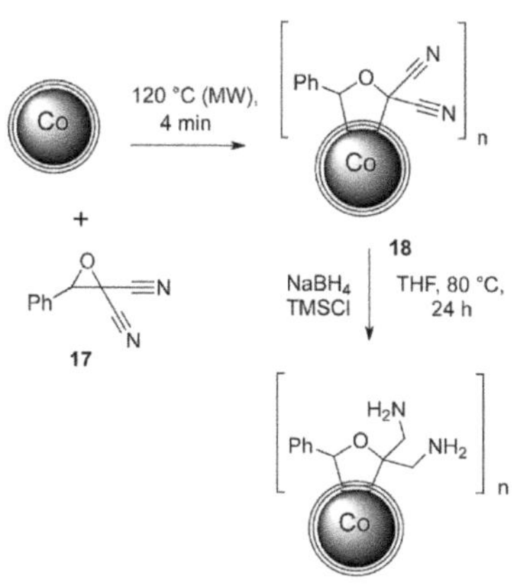

Scheme 4 Functionalization of carbon-coated nanoparticles *via* [3 + 2]-cycloaddition with benzylidenmalononitrile epoxide (**19**).

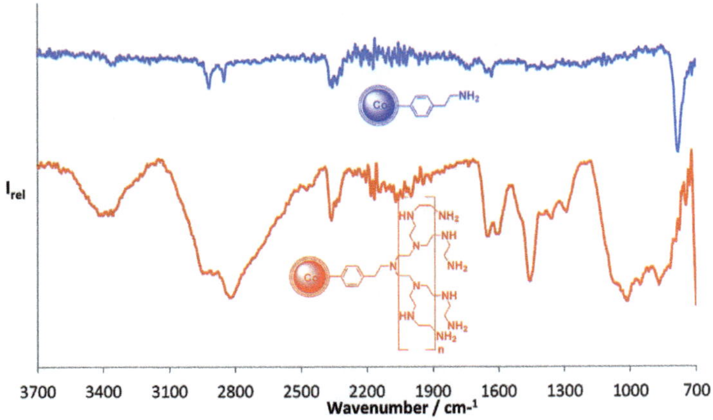

Fig. 3 Structure confirmation by IR spectra of C/Co nanoparticles functionalized with phenethylamine **21** (top) and polyethyleneimine (PEI) **23** (bottom).

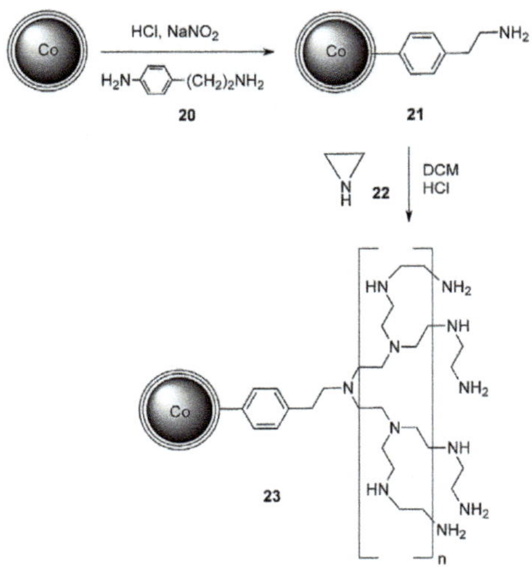

Scheme 5 Synthesis of PEI@C@Co nanoparticles.

protein adsorption and decreases platelet and cell adhesion.[28–30] Studies have shown that an average molecular weight of 2000 g mol^{-1} is enough for this cell-repellent effect.[31] PEG features some unique properties, such as excluded volume on the surface and flexible hydrophilic chain motion against proteins and cells, along with its non-toxicity and non-immunogenicity.[32,33] Therefore, PEG-based coatings have already been used to stabilize dispersions of CNTs,[34] as well as gold[35] and magnetite[36] nanoparticles with excellent results.

The nanoparticles thus obtained were tested for their dispersibility in water (Table 1 and Fig. 2), comparing the type of functional groups attached, and the strategy used for the functionalization of the nanoparticles (covalent or non-covalent ligation).

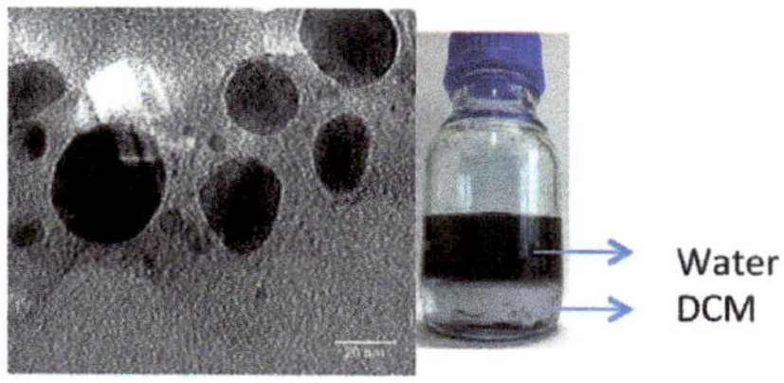

Fig. 4 TEM of PEI-functionalized carbon-coated cobalt particles **23** (left) and dispersibility in water and between water and dichloromethane (right). The apparent separation of the nanoparticles **23** suggests that the PEI has been reliably grown on the carbon shell of the nanoparticles.

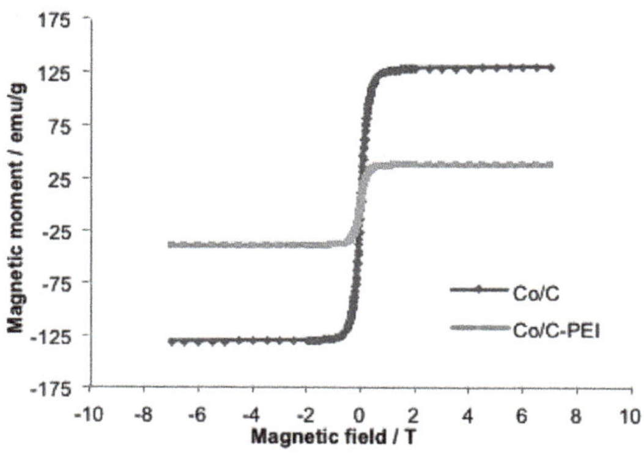

Fig. 5 Magnetization (SQUID measurement at room temp.) for both Co/C and Co/C-PEI **23** particles.

Comparing **4** and **5**, which have approximately the same number of terminal groups attached to each nanoparticle,[37] revealed that the positively charged ammonium groups conferred a better dispersibility than hydroxy groups. However, these differences are small, and nanoparticles **6** which already had an increased number of hydroxy groups due to the terminating glucose moieties led to equal, and in the case of the non-covalent functionalization (**9** *vs.* **10**) to even better dispersibility. Comparing the same terminal groups, in general the non-covalent ligation strategy led to decreased dispersibility of the nanoparticles. Differences were especially pronounced for the PEG-modified nanoparticles **12** and **16**, for which the former could be considerably better dispersed in water than the latter despite the same degree of loading (0.02 mmol PEG_{2000}-OMe per gram nanomaterial) and magnetization (**12:** 134 emu g^{-1}; **16:** 131 emu g^{-1}). Thus, from all of the nanoparticles evaluated at this point, **12** showed the best dispersibility but its stability was nevertheless still moderate in water (60 min). In dichloromethane, dispersions of **12** were stable for a several days.

These results clearly point to the necessity for an exceptionally high loading of polar functional groups to break up the high tendency for agglomeration of the cobalt nanoparticles. We therefore investigated two different approaches to arrive at highly amine-loaded carbon@metal nanoparticles. Cravotto *et al.* recently reported the functionalization of single-walled carbon nanotubes through 1,3-

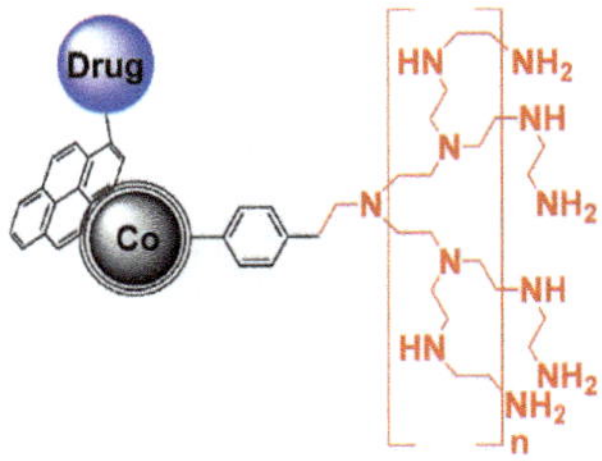

Fig. 6 Potential orthogonal functionalization of carbon-coated nanoparticles.

cycloaddition of the carbonyl ylide obtained by thermal ring opening from the benzylidenmalononitrile epoxide (**17**).[38] We were pleased to see that this methodology was applicable to carbon-coated cobalt nanoparticles, giving rise to **18** with a loading based on an attached **17** of 0.41 mmol g^{-1}, which is about 10–20 times higher than the functionalization on the surface of the carbon shell that was achieved by tagging with phenyl radicals or pyrene moieties described above (Scheme 4). The reduction of the nitrile groups by trimethylsilane resulted in a quantitative conversion to yield **19** as judged by IR spectroscopy as used to monitor the vanishing nitrile band. Unfortunately, despite a loading of those nanoparticles of around 0.8 mmol g^{-1} amine functionalities and a significantly reduced magnetization (104 emu g^{-1}), their dispersibility in water was much lower than for the particles functionalized with dendrimers or PEG_{2000}-OMe, which might be due to the close proximity of the amine groups to the carbon shell.

We therefore attempted to graft a polyethyleneimine (PEI) onto the nanoparticles by polymerizing aziridine **22** with an appropriate nanoparticle-bound amine precursor (Fig. 3). For our purposes this best turned out to be the phenylethyleneamine-functionalized particles **21** that could be obtained by reacting carbon-coated nanoparticles and the diazonium salt generated from the aniline **20**. Using 1000 equivalents of aziridine **22**, PEI@C@Co nanoparticles **23** were obtained (Scheme 5) with a loading of 10–14 mmol amino functionalities per gram of nanomaterial based on the nitrogen content determined by elemental analysis or by TGA measurements.

TEM analysis (Fig. 4) showed that **23** consisted of discrete particles with no indication that cross-linking between the particles had taken place. The magnetization of **23** was considerably reduced from 128 to 39 emu g^{-1} compared to the magnetization of the particles before polymerization (Fig. 5), which is consistent with the high mass proportion of the PEI-polymer (60 wt% by TGA). Indeed, this nanomaterial proved to form stable dispersions in water over days with no tendency for agglomeration or sedimentation, even in a biphasic system with dichloromethane present as a second layer, contrasting the PEG-functionalized C/Co particles **12** described above that formed much more stable dispersions in dichloromethane compared to water. The particles can be collected using a magnet and redispersed as long as they stay wet; however, once the solvent had been removed and the nanoparticles dried under vacuum, dispersing **23** again in water or acid was not possible, for reasons that are not yet understood and which are under investigation in our laboratory.

Conclusions

In conclusion, we have been able to functionalize carbon-coated cobalt nanoparticles with polyhydroxy-, polyamine- or PEG_{2000}-functionalized dendrons or polymers either through a covalent or non-covalent π-stacking ligation strategy. Due to the high magnetization and the non-polar shell of the parent material, a high loading with polar groups is necessary to generate dispersions that stable in water, a prerequisite for biomedical applications. Dendrons with a loading of up to 1.3 mmol g^{-1} amino or hydroxyl groups being displayed on the surface of the particles turned out to be insufficient to break-up the agglomeration of the highly magnetic core material. However, a covalently attached polyethyleneimine (PEI) polymer with a loading of approximately 10–14 mmol amino groups per gram of nanomaterial resulted in nanoparticles that could form stable dispersions in water. This material might offer the possibility of being further developed as a carrier for biologically relevant molecules by attaching them through the orthogonal, non-covalent functionalization strategy *via* π-stacking either before or after generating the PEI-polymer on the nanoparticles by covalent ligation (Fig. 6).

Acknowledgements

This work was supported by PITN-GA-2012 290248_Magneticfun, the Studienstiftung des Deutschen Volkes and the Hans Seidel Stiftung.

References

1 T. L. Doane and C. Burda, *Chem. Soc. Rev.*, 2012, **41**, 2885.
2 M. Elsabahy and K. L. Wooley, *Chem. Soc. Rev.*, 2012, **41**, 2545.
3 E. C. Dreaden, A. M. Alkilany, X. Huang, C. J. Murphy and M. A. El-Sayed, *Chem. Soc. Rev.*, 2012, **41**, 2740.
4 M. Swierczewska, G. Liu, S. Lee and X. Chen, *Chem. Soc. Rev.*, 2012, **41**, 2641.
5 M. Perfezou, A. Turner and A. Merkoçi, *Chem. Soc. Rev.*, 2012, **41**, 2606.
6 Z. Li, J. C. Barnes, A. Bosoy, J. F. Stoddart and J. I. Zink, *Chem. Soc. Rev.*, 2012, **41**, 2590.
7 A. Akbarzadeh, M. Samiei and S. Davaran, *Nanoscale Res. Lett.*, 2012, **7**, 144.
8 J. Gao, H. Gu and B. Xu, *Acc. Chem. Res.*, 2009, **42**, 1097.
9 Y. Pan, X. Du, F. Zhao and B. Xu, *Chem. Soc. Rev.*, 2012, **41**, 2912.
10 M. Bañobre-López, A. Teijeiro and J. Rivas, *Rep. Pract. Oncol. Radiother.*, 2013, **18**, 397.
11 N. Lee and T. Hyeon, *Chem. Soc. Rev.*, 2012, **41**, 2575.
12 M. Hofmann-Amtenbrink, B. von Rechenberg and H. Hofmann, in *Nanostructured Materials for Biomedical Applications*, ed. M. C. Tan, Transworld Research Network, 2009, p. 119.
13 Q. M. Kainz and O. Reiser, *Acc. Chem. Res.*, 2014, **47**, 667.
14 R. N. Grass, E. K. Athanassiou and W. J. Stark, *Angew. Chem., Int. Ed.*, 2007, **46**, 4909.
15 A.-H. Lu, E. L. Salabas and F. Schüth, *Angew. Chem., Int. Ed.*, 2007, **46**, 1222.
16 M. Zeltner, R. N. Grass, A. Schaetz, S. B. Bubenhofer, N. A. Luechinger and W. J. Stark, *J. Mater. Chem.*, 2012, **22**, 12064.

17 A. Schätz, R. N. Grass, W. J. Stark and O. Reiser, *Chem.–Eur. J.*, 2008, **14**, 8262.
18 S. Wittmann, A. Schätz, R. N. Grass, W. J. Stark and O. Reiser, *Angew. Chem., Int. Ed.*, 2010, **49**, 1867.
19 M. Keller, V. Collière, O. Reiser, A.-M. Caminade, J.-P. Majoral and A. Ouali, *Angew. Chem., Int. Ed.*, 2013, **52**, 3626.
20 M. Malkoch, K. Schleicher, E. Drockenmuller, C. J. Hawker, T. P. Russell, P. Wu and V. V. Fokin, *Macromolecules*, 2005, **38**, 3663.
21 P. Wu, M. Malkoch, J. N. Hunt, R. Vestberg, E. Kaltgrad, M. G. Finn, V. V. Fokin, K. B. Sharpless and C. J. Hawker, *Chem. Commun.*, 2005, 5775.
22 B. Li, A. L. Martina and R. E. Gillies, *Chem. Commun.*, 2007, 5217.
23 Q. M. Kainz, A. Schatz, A. Zopfl, W. J. Stark and O. Reiser, *Chem. Mater.*, 2011, **23**, 3606.
24 X.-L. Sun, C. Haller, X. Wu, V. P. Conticello and E. L. Chaikof, *J. Proteome Res.*, 2005, **4**, 2355.
25 P. Wu, X. Chen, N. Hu, U. C. Tam, O. Blixt, A. Zettl and C. R. Bertozzi, *Angew. Chem., Int. Ed.*, 2008, **47**, 5022.
26 H. Wenker, *J. Am. Chem. Soc.*, 1935, **57**, 2328.
27 The maximum loading of the carbon surface of the nanoparticles with pyrene molecules was estimated by grafting 4-nitrophenyl-4-(pyren-1-yl)butanoate onto C/Co nanoparticles. After isolation of the nanoparticles, the nitrophenol ester was hydrolyzed by stirring in 10% NaOH for 24 h. By comparing the UV/Vis absorbance at 402 nm with reference solutions a loading of 0.2 mmol g^{-1} was determined.
28 H.-S. Shin, K. Park, J. H. Kim, J.-J. Kim and D. K. Han, *J. Bioact. Compat. Polym.*, 2009, **24**, 316.
29 F. Zhang, E. T. Kang, K. G. Neoh and W. Huang, *J. Biomater. Sci., Polym. Ed.*, 2001, **12**, 515.
30 D. K. Han and J. A. Hubbell, *Macromolecules*, 1997, **30**, 6077.
31 D. Bozukova, C. Pagnoulle, M.-C. De Pauw-Gillet, S. Desbief, R. Lazzaroni, N. Ruth, R. Jérôme and C. Jérôme, *Biomacromolecules*, 2007, **8**, 2379.
32 M. Amiji and K. Park, *J. Biomater. Sci., Polym. Ed.*, 1993, **4**, 217.
33 F. Fuerteges and A. Abuchowski, *J. Controlled Release*, 1990, **11**, 139.
34 J. Liu, O. Bibari, P. Mailley, J. Dijon, E. Rouviere, F. Sauter-Starace, P. Caillat, F. Vinet and G. Marchand, *New J. Chem.*, 2009, **33**, 1017.
35 H. Furusho, K. Kitano, S. Hamaguchi and Y. Nagasaki, *Chem. Mater.*, 2009, **21**, 3526.
36 N. Kohler, C. Sun, A. Fichtenholtz, J. Gunn, C. Fang and M. Zhang, *Small*, 2006, **2**, 785.
37 The lower loading of **5** in comparison with **4** is due to the increased molecular weight of the molecules attached to the nanoparticles; however, as judged by the completely disappeared IR band of the azide, TGA and elemental analysis, conversion of **3** to both **4** and **5** proceeded quantitatively, thus the number of terminal groups for each nanoparticle remains approximately the same.
38 S. Tagliapietra, G. Cravotto, E. Gaudino, S. Visentin and V. Mussi, *Synlett*, 2012, **23**, 1459; for the underlying synthetic process, see: G. Bentabed, A. Derdour and H. Benhaoua, *Synth. Commun.*, 2003, **33**, 1861.

Faraday Discussions

PAPER

Prolonging the circulatory retention of SPIONs using dextran sulfate: *in vivo* tracking achieved by functionalisation with near-infrared dyes

Maha R. A. Abdollah,[a] Tammy Kalber,[b] Berend Tolner,[a] Paul Southern,[c] Joseph C. Bear,[d] Mathew Robson,[a] R. Barbara Pedley,[a] Ivan P. Parkin,[d] Quentin A. Pankhurst,[c] Paul Mulholland[a] and Kerry Chester*[a]

Received 20th May 2014, Accepted 25th June 2014
DOI: 10.1039/c4fd00114a

The rapid reticuloendothelial system (RES) mediated clearance of superparamagnetic iron oxide nanoparticles (SPIONs) from circulation is considered a major limitation of their clinical utility. We aimed to address this by using dextran sulfate 500 (DSO4 500), a Kupffer cell blocking agent, to prolong SPIONs circulatory time. Blood concentrations of SPIONs are difficult to quantify due to the presence of haemoglobin. We therefore developed methods to functionalise SPIONs with near-infrared (NIR) dyes in order to trace their biodistribution. Two SPIONs were investigated: Nanomag®-D-spio-NH_2 and Ferucarbotran. Nanomag®-D-spio-NH_2 was functionalised using NHS (*N*-hydroxysuccinimide) ester NIR dye and Ferucarbotran was labelled using periodate oxidation followed by reductive amination or a combination of EDC (ethyl(dimethylaminopropyl) carbodiimide)/NHS and click chemistries. Stability after conjugation was confirmed by dynamic light scattering (DLS), superconducting quantum interference device (SQUID) and transmission electron microscopy (TEM). *In vivo* experiments with the functionalised SPIONs showed a significant improvement in SPIONs blood concentrations in mice pre-treated with dextran sulfate sodium salt 500 (DSO4 500).

Introduction

Superparamagnetic iron oxide nanoparticles (SPIONs) provide a versatile platform for biomedical applications[1] due to their inducible magnetic properties (superparamagnetism).[2] Potential applications include magnetic resonance

[a]*UCL Cancer Institute, University College London (UCL), Paul O'Gorman Building, 72 Huntley Street, London, WC1E 6BT, UK. E-mail: k.chester@ucl.ac.uk; Tel: +44 20 7679 0700*
[b]*Centre for Advanced Biomedical Imaging, UCL, London, UK*
[c]*UCL Healthcare and Biomagnetics Laboratory, The Royal Institution of Great Britain, London, UK*
[d]*Department of Chemistry, UCL, London, UK*

imaging (MRI),[3–7] hyperthermia,[8–11] drug delivery,[12–14] stem cell labelling and tracking,[15–17] tissue repair and cell separation.[12] To date, the most established clinical use of SPIONs is for MRI contrast enhancement to detect focal liver and spleen cancerous lesions with SPIONs such as Endorem® (Ferridex® in the US) and Resovist® (Ferucarbotran).[6,18]

When administered intravenously, SPIONs are rapidly cleared from the circulation by the reticuloendothelial system (RES),[19] with uptake dependent on factors such as the hydrodynamic diameter, electrical charge and surface coating.[20] Thus, in patients, approximately 80% of the injected dose of Ferucarbotran rapidly accumulates in the liver within minutes, resulting in a distribution half-life of only 3.9 to 5.8 minutes and an elimination half-life of 2.4–3.6 hours.[6] The preferential uptake of SPIONs by the liver and spleen RES is the property exploited for MRI, because it reveals liver tumour deposits as negative images.[6] However, the rapid clearance of SPIONs from circulation is considered to be one of the main challenges for applications other than those dependent on RES targeting. We hypothesized that the uptake of SPIONs by macrophages could be blocked by DSO4 500, enabling further development of SPIONs and expanding their clinical potential for exciting new applications, such as magnetic hyperthermia in cancer.[9,11]

We aimed to investigate the possibility of using DSO4 500 to block uptake of Ferucarbotran by RES *in vivo* and thus increase its blood residence time. This would necessitate tracing the blood concentrations of SPIONs, which can be difficult because haemoglobin interferes with iron quantification assays. Alternative methods, such as incorporation of radioactive iron isotopes (*e.g.* ^{59}Fe) during the synthesis process of SPIONs,[21] require bespoke manufacture. Other approaches utilise the ability of SPIONs to shorten the proton relaxation time of water by MR[3] or magnetic susceptibility measurements (MSM) *via* a magneto-susceptometer.[22]

We designed a new method to trace blood concentrations of SPIONs by labelling them with near-infrared (NIR) dyes. NIR labelling has been used extensively for non-invasive *in vivo* imaging.[23] The molar extinction coefficients of haemoglobin and oxyhaemoglobin are at their lowest in the NIR range, especially around wavelengths of 800 nm,[24] which is predicted to give the least background if NIR dye-labelled SPIONs are to be measured in whole blood.

Two dextran-coated SPIONs were investigated: Ferucarbotran and Nanomag®-D-spio amine. Nanomag®-D-spio amine are dextran-coated, amine-functionalised nanoparticles with hydrodynamic diameter in the range of 50–120 nm.[25] The presence of amine (NH_2) groups on the SPIONs surface is ideal for conjugation to NHS ester dyes, therefore Nanomag®-D-spio amine was first investigated as a proof-of-concept. However, our main SPION of interest is Ferucarbotran, the active ingredient of Resovist® (SHU 555A), a clinical MRI contrast agent[6] previously marketed by Schering AG, Berlin, Germany and currently produced by Fujifilm RI Pharma Co., Tokyo, Japan. Resovist® has an average hydrodynamic diameter of 60 nm and consists of a mixture of magnetite-Fe_3O_4 and maghemite-γFe_2O_3 crystals embedded in a biocompatible carboxydextran shell that provides stability to the particles, both in solution and *in vivo*.[2,6] The well-documented safety profile of Ferucarbotran in pre-clinical studies[4] and in patients[3,6,26–29] make it an attractive agent for development for therapeutic use. In particular, when placed in an alternating magnetic field, Ferucarbotran can be induced to heat by a

Table 1 List of nanoparticles investigated

SPION	Supplier	Size range	Coating	Functional groups	Iron concentration
Ferucarbotran (FC)	Meito Sangyo Co. LTD., Japan	45–65 nm	Carboxydextran	OH, COOH (hydroxyl and carboxylic)	55.6–56.2 mg ml^{-1}
Nanomag®-D-spio-amine (NM)	Micromod Partikeltechnologie GmbH, Germany	50–120 nm	Dextran	NH_2 (amine)	2.4 mg ml^{-1}

process known as magnetic field hyperthermia. Temperatures as high as 72.4 °C *in vitro* and 45–59.9 °C *in vivo* have been achieved,[30–34] indicating that Ferucarbotran has potential for localised hyperthermic cancer treatment. The conjugation of carboxydextran-coated Ferucarbotran has proved to be challenging and thus a combination of EDC/NHS and click chemistry was devised in order to functionalise Ferucarbotran with a NIR dye.

Materials and methods

All chemicals were purchased from Sigma Aldrich unless otherwise specified.

SPIONs

Details of the SPIONs investigated, as stated by the manufacturers, are shown in Table 1 and a schematic presentation of a SPION is shown in Fig. 1.

Characterisation of SPIONs

DLS: SPIONs were characterised before and after functionalisation using a Zetasizer Nano ZS90 (Malvern, Worcestershire, UK) to identify their hydrodynamic diameter and zeta potential. SPIONs solutions were prepared in sterile-filtered 5 mM NaCl.

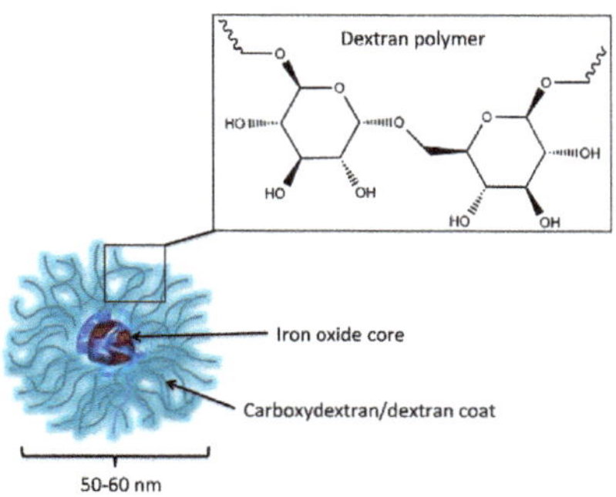

Fig. 1 Schematic presentation of a SPION illustrating the iron oxide core embedded in dextran (Nanomag®-D-spio-amine) or carboxydextran (Ferucarbotran) coating. Inset: the chemical structure of dextran polymer, which is made up of repeated glucose unites.

Table 2 List of near infrared (NIR) dyes used

Product	Supplier	Functional group	Reactivity
DyLight® 800 NHS ester	Thermo Scientific, Pierce Biotechnology	NHS ester	Amine
iFluor™ 780 amine	Stratech Scientific Ltd.	Amine	Carbonyl (aldehydes)
IRDye® 800 CW azide	LI-COR Biosciences	Azide	Copper-free click reaction with dibenzocyclooctyne (DBCO)

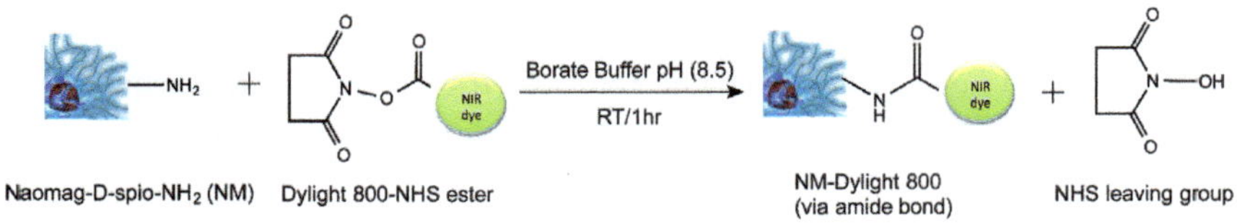

Fig. 2 Schematic presentation of the chemical reaction between NM and DyLight® 800 NHS ester dye. The NHS ester attacks the amine groups on the surface of the SPIONs to form a stable amide bond. The reaction was allowed to proceed for 1 hour at RT.

TEM: SPIONs samples were diluted in distilled water and visualised with a Jeol 2100 HRTEM with a LaB_6 source operating at an acceleration voltage of 200 kV with an Oxford Instrument UTW EDX detector running AZTEC software. Micrographs were taken in a Gatan-Orius charge coupled device (CCD). Core sizes were measured from TEM images using ImageJ software.

SQUID: Samples were prepared for SQUID magnetometry by pipetting 5 µl of solution into a polycarbonate powder capsule filled with cotton wool to absorb the liquid. The sample was then left to dry for 2 hours before mounting onto a brass SQUID-VSM rod. A Quantum Design MPMS SQUID VSM Evercool system (Quantum Design, San Diego, USA) was used to measure the magnetic moment as a function of field between ±7 T at 300 K at atmospheric pressure.

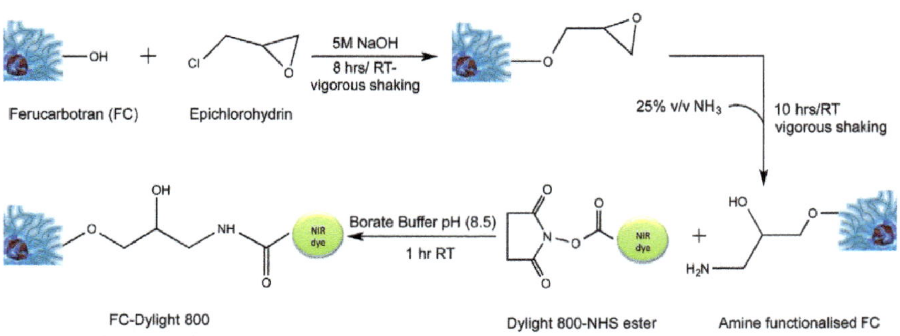

Fig. 3 Schematic presentation of the amination of FC using epichlorohydrin followed by reaction with DyLight® 800 NHS ester NIR dye.

NIR dyes

Three NIR dyes were used to functionalise the SPIONs (Table 2).

RES blockers

DSO4 500 was used as an RES blocker in doses ranging from 3 to 30 mg kg^{-1}. DSO4 500 has an average molecular weight of >500 000 Da (Sigma Aldrich data sheet, product code: D6001).

Conjugation of NM to DyLight® 800 NHS ester dye (Fig. 2): The particles were purchased as a suspension in water, which is incompatible with the labelling protocol. Consequently, water was exchanged with 50 mM sodium borate buffer (pH 8.5) using a PD-10 desalting column (GE Healthcare, UK).

500 µl of 50 mM sodium borate buffer was added to 50 µg of the lyophilized dye. The resultant dye solution was subsequently added to 1 ml of SPIONs (2.4 mg Fe) solution. The reaction was allowed to proceed for 1 hour at room temperature (RT) on a roto-torque. The solution was purified by applying the sample to a PD-10 column equilibrated with PBS.

Conjugation of FC to DyLight® 800 NHS ester using epichlorohydrin (Fig. 3): The protocol was derived from the procedures described by Pittet *et al.* 2006[35] and Maxwell *et al.* 2007.[36] In a conical flask, 200 µl of FC (11 mg Fe) was diluted with 800 µl of distilled water. 1.6 ml of 5 M sodium hydroxide (NaOH) was then added and the solution was allowed to react for 15 minutes at RT with vigorous agitation on a shaking platform, before 0.7 ml of epichlorohydrin solution was added and the solution was left to react on the shaking platform at RT for 8 hours. Next, 0.6 ml of 25% v/v of ammonia solution was added and the solution was left at RT for 10 hours while shaking. Finally, the solution was extensively dialysed against distilled water to remove excess epichlorohydrin using a dialysis cassette with 20K

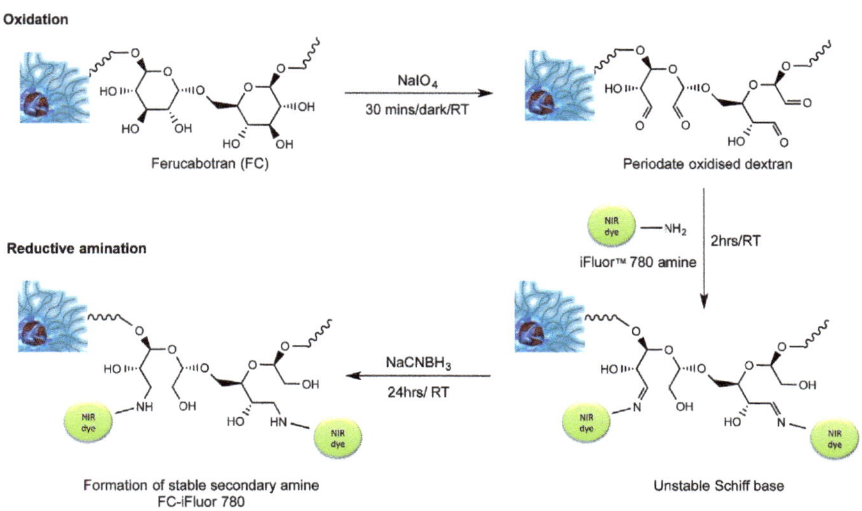

Fig. 4 A schematic presentation of the 2-step reaction to conjugate FC to iFluor 780 amine. The reaction has two main steps: first, oxidation of the hydroxyl groups on the dextran coat of FC with sodium periodate, followed by reductive amination using $NaCNBH_3$ to form a stable amine conjugate.

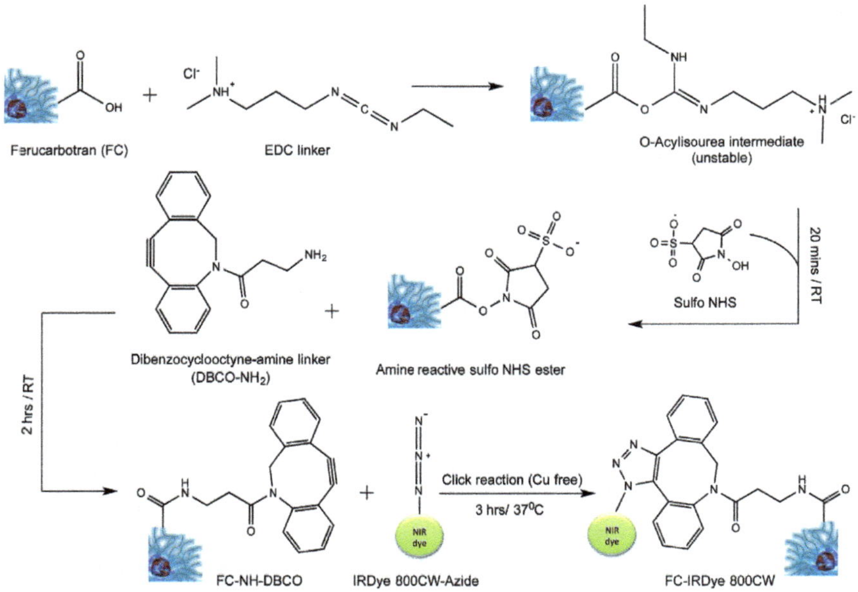

Fig. 5 Schematic presentation of the labelling reaction of FC with IRDye® 800CW-Azide. This method combines the activation of the COOH groups present on the carboxydextran coat of FC using EDC/NHS to react with a DBCO-amine linker. The conjugate is then "clicked" with azide functionalised NIR dye (IRDye® 800CW-Azide).

cut off (Thermo Fisher Scientific). The aminated FC was subsequently labelled with DyLight® 800 NHS ester using the protocol described above for NM.

Conjugation of FC to iFluor™ 780 amine using periodate oxidation followed by reductive amination (Fig. 4): The protocol was derived from Vigor *et al.* 2012.[37]

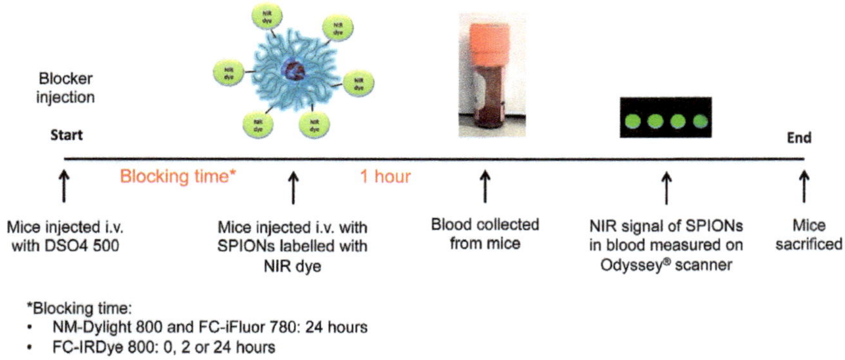

Fig. 6 Experimental design of the *in vivo* blocking experiments. DSO4 500 was injected intravenously then a certain blocking period was allowed, ranging from 0 to 24 hours depending on the SPIONs investigated. After this, NIR dye labelled SPIONs (either NM-DyLight 800, FC-iFluor 780 or FC-IRDye 800) were injected intravenously. 1 hour following SPIONs injection blood was collected from mice and measured on Odyssey® scanner.

FC is suspended in water, which was exchanged with 50 mM sodium borate buffer (pH 8.5) using a PD-10 desalting column. FC (200 μl, 11 mg Fe) was added to a 10 mg ml^{-1} solution of sodium periodate ($NaIO_4$) in 50 mM sodium borate buffer (pH 8.5). The reaction was allowed to proceed for 25–30 minutes in the dark on a roto-torque and then terminated by running the sample on a PD-10 column equilibrated with 50 mM borate buffer (pH 8.5). 100 μl of 0.1 mg ml^{-1} of iFluor™ 780 amine NIR dye was added to the reaction mixture and left to react for 2 hours at RT on a roto-torque. Finally, 10 μl of 5 M sodium cyanoborohydride ($NaCNBH_3$) in 1 M NaOH solution was added per 1 ml of reaction mixture. The reaction was left on roto-torque for 24 hours then purified by passing it twice through a PD-10 column equilibrated with PBS.

Conjugation of FC to IRdye® 800CW azide using EDC/NHS and click chemistries (Fig. 5): FC (88.6 μl, 5 mg Fe) was incubated with 200 μl of EDC/sulfo-NHS (1-ethyl-3-[3-dimethylaminopropyl]carbodiimide-hydrochloride)/(*N*-hydroxysulfosuccinimide; Thermo Fisher Scientific) activation buffer (1.21 mg EDC and 8.8 μl of 230 mM sulfo-NHS prepared in 0.1 M MES buffer, pH = 6). The reaction was allowed to proceed for 20 minutes at RT (on a rotator) and was terminated by application to a PD-10 column equilibrated with PBS (pH = 7.4). Next, 200 μg of dibenzocyclooctyne-amine ($DBCO\text{-}NH_2$) linker in 1 ml PBS was added to the reaction mixture and incubated for 2 hours at RT followed by 100 μl ml^{-1} of 25 mM glycine for 30 minutes to block the remaining reactive sites. The sample (FC–NH–DBCO conjugate) was buffer exchanged using a PBS-equilibrated PD-10 column to remove unreacted linker. Finally, 2.5 μl of 20 mg ml^{-1} (~50 μg) of IRDye® 800 CW azide was incubated with the FC–NH–DBCO conjugate for 3 hours at 37 °C in a water bath. The resultant NIR-conjugated SPIONs were extensively purified by passing them at least 3× through PD-10 columns equilibrated with PBS, to remove any dye that was unspecifically attached to the dextran coat of FC.

Following all SPIONs conjugations, the integrated signal intensity of a serial dilution of the conjugates of known iron concentration was measured at 800 nm on Odyssey® infrared scanner (LI-COR Biosciences, Lincoln, Nebraska, USA) to plot a standard curve. Integrated signal intensity (also known as pixel volume) is defined by the Odyssey® infrared scanner software as the sum of the intensity values for all pixels enclosed by a shape, multiplied by the area of the shape (counts mm^2).

In vivo blocking of the uptake of SPIONs by the RES

All *in vivo* experiments were performed using mouse models in compliance with licenses issued under the U.K. Animals (Scientific Procedures) Act 1986 after local ethical committee review. All mice were 6–8 week old female BALB/c (Charles River Laboratories, UK). The mice had an average weight of 20 g at the start of the experiments. All labelled SPIONs were injected at a dose of 10 μmol Fe per kg. A general layout of the experimental design is shown in Fig. 6. SPIONs and DSO4 500 were prepared in PBS and sterile filtered (using 0.22 μm syringe filter) in a cell culture hood prior to injections to minimise bacterial contamination. All injections were administered intravenously through tail veins.

For Nanomag-D-spio-DyLight 800 (NM-DyLight 800), DSO4 500 was injected at doses ranging from 3 to 30 mg kg^{-1}. A period of 24 hours was given to allow for sufficient blocking of the liver by DSO4 500 then NM-DyLight 800 was injected.

Table 3 Characterisation data of SPIONs

	Hydrodynamic diameter		Zeta potential	
Sample	Z-average (nm)	PDI	Mean (mV)	Zeta deviation (mV)
FC batch A[a]	59.0	0.22	−28.1	3.41
FC-IRDye 800	64.8	0.24	−33.7	13.9
FC batch B[a]	56.7	0.20	−29.4	4.02
FC-iFluor 780	82.8	0.36	−43.9	5.56
NM	75.4	0.19	−4.21	9.69
NM-DyLight 800	74.3	0.21	−3.82	7.00

[a] 2 different batches of FC were used. Batch A was used in FC-IRDye 800 and batch B in FC-iFluor 780. NM = Nanomag®-D-spio-NH_2 (unmodified).

For Ferucarbotran-iFluor™ 780 (FC-iFluor 780), 30 mg kg^{-1} of DSO4 500 was injected and after 24 hours FC-iFluor 780 was injected.

For Ferucarbotran-IRDye® 800CW (FC-IRDye 800), 30 mg kg^{-1} of DSO4 500 was injected. A period of 0, 2 or 24 hours was given before FC-IRDye 800 was administered.

One hour following the injection of the dye labelled SPIONs, the mice were anaesthetized using a 1% isoflurane (Ivax Pharmaceuticals, UK) oxygen mixture. Blood was collected by cardiac puncture in EDTA-coated tubes (Teklab or BD Vacutainer®) and the mice were sacrificed by cervical dislocation. For SPIONs quantification, a 100 μl aliquot of each blood sample was transferred into a 96-well plate (Corning) and then the integrated signal intensity was measured at 800 nm on an Odyssey® infrared scanner.

Statistical analysis

For NM-DyLight 800 and FC-IRDye 800 *in vivo* experiments, the effect of blocking was assessed using a total of 8 mice per group in 4 independent experiments (2 mice per group for NM-DyLight 800 and 1–3 mice per group for FC-IRDye 800). An ANOVA test was used to assess variance associated with the different experiments, and statistical significance between groups (blocked *versus* unblocked) was assessed using a Tukey HSD (honest significant difference) test on the fitted model. For the FC-IRDye 800 final experiment (Fig. 17), Shapiro–Wilk test confirmed that the data was normally distributed, and therefore a Student's unpaired 2-tailed t-test was used to assess statistical significance of the data. p values less than 0.05 were considered statistically significant.

Results

Characterisation of SPIONs

Unmodified and dye-conjugated SPIONs were characterised using DLS to monitor changes in their overall size and charge due to functionalisation. In general, only minor differences were observed between functionalised and unmodified SPIONs

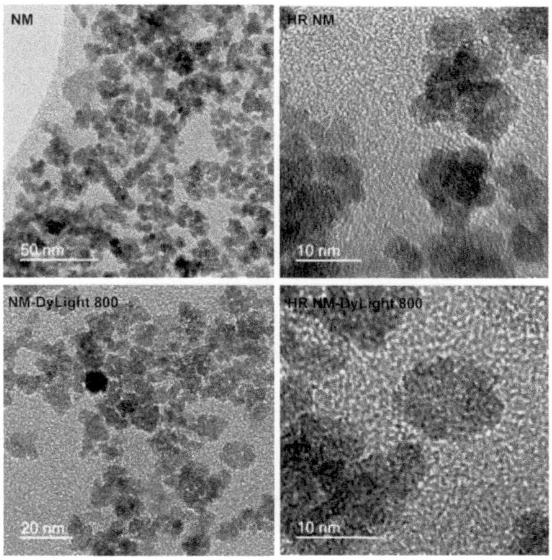

Fig. 7 TEM images of NM before and after conjugation to NIR dye. NM appears unaffected by conjugation and HR images (right) are showing lattice planes of iron oxide cores of 6.019 (±1.82) nm in diameter.

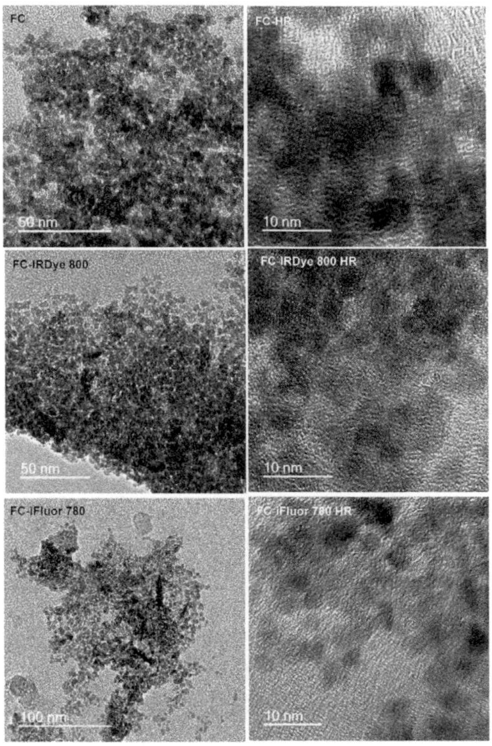

Fig. 8 TEM images of FC before and after conjugation. TEM images have revealed no changes in the iron oxide cores following conjugation. HRTEM images (right) of FC show iron oxide lattice planes with core sizes of 3.75 (±0.834) nm in diameter.

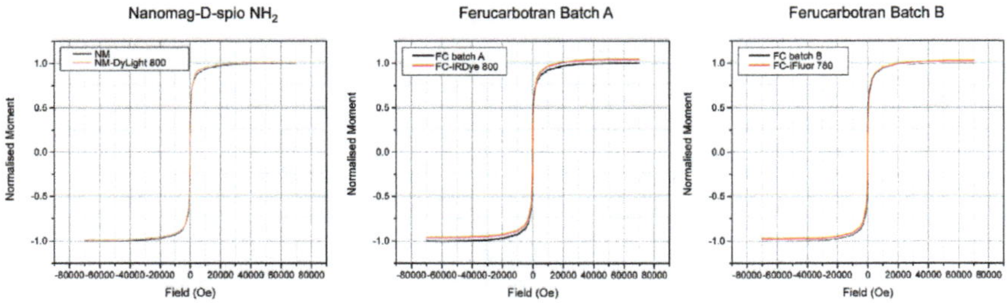

Fig. 9 Magnetic hysteresis loops of SPIONs as measured by SQUID magnetometry showing the normalised magnetisation *versus* field curves. No difference was observed in all the samples measured.

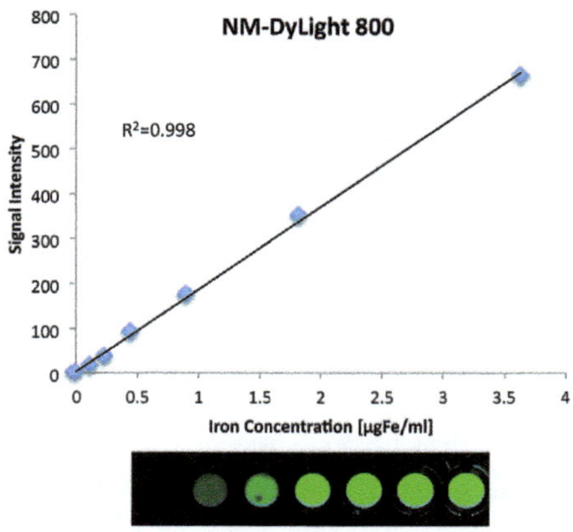

Fig. 10 A linear increase (R^2 = 0.998) in the signal intensity of NM-DyLight 800 was observed as the iron concentration increased. Images of individual wells illustrating florescence intensity are shown below the graph.

(Table 3). The exception, FC-iFluor 780, showed a 26.1 nm increase in diameter, an increase in the zeta potential of −14.5 mV, and an increase in its polydispersity index (PDI) to 0.36. Precipitation in SPIONs solution was observed by visual inspection following the reaction. These changes may be an indication of destabilisation caused by the conjugation reaction.

High resolution (HR) TEM was used to investigate the SPIONs iron oxide cores. All samples of FC and NM appeared highly crystalline in this assessment and the iron oxide cores appeared to be unaffected by the conjugation process. Average core sizes were 3.75 (±0.83) nm for FC and 6.02 (±1.82) nm for NM (Fig. 7 and 8).

SQUID magnetometry was used to test the effect of conjugation on the superparamagnetic properties of both SPIONs. The normalised magnetisation *versus* field curves show little or no difference between the unmodified SPIONs solution and the conjugated samples, all showing hysteretic behaviour indicating that the magnetic particles are superparamagnetic at 300 K (Fig. 9).

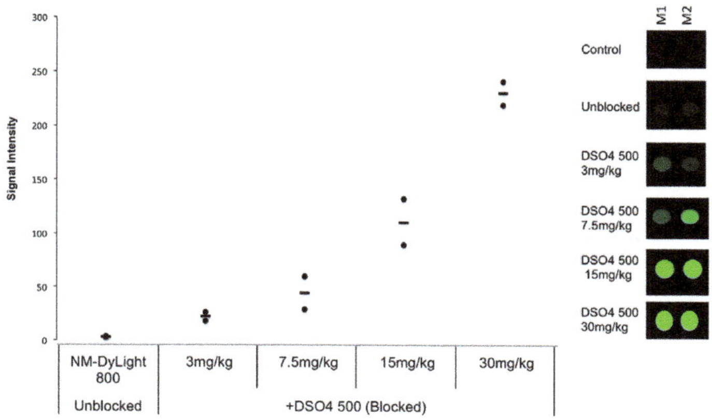

Fig. 11 Dot plot showing the mean signal intensity (bar) ($n = 2$) for mice treated with NM-DyLight 800 with and without pre-treatment with different doses of DSO4 500. A dose-dependent increase in the signal intensity was observed as the dose of DSO4 500 increased. Each mouse is represented by a circle (●). Right: Images of individual wells illustrating florescence intensity are also shown. M1 = mouse 1, M2 = mouse 2.

Conjugation of NM to NIR dye

NM SPIONs were used as a proof-of-concept because they could be readily functionalised with a near infrared dye (DyLight® 800 NHS ester) *via* available amines on the SPIONs coating. Following conjugation, the samples were measured on Odyssey® scanner and a linear increase in the signal intensity was observed as the iron concentration increased (Fig. 10).

The NIR-functionalised SPIONs were used in a pilot experiment (2 mice per group) to examine the effects of pre-treatment with DSO4 500 prior to administration of SPIONs. Four DSO4 500 doses were tested: 3, 7.5, 15 and 30 mg kg^{-1}. Results showed that the NIR signal intensity of the NIR-labelled SPIONs was strongly enhanced in a dose dependent effect (Fig. 11). The highest dose of the

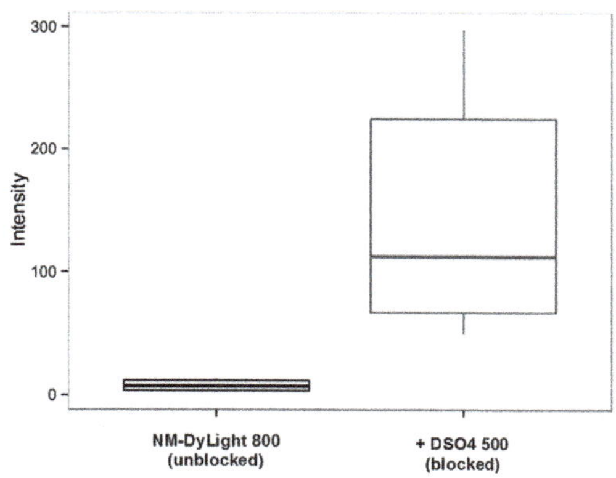

Fig. 12 Box plot of all NM-DyLight 800 experiments showing a statistically significant difference between the signal intensity of NM-DyLight 800 in mice pre-treated with the blocker compared to unblocked control. Median signal intensity of blocked group was 112 compared to 7.17 for unblocked mice ($p = 0.0016$).

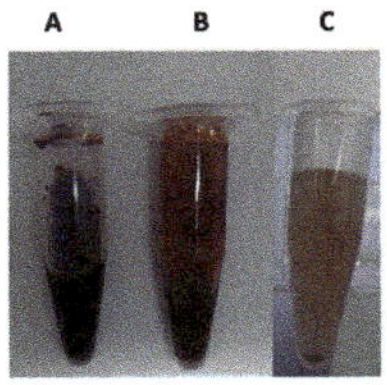

Fig. 13 Visual appearance of 3 different FC solutions. Unmodified FC (A) is a dark brown translucent solution while chemically modified FC, if unstable, can appear as either cloudy (B) or precipitate entirely (C).

blocker (30 mg kg^{-1}) was further investigated in four independent 2-mice experiments with different batches of the conjugate. A significant difference was observed ($p = 0.0016$, Tukey HSD test) in the mice pretreated with the blocker DSO4 500 compared to the unblocked control. There was a 15.6 times improvement in the median signal intensity of blocked mice compared to unblocked group (Fig. 12).

Conjugation of FC to NIR dyes

FC is coated with carboxydextran and therefore functionalisation of either hydroxyl or carboxylic groups is possible. Nevertheless, labelling FC has proved to be very challenging, due to the destabilising effects of chemical modification, leading to visibly detectable aggregation and precipitation of nanoparticles (Fig. 13B and C).

We investigated 3 different labelling strategies for FC functionalisation. First, amination of the hydroxyl groups using epichlorohydrin was employed.[35,36] This method, which uses harsh alkali conditions, led to rapid SPIONs destabilisation, with the solution appearing cloudy immediately after the completion of the

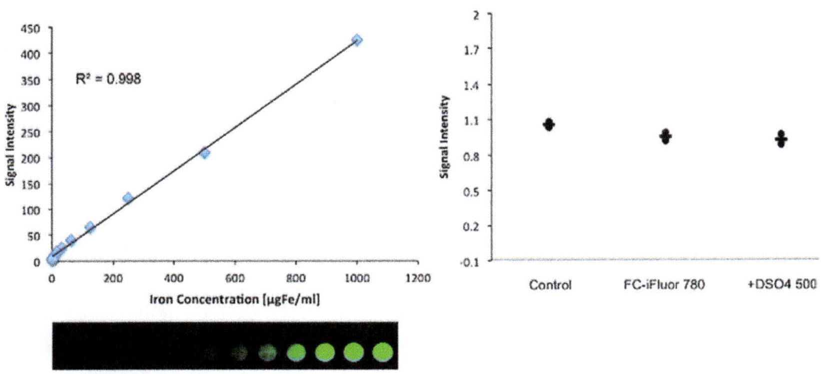

Fig. 14 Results of conjugation of FC to iFluor 780 amine. Left: A linear increase ($R^2 = 0.998$) in the signal of the conjugate was seen as the iron concentration increased *in vitro*, images of individual wells illustrating florescence intensity are shown below the graph. Right: dot plot showing the mean signal intensity (bar) ($n = 2$) for mice treated with FC-iFluor 780 with and without pre-treatment with DSO4 500. No signal higher than that of the control blood (untreated mice) was seen in all the treatment groups. Each mouse is represented by a circle (●).

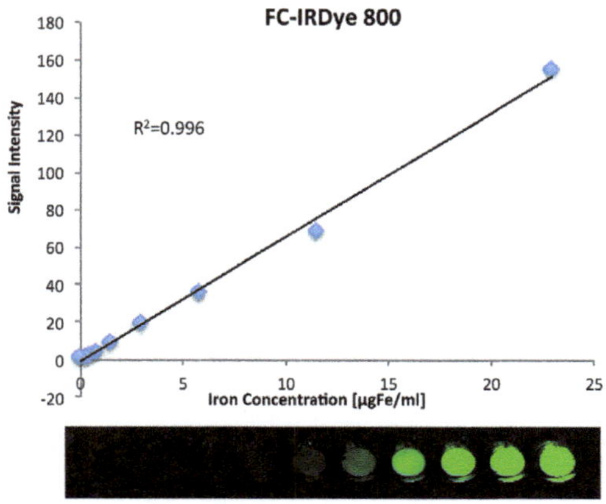

Fig. 15 A linear relationship ($R^2 = 0.99$) was observed between the iron concentration of FC-IRDye 800 and the signal intensity as measured on Odyssey® scanner. Below: images of individual wells illustrating florescence intensity.

reaction. The second method combined periodate oxidation of the hydroxyl groups of the dextran into aldehydes, followed by reaction with amine-functionalised NIR dye, iFluor 780 amine, and stabilisation by reductive amination using cyanogen bromide. This method resulted in only minor precipitation and the measured NIR signal was linear (Fig. 14 left). Despite this, the conjugate did not appear to be stable *in vivo*, as, when evaluated in a small (2 mice per group) pilot experiment, no signal higher than that of the untreated control was detected (Fig. 14 right).

The third method investigated for functionalising FC was to employ click chemistry attachment to carboxylic groups present in the carboxydextran coat. The method combines the use of EDC/NHS to conjugate Ferucarbotran to a DBCO-amine linker, which is subsequently "clicked" with an IRDye® 800CW

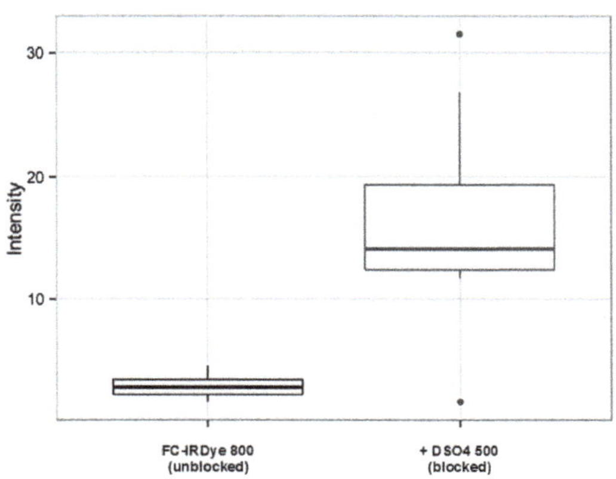

Fig. 16 Box plot of FC-IRDye 800 treated mice with and without blocking showing a significant increase in the signal of FC-IRDye 800 in the blood of blocked mice compared to unblocked control ($p = 0.0039$). Median signal intensity of blocked group is 14.13 compared to 2.76 in unblocked control.

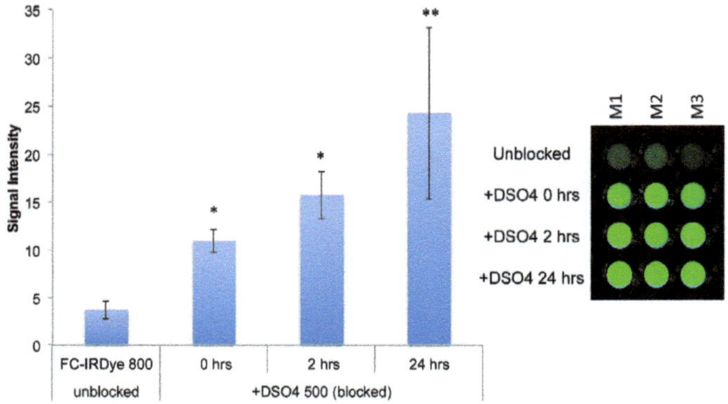

Fig. 17 Signal of FC-IRDye 800 remaining in the blood of mice pre-treated with DSO4 500 with different blocking times. All blocking times tested have shown an improvement in the blood concentration of FC-IRDye 800 with the highest observed with 24 hour blocking. * $p = 0.001$ ** $p = 0.01$ compared to unblocked control, bars represent mean values and error bars are for standard deviation. Right: images of individual wells illustrating florescence intensity. M1, M2 and M3 stand for mouse number.

azide dye. This reaction maintained the stability of the particles with no visible changes in the conjugate solution, as well as only a slight increase in the hydrodynamic diameter after the conjugation. To confirm the success of labelling, a serial dilution of the conjugate (FC-IRDye 800) was measured on Odyssey® scanner and a linear relationship was observed between the iron concentration and the signal intensity (Fig. 15).

The conjugate was subsequently tested *in vivo*. Fig. 16 shows a box plot of 4 different experiments performed with 4 different batches of FC-IRDye 800. In mice pre-treated with DSO4 500 24 hours prior to FC-IRDye 800 injection, a statistically significant ($p = 0.0039$, Tukey HSD test) 5-fold increase in the blood signal was seen in the blocked group.

After confirming that prior blocking with DSO4 500 for 24 hours can increase the blood concentration of FC-IRDye 800, we also investigated this blocking effect at 3 different blocking times: either FC-IRDye 800 and DSO4 500 were given simultaneously (0 hours), or the blocker was given either 2 hours or 24 hours prior to FC-IRDye 800. Results in Fig. 17 show that administration of DSO4 500 increased the blood concentration of FC-IRDye 800 at all the tested blocking times and the highest effect was seen at 24 hours.

Discussion

The rapid clearance of the SPIONs by the RES is advantageous when SPIONs are used as liver contrast agents (*e.g.* Resovist® and Endorem®), because it results in localisation of the particles within the Kupffer cells in healthy liver tissues, causing a signal loss in MRI, compared to tumour cells that have lower phagocytic function.[38] This enables the differentiation of cancerous lesions from normal tissues and consequently their improved detection.[38] However, if SPIONs are to be used in other clinical applications such as hyperthermia or as drug carriers, the rapid RES-mediated clearance becomes a problem.

A number of approaches have been taken to increase the circulating half-life of SPIONs by modifying their surface or using a smaller size. For example, PEG coating of the particles has been used to prevent their adsorption and agglomeration.[12,39] Others have used stealth surface modification with folic acid and *o*-carboxymethyl chitosans (OCMCS) to evade the RES, as well as to target the SPIONs to tumours expressing folate receptor.[40] In this study, a 27% reduction in the T2-weighted MRI signal intensity was observed in mice bearing folate receptor positive tumours and receiving the targeted/stealth SPIONs.[40] Decreasing the hydrodynamic diameter of the particles by using ultra small SPIONs (USPIO) (with diameters less than 50 nm) has also been shown to increase their half-life.[20]

For our study, we took an alternative approach to blocking the RES, allowing use of clinically validated SPIONs such as Ferucarbotran. DSO4 500 was chosen as it is known to block liver Kupffer cells *in vivo*,[41–49] and has been reported to block SPIONs uptake by macrophages *in vitro*.[50,51] To prove the hypothesis that DSO4 500 can block liver uptake of SPIONs *in vivo*, SPIONs were labelled with NIR dye to obtain a simple and quantitative means to measure their concentration in the blood.

NM is readily labelled with an NHS ester NIR dye and thus it was used to establish experimental conditions. Labelling FC was challenging, because the functional groups available on the carboxydextran coating do not allow a straightforward method of conjugation. Furthermore, the majority of commercially available NIR dyes are functionalised with either NHS ester or maleimide, to react with amines and sulfhydryl groups, respectively. These moieties are abundantly present in biomolecules, but not on FC, where the available chemical groups are either hydroxyl or carboxylic.

First, we investigated labelling hydroxyl groups using either epichlorohydrin or periodate oxidation, but neither of these proved suitable for *in vivo* use. Epichlorohydrin chemistry resulted in particle aggregation and SPIONs labelled by periodate oxidation were not detectable *in vivo*. The latter might be related to physical changes leading to rapid clearance *in vivo*. Another explanation could be the low labelling efficiency; however, attempts to optimise labelling conditions resulted in SPIONs aggregation. This could have been aggravated by breaking the hexose rings of dextran during the reaction to form aldehydes.

Our most successful functionalisation of FC was achieved *via* carboxylic groups, using a combination of the standard EDC/NHS reaction to a dibenzocyclooctyne-amine linker followed by a copper-free click reaction with azide dye. This reaction is performed in mild conditions, which helps to maintain the stability of the particles.

The effect of conjugation on the superparamagnetic properties of SPIONs was evaluated to ensure that they were not affected during the conjugation process. No major differences were observed in the hysteresis loops of SPIONs as measured on SQUID. In addition, successfully conjugated SPIONs (NM-DyLight 800 and FC-IRDye 800) showed no significant changes in the hydrodynamic diameter or PDI compared to unmodified SPIONs.

In vivo, pre-treatment with DSO4 500 showed a median 15-fold improvement in the signal intensity of NM-DyLight 800, compared to a median 5-fold improvement with FC-IRDye 800. Thus, in our pilot experimental conditions, DSO4 500 appeared to be more effective in reducing the clearance of NM than of FC (Fig. 12 and 16). Consequently, our future work with NIR-functionalised SPIONs will focus on optimising the experimental conditions required to maximise the pharmacokinetic profiles for individual nanoparticles.

The work presented in this paper provides methods to functionalise SPIONs with dyes that allow sensitive measurements of their blood concentrations. The labelled SPIONs have been used *in vivo* to demonstrate that RES blockers, such as DSO4 500, can prolong SPIONs half-life, potentially widening their clinical applications.

Acknowledgements

The authors would like to thank Mr Ankur Ravinarayana Chakravarthy, MSc., for his insightful help in the statistical analysis of the data. The authors would also like to thank Meito Sangyo Co. LTD., Japan for helpful discussions and for providing Ferucarbotran. This project is funded by the EU Framework 7 Programme, Dartrix project contract no. 234870; Celia Abrahams and The Mothers and Daughters Committee; Cancer Research UK (CR-UK); Department of Health (ECMC, Experimental Cancer Medicine Network Centre); NIHR University College London Hospitals Biomedical Research Centre. Dr Tammy Kalber is an EPSRC Fellow in Stem Cell Imaging (EP/L006472/1).

References

1 K. M. Krishnan, *IEEE Trans. Magn.*, 2010, **46**, 2523–2558.

2 A. K. Gupta, R. R. Naregalkar, V. D. Vaidya and M. Gupta, *Nanomedicine*, 2007, **2**, 23–39.

3 B. Hamm, T. Staks, M. Taupitz, R. Maibauer, A. Speidel, A. Huppertz, T. Frenzel, R. Lawaczeck, K. J. Wolf and L. Lange, *J. Magn. Reson. Imaging*, 1994, **4**, 659–668.

4 R. Lawaczeck, H. Bauer, T. Frenzel, M. Hasegawa, Y. Ito, K. Kito, N. Miwa, H. Tsutsui, H. Vogler and H. J. Weinmann, *Acta Radiol.*, 1997, **38**, 584.

5 B. Bonnemain, *J. Drug Targeting*, 1998, **6**, 167–174.

6 P. Reimer and T. Balzer, *Eur. J. Radiol.*, 2003, **13**, 1266–1276.

7 C. Sun, J. S. H. Lee and M. Zhang, *Adv. Drug Delivery Rev.*, 2008, **60**, 1252–1265.

8 M. Plotkin, U. Gneveckow, K. Meier-Hauff, H. Amthauer, A. Feußner, T. Denecke, M. Gutberlet, A. Jordan, R. Felix and P. Wust, *Int. J. Hyperthermia*, 2006, **22**, 319–325.

9 G. L. DeNardo and S. J. DeNardo, *Cancer Biother. Radiopharm.*, 2008, **23**, 671–680.

10 K. Maier-Hauff, F. Ulrich, D. Nestler, H. Niehoff, B. Thiesen, H. Orawa, V. Budach and A. Jordan, *J. Neuro-Oncol.*, 2011, 1–8.

11 C. Vauthier, N. Tsapis and P. Couvreur, *Nanomedicine*, 2011, **6**, 99–109.

12 A. K. Gupta and M. Gupta, *Biomaterials*, 2005, **26**, 3995–4021.

13 J. Dobson, *Drug Dev. Res.*, 2006, **67**, 55–60.

14 W. H. De Jong and P. J. A. Borm, *Int. J. Nanomed.*, 2008, **3**, 133.

15 B. H. Park, J. C. Jung, G. H. Lee, T. J. Kim, Y. J. Lee, J. Y. Kim, Y. W. Kim, J. H. Jeong and Y. Chang, *Colloids Surf., A*, 2008, **313**, 145–149.

16 V. Mailänder, M. R. Lorenz, V. Holzapfel, A. Musyanovych, K. Fuchs, M. Wiesneth, P. Walther, K. Landfester and H. Schrezenmeier, *Mol. Imaging Biol.*, 2008, **10**, 138–146.

17 M. R. Loebinger, P. G. Kyrtatos, M. Turmaine, A. N. Price, Q. Pankhurst, M. F. Lythgoe and S. M. Janes, *Cancer Res.*, 2009, **69**, 8862.
18 A. Chachuat and B. Bonnemain, *Radiologe.*, 1995, **35**, S274.
19 D. Simberg, J.-H. Park, P. P. Karmali, W.-M. Zhang, S. Merkulov, K. McCrae, S. N. Bhatia, M. Sailor and E. Ruoslahti, *Biomaterials*, 2009, **30**, 3926–3933.
20 C. Chouly, D. Pouliquen, I. Lucet, J. Jeune and P. Jallet, *J. Microencapsulation*, 1996, **13**, 245–255.
21 S. Majumdar, S. Zoghbi and J. Gore, *Invest.Radiol.*, 1990, **25**, 771–777.
22 L. Maurizi, U. Sakulkhu, A. Gramoun, J.-P. Vallee and H. Hofmann, *Analyst*, 2014, **139**, 1184–1191.
23 V. R. Kondepati, H. M. Heise and J. Backhaus, *Anal. Bioanal. Chem.*, 2008, **390**, 125–139.
24 J. G. Kim, M. Xia and H. Liu, *IEEE Eng. Med. Biol. Mag.*, 2005, **24**, 118–121.
25 Micromod-Partikeltechnologie-GmbH, Nanomag-D-spio Technical Data Sheet, 2012.
26 D. T. Kehagias, A. D. Gouliamos, V. Smyrniotis and L. J. Vlahos, *J. Magn. Reson. Imaging*, 2001, **14**, 595–601.
27 P. Reimer, E. J. Rummeny, H. E. Daldrup, T. Balzer, B. Tombach, T. Berns and P. E. Peters, *Radiology*, 1995, **195**, 489–496.
28 K. Shamsi, T. Balzer, S. Saini, P. R. Ros, R. C. Nelson, E. C. Carter, S. Tollerfield and H. P. Niendorf, *Radiology*, 1998, **206**, 365–371.
29 T. Balzer, E. C. Carter, K. Shamsi and H. P. Niendorf, *Academic Radiology*, August 1996, vol. 3 suppl 2, pp. S417–S419.
30 T. Hao-Yu, L. Chen-Yi, S. Ying-Hsia, L. Xi-Zhang and L. Gwo-Bin, *Nano/Micro Engineered and Molecular Systems, NEMS '07, 2nd IEEE International Conference on, 2007*, 2007.
31 H. Y. Tseng, G. B. Lee, C. Y. Lee, Y. H. Shin and X. Z. Lin, *IET Nanobiotechnol.*, 2009, **3**, 46–54.
32 S. Takamatsu, O. Matsui, T. Gabata, S. Kobayashi, M. Okuda, T. Ougi, Y. Ikehata, I. Nagano and H. Nagae, *Radiat. Med.*, 2008, **26**, 179–187.
33 T. Araya, K. Kasahara, S. Nishikawa, H. Kimura, T. Sone, H. Nagae, Y. Ikehata, I. Nagano and M. Fujimura, *OncoTargets and Therapy*, 2013, **6**, 237.
34 K. Yamada, T. Oda, S. Hashimoto, T. Enomoto, N. Ohkohchi, H. Ikeda, H. Yanagihara, M. Kishimoto, E. Kita and A. Tasaki, *Int. J. Hyperthermia*, 2010, **26**, 465–474.
35 M. J. Pittet, F. K. Swirski, F. Reynolds, L. Josephson and R. Weissleder, *Nat. Protoc.*, 2006, **1**, 73–79.
36 D. J. Maxwell, J. Bonde, D. A. Hess, S. A. Hohm, R. Lahey, P. Zhou, M. H. Creer, D. Piwnica-Worms and J. A. Nolta, *Stem Cells*, 2007, **26**, 517–524.
37 K. L. Vigor, P. G. Kyrtatos, S. Minogue, K. T. Al-Jamal, H. Kogelberg, B. Tolner, K. Kostarelos, R. H. Begent, Q. A. Pankhurst and M. F. Lythgoe, *Biomaterials*, 2010, **31**, 1307–1315.
38 G. M. Lanza, P. M. Winter, S. D. Caruthers, A. M. Morawski, A. H. Schmieder, K. C. Crowder and S. A. Wickline, *J. Nucl. Cardiol.*, 2004, **11**, 733–743.
39 V. P. Torchilin and V. S. Trubetskoy, *Adv. Drug Delivery Rev.*, 1995, **16**, 141–155.
40 C. Fan, W. Gao, Z. Chen, H. Fan, M. Li, F. Deng and Z. Chen, *Int. J. Pharm.*, 2011, **404**, 180–190.
41 J. W. Bradfield, R. L. Souhami and I. E. Addison, *Immunology*, 1974, **26**, 383–392.

42 K. R. Patel, M. P. Li and J. D. Baldeschwieler, *Proc. Natl. Acad. Sci. U. S. A.*, 1983, **80**, 6518.
43 L. Illum, N. W. Thomas and S. S. Davis, *J. Pharm. Sci.*, 1986, **75**, 16–22.
44 N. Das, M. K. Basu and M. K. Das, *Biosci. Rep.*, 1987, **7**, 465–470.
45 H. Sands and P. L. Jones, *J. Nucl. Med.*, 1987, **28**, 390.
46 R. W. Jansen, G. Molema, T. L. Ching, R. Oosting, G. Harms, F. Moolenaar, M. J. Hardonk and D. K. Meijer, *J. Biol. Chem.*, 1991, **266**, 3343.
47 D. Liu, A. Mori and L. Huang, *Biochim. Biophys. Acta, Biomembr.*, 1992, **1104**, 95–101.
48 T. Yoshinobu, F. Takuya, F. Hisao, N. Makiya, S. Hitoshi and H. Mitsuru, *Int. J. Pharm.*, 1994, **105**, 19–29.
49 M. Fujiwara, J. D. Baldeschwieler and R. H. Grubbs, *Biochim. Biophys. Acta, Biomembr.*, 1996, **1278**, 59–67.
50 O. Lunov, V. Zablotskii, T. Syrovets, C. Röcker, K. Tron, G. U. Nienhaus and T. Simmet, *Biomaterials*, 2011, **32**, 547–555.
51 C. Y. Yang, M. F. Tai, C. P. Lin, C. W. Lu, J. L. Wang, J. K. Hsiao and H. M. Liu, *PLoS One*, 2011, **6**, e25524.

Faraday Discussions

PAPER

Final step gallium-68 radiolabelling of silica-coated iron oxide nanorods as potential PET/MR multimodal imaging agents†

Benjamin P. Burke,[ab] Neazar Baghdadi,[a] Gonçalo S. Clemente,[b] Nathalie Camus,[c] Amaury Guillou,[c] Alicja E. Kownacka,[a] Juozas Domarkas,[ab] Zakaria Halime,[c] Raphaël Tripier[c] and Stephen J. Archibald*[ab]

Received 24th June 2014, Accepted 24th July 2014
DOI: 10.1039/c4fd00137k

The investigation of iron oxide-based positron emission tomography/magnetic resonance (PET/MR) multimodal imaging agents is an expanding field in which a variety of nanoparticle sizes, shapes, surface coatings and radioisotopes are open for exploration. This study develops iron oxide nanorods which are coated with various mixtures of poly(ethylene glycol) (PEG) and macrocyclic ligand (DO3A) *via* the formation of a silica layer on the surface. Gallium-68 radiolabelling of the nanorods was carried out in high radiochemical yields (RCY) and their stability in human serum was demonstrated for all constructs, even in the absence of the macrocyclic chelating unit. Further studies were carried out in an attempt to determine the appropriate amount of PEG coating to give optimal properties for future *in vivo* studies.

1. Introduction

Medical imaging is the imaging of the intact human body for clinical reasons such as diagnosis or to study disease pathology. Positron emission tomography (PET) is a nuclear imaging technique which relies on the radioactive decay of a positron-emitting radionuclide to give a functional image with high sensitivity and accurate quantification.[1–3] However, as a molecular imaging technique it does not provide general anatomical information therefore is often combined with computed tomography (CT) or magnetic resonance imaging (MRI). Although the

[a]*Department of Chemistry, University of Hull, Cottingham Road, Hull, HU6 7RX, UK. E-mail: s.j.archibald@hull.ac.uk*
[b]*Positron Emission Tomography Research Centre, University of Hull, Cottingham Road, Hull, HU6 7RX, UK*
[c]*Université de Brest, UMR-CNRS 6521/SFR148 ScInBioS, UFR Sciences et Techniques, 6 Avenue Victor le Gorgeu, C.S. 93837, 29238 Brest, France*

† Electronic supplementary information (ESI) available. See DOI: 10.1039/c4fd00137k

combination of PET with CT is more prevalent, more recently the technical limitations of combining PET with MRI have been mitigated and its usage has been increasing because of the advantages of PET/MRI over PET/CT, such as increased soft tissue contrast, a lower radiation dose needed and the availability of MRI contrast agents.[4–8] Super-paramagnetic iron oxide nanoparticles (SPIONs) are used in MRI to reduce the amount of time taken for nuclei to transfer their spin to neighbouring nuclei, resulting in the loss of transverse magnetisation (T_2 relaxation), darkening regions in which the SPIONs are accumulated.[9–11] Iron oxide derivatives are the most commonly used T_2 MRI contrast agents due to their low toxicity[12,13] and modifiable surface.[11,14] Nanorods (NRs) are more unusual in the field of iron oxide MRI contrast agents, which is dominated by spherical nanoparticles.[15–17] NRs, however, have been shown to have similar and sometimes improved MRI T_2 relaxivities.[18–20]

Gallium-68 ($t_{1/2} = 68$ min) is an attractive radioisotope for PET due to its high proportion of positron decay (89%) and its availability in a generator-based system from a parent isotope with a long half-life (^{68}Ge $t_{1/2} = 271$ days), eliminating the need for an on-site cyclotron.[21–25] ^{68}Ga is exclusively in the oxidation state +3 in aqueous solution at physiological pH and most synthetic procedures are carried out in the presence of weakly coordinating ligands such as citrate, acetate or oxalate at low pH to avoid the formation of insoluble $Ga(OH)_3$ and soluble $Ga(OH)_4^-$, both of which dramatically reduce the kinetics of complex formation.[3,4,24,26] Gallium(III) generally forms six-coordinate complexes and is classified as a hard Lewis acid, therefore it binds to borderline and hard Lewis base donor atoms such as nitrogen and oxygen which are present in routinely used chelators for biomedical imaging.[3,27–29] Incorporation of ^{68}Ga into a targeting vector is generally undertaken with the use of a bifunctional chelator (BFC), a molecule which forms complexes that are stable *in vivo* with a reactive terminating group for conjugation to a bioactive molecule, protein or peptide.[2,3,30] Macrocyclic BFCs are generally more kinetically inert and thermodynamically stable than their acyclic counterparts but generally have slower metal ion binding kinetics,[25,31] with 1,4,7,10-tetraazacyclododecane-1,4,7,10-tetraacetic acid (DOTA) being the most employed BFC due to its ease of synthesis and adequate complex stability for a range of metal ions used in a wide range of applications.[22,32,33]

The introduction of clinical PET/MRI scanners allows the use of multimodal contrast agents that are responsive in both imaging techniques. This can be carried out by using either a cocktail of imaging agents or by developing a dual modality contrast agent where both components are linked in one construct.[16,34–37] A general comment on this approach is the large difference in sensitivities of the two techniques, with MRI generally requiring at least 10^6 higher concentrations to achieve the necessary contrast. This discrepancy is not a significant concern when adding a radioisotope to an MRI contrast agent as this can be achieved by radiolabelling at very low specific activities (radioactivity/moles of MRI contrast agent).[38–41] There may be issues in low receptor density targeted imaging in comparison to PET imaging alone.

Herein we describe the synthesis of various silica-coated NRs, together with the final step ^{68}Ga radiolabelling followed by study of their characteristics that are relevant to their potential applications as PET/MR imaging agents.

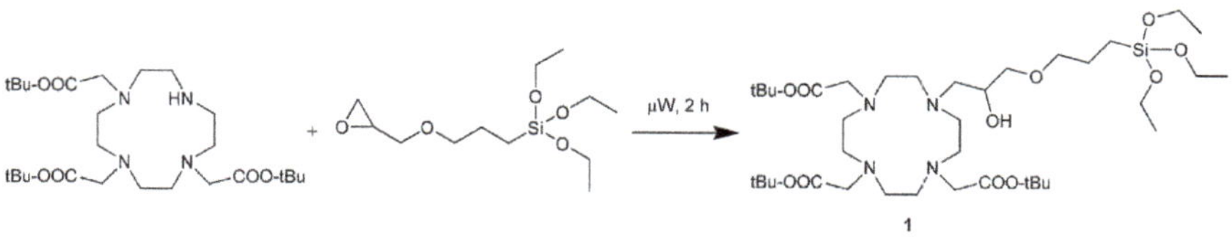

Scheme 1 Synthesis of siloxane-derivatised *t*Bu-DO3A (**1**).

2. Results

2.1. Formation of siloxane-coated nanorods with various DO3A : PEG ratios

In order to form NRs coated with macrocyclic ligands, a siloxane approach was used to form a shell of silica: 1,4,7-tris(*tert*-butoxycarboxymethyl)-1,4,7,10-tetraazadodecane (*t*Bu-DO3A) was heated *via* microwave irradiation with a 1.1 excess of (3-glycidyloxypropyl)triethoxysilane (GPTES) in chloroform to form a siloxane-terminated DO3A derivative (**1**), see Scheme 1. The synthesis and isolation of **1** were confirmed by ^{1}H NMR and MS.

In order to compare the gallium-68 radiolabelling characteristics and the chemical stability properties (*vide infra*), a range of three nanorod coatings was synthesised with varying amounts of DO3A and a siloxane PEG derivative with a molecular weight range of 459–591 g mol^{-1}. Iron oxide nanorods were synthesised *via* modified literature methods[42,43] (**2**) and were coated with either 100% Si-*t*Bu-DO3A (**1**) to form **3**, or with a mixture of 50% Si-*t*Bu-DO3A (**1**)/50% Si-PEG to form **4**, or 100% Si-PEG to form **5**, see Scheme 2. Isolation of the nanorods was carried out using a strong rare earth magnet followed by successive washings in order to remove both weakly magnetic NRs and unreacted starting materials. In order to form effective coordination bonds with transition metals, the *tert*-butyl esters must be cleaved to form carboxylic acids; this was carried out using trifluoroacetic acid (TFA) in dichloromethane (DCM) as this method is routinely used for analogous macrocyclic ester cleavage reactions.[32,44] This reaction was carried out on both the mixed *t*Bu-DO3A/PEG (**4**) and *t*Bu-DO3A (**3**) NRs to form **6**

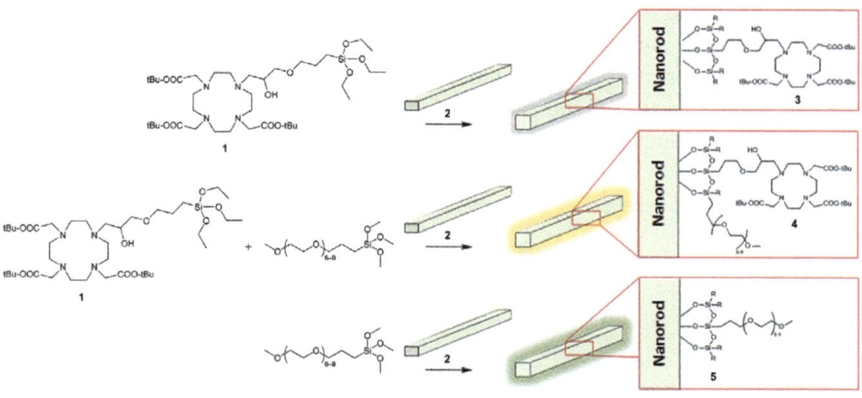

Scheme 2 Synthesis of siloxane-coated iron oxide nanorods: 100% *t*Bu-DO3A (**3**), 50% *t*Bu-DO3A/50% PEG (**4**) and 100% PEG (**5**).

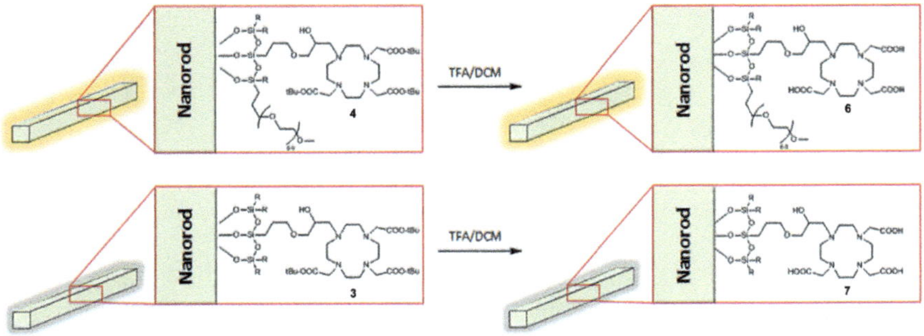

Scheme 3 Ester hydrolysis of *t*Bu-DO3A derivatives to give 50% DO3A/50% PEG (**6**) and 100% DO3A (**7**).

and **7** respectively, see Scheme 3, where the products were isolated with a similar washing procedure to that used previously. Successful hydrolysis was challenging to confirm but was indicated by the increase in aqueous suspension stability of the NRs, as was expected from the increased hydrophilic character conferred by the free carboxylic acid groups.

2.2. Surface coating characteristics and analysis

Transmission electron microscopy (TEM) was carried out on the uncoated NRs (**2**) and the final constructs (**5–7**) to assess how the coating procedure and hydrolysis (to form **6** and **7**) affected their size and shape, see Fig. 1. It can be observed that the form is almost universally rod-like, with lengths generally between 60 and 80 nm, with particle shapes both internally consistent and showing little variation between samples, indicating that the coating and hydrolysis steps have a low impact on the nanorod size and shape.

Fourier transform infra-red (FTIR) spectroscopy was carried out to verify both the successful surface coating and to study the difference between the variable DO3A/PEG coatings, see Fig. S1.† It can be observed that all samples have peaks between 600 and 1000 cm^{-1} which can be attributed to Fe–O, Si–O–C and Si–O–Fe,[45,46] and it should also be noted that there is an absence of a peak at 3400 cm^{-1}, usually attributed to Fe–O–H bonds. **5** has strong peaks at 2873 and 1088 cm^{-1} due to the large numbers of C–H and C–O–C respectively.[46] Introduction of a macrocycle to form **6** seems to show very little difference when compared to **5**, whilst **7** has a peak at 1200 cm^{-1} which is attributed to the macrocyclic C–N stretching frequency and an absence of PEG-related peaks.

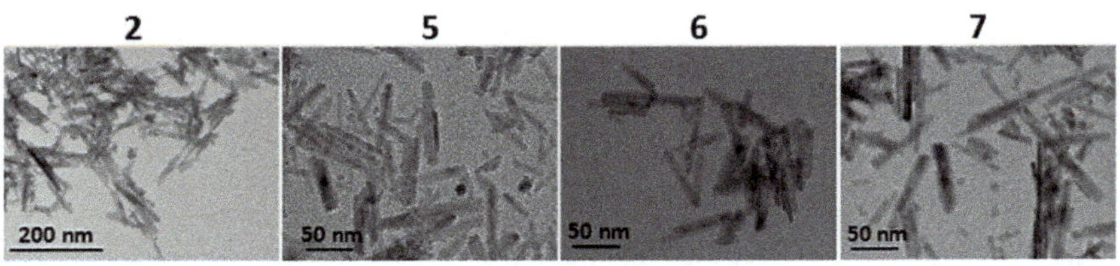

Fig. 1 TEM images of **2** and **5–7**.

2.3. ^{68}Ga radiolabelling and radiochemical stability of NRs

In aqueous solutions, gallium is found almost exclusively in the oxidation state +3, and is classified as a hard Lewis acid. This property allows gallium(III) to bind to highly ionic hard base ligand donors such as the carboxylic acids or nitrogen atoms present in the DO3A macrocycles. Gallium chemistry is highly influenced by pH, with the optimal coordination conditions situated between pH 3 and 5, since a more acidic environment can inhibit the reaction due to protonation of donor atoms and a more basic milieu will favour hydrolysis and lead to gallium hydroxide formation.[47]

To carry out the radiolabelling, a sample of $^{68}GaCl_3$ (~150 MBq) was completely dried at 90 °C under compressed air flow. A minimum volume of 35 μl of an NR suspension (10 mM of effective iron concentration) in 0.2 M ammonium acetate buffer, pH 5, was then added to the dry ^{68}Ga and left to react for 15 min at 90 °C. The three different sets of NRs were radiolabelled to determine the impact of the different ratio between the PEGylated arm and the macrocycle when preparing the coating (0% (**5**), 50% (**6**) and 100% (**7**) DO3A), all of which gave full ^{68}Ga incorporation into the NR within 5 min. At the end of the reaction the solution was diluted to 95% phosphate buffer saline (PBS), pH 7.4, in order to produce a bimodal radiotracer preparation suitable for administration under physiological conditions.

Radiolabelling yields and reaction kinetics were followed by radio-thin layer chromatography (radioTLC) using thin layer silica gel chromatography plates (TLC-SG) developed in an aqueous 0.1 M citric acid solution. In this system the radiolabelled NRs remain at the application point ($R_f = 0$) whilst non-chelated ^{68}Ga follows the solvent front ($R_f = 1$).

The stability of the radiolabelled NRs in the final 95% PBS solution was followed by radioTLC after leaving the formulation in a glass vial at room temperature for 3 h. The strength and stability of the DO3A coordination with gallium(III) was assessed by having the radiolabelled NRs in competition with a 100-fold excess of ethylenediaminetetraacetic acid (EDTA) solution for 1 h at room temperature. The amount of *trans*-chelated ^{68}Ga was then measured by radioTLC. The stability of the radiolabelling product for future *in vivo* studies was carried out in serum. Approximately 74 MBq of the final physiological solution of NRs were incubated in human serum (500 μL) at 37 °C for 3 h and then analysed by radioTLC. In all of the stability tests, the three types of radiolabelled NRs were shown to be highly stable, with no noticeable difference when various ratios of PEG/DO3A were used in the preparation mixture.

In general it was seen that all NRs are easily radiolabelled to completion with ^{68}Ga at 90 °C and have high stability under the conditions tested. No significant variation in the results from the radiolabelling process or serum stability was observed over the different coating ratios tested.

2.4. Characterisation of nanorods

2.4.1. Precipitation stability in solution. It was expected that introducing PEG onto the nanorod surface would cause an increase in the aqueous suspension stability (*i.e.* avoiding aggregation/precipitation). In order to test this, qualitative observational studies were carried out on **5**, **6** and **7** suspended in aqueous solutions of variable pH and ionic strength, see Fig. 2. In both experiments, the

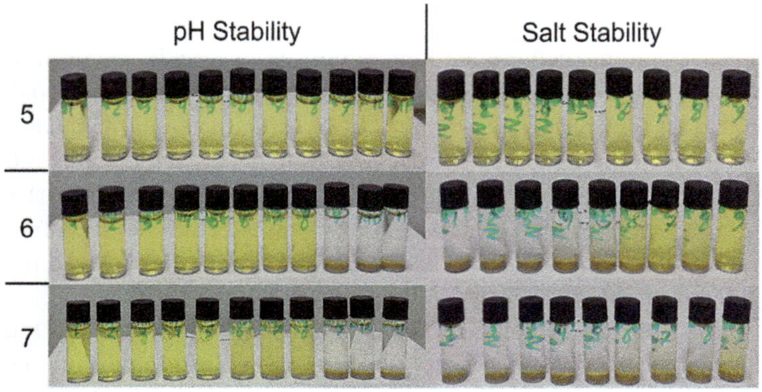

Fig. 2 Suspension stability of **5–7** to precipitation within a range of pH 1–11 (left) and increasing NaCl concentration (right) from 2 mOsm L^{-1} (far right) to 200 mOsm L^{-1}.

iron concentration was limited to 7.5 nM as this is the highest concentration at which the NRs can stably be suspended in pure water. In the pH studies, the NRs were suspended in pure water and the pH was adjusted to the correct value by the addition of 0.1 M NaOH or 0.1 M HCl before being left to stand for 10 h before making an observation. In the ionic strength experiments, the NRs were suspended in a range of concentrations of NaCl and left to stand for 3 days before observation. An increase in suspension stability was observed which correlated with an increased PEG coating, with **5** being stable at most pH values and to a salt concentration of 200 mOsm L^{-1}.

2.4.2. Nanoparticle tracking analysis (NTA). Nanoparticle tracking analysis (NTA, NanoSight®) was carried out to determine the mean and modal sizes of the NRs. Samples were suspended in water at a 0.1 mM iron concentration. During the ^{68}Ga radiolabelling procedure, the NRs were subjected to a high salt concentration (0.2 M ammonium acetate, pH 5) along with high temperatures (90 °C) and potential radiation effects. NTA of decayed samples was of interest in order to investigate how the radiolabelling process influences the NR aggregation.

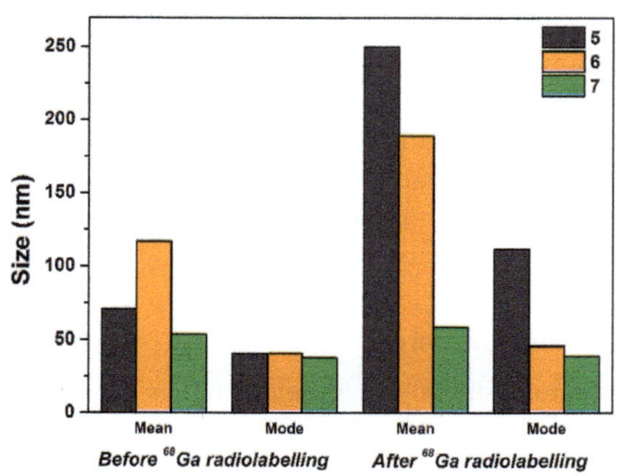

Fig. 3 Mean and mode size of **5–7** before and after ^{68}Ga radiolabelling using NTA measurements.

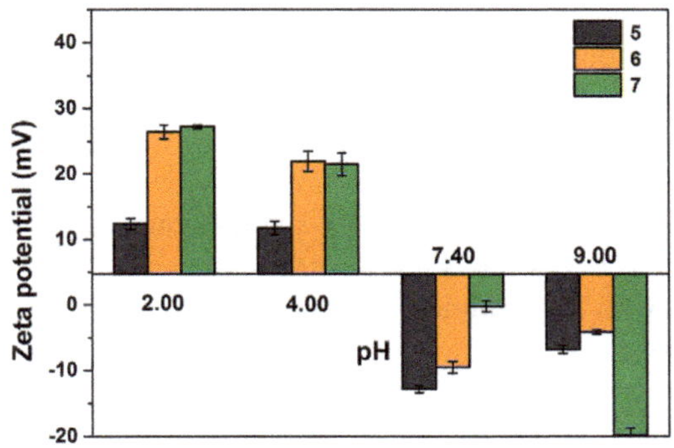

Fig. 4 Evolution of the zeta potential for **5–7** measured in buffers with different pH values.

Increasing amounts of PEG in the coating reaction mixture caused an increase in the mean and modal hydrodynamic size of the nanoparticles, see Fig. 3 and S2.† The radiolabelling procedure had a significant effect on the hydrodynamic size of the PEG-coated NRs, with no observable effect when PEG was absent from the coating reaction mixture.

2.4.3. Zeta potential. The surface charge of the iron oxide nanorods with various coatings has been investigated by measuring the zeta potential. Analysis of the zeta potential for the modified iron oxide NRs is indicative of both the suspension stability of the system and the presence of the coating molecules. The experiments were performed in various buffers to give consistent pH values (2.0; 4.0; 7.4; 9.0) with a fixed concentration of iron oxide (7.5 mM). The data were acquired in triplicate and the standard deviation was calculated for each sample, see Fig. 4. At pH 7.4, increasing the PEG coating causes an increase in negative values which translates to an increase in the suspension stability of these constructs at physiological pH.

3. Discussion

Medical imaging is increasingly moving towards multi-modality approaches, with the combination of the independent strengths of individual techniques covering their specific shortcomings. Most commonly this is carried out by combining a functional technique such as PET or SPECT (single-photon emission computed tomography) with a structural technique such as CT. More recently, the use of MRI has been used in place of CT to provide PET/MRI and SPECT/MRI, capable of increased soft tissue contrast and resulting in a lower radiation exposure. In addition, using MRI allows researchers or clinicians to take advantage of the wealth of MRI contrast agents that are already available to give clinically relevant information for a better diagnosis or monitoring of disease progression. Iron oxide-based nanoparticles have been used routinely as T_2 MRI contrast agents, and have now been combined with PET radioisotopes to form multi-modal imaging agents.[48–50] PET/MR imaging agents still need more development towards clinical applications and more research is needed to explore the optimal size, shape and surface coating for successful *in vivo* applications. Nanoparticles coated with silica are gaining interest due to their biocompatibility and ease of covalent functionalisation.[51–53] The only requirement for a potential coating is a terminal siloxane group which would react with the Fe–O–H surface. In order to radiolabel NRs with the PET radioisotope gallium-68, we modified the tetraazamacrocyclic chelator DO3A with a siloxane derivative (GPTES) *via* an epoxide ring-opening reaction. Tetraazamacrocycles have been previously siloxane functionalised in a similar manner for similar applications using ^{64}Cu with nanoparticles.[48] Siloxane DO3A was subsequently attached to the iron oxide surface both alone and with an equimolar amount of a siloxane-functionalised PEG derivative. As a control, 100% siloxane PEG-coated NRs were produced in order to investigate how the addition of PEG affects the chemical and radiochemical characteristics. An increased content of PEG in the coating was initially expected to lead to a higher aqueous suspension stability of the construct but was thought to reduce its radiochemical complexation characteristics with limited or no binding of gallium-68.

In order to characterise how the surface coating affected the nanoconstructs and to validate our attempted coating methodologies, a range of analyses was carried out. TEM images of all three constructs were similar to those of uncoated NRs (**2**), with iron oxide cores present in a nanorod shape. Infrared (IR) studies were very clear and distinct for **5** and **7** when the surface was uniformly coated with PEG and DO3A respectively. However, **6** shows little evidence of the characteristic C–N shift, which could be due to the relative ratios of C–N peaks in comparison to C–O–C peaks per molecule.

Once the coated constructs had been synthesised and their structures confirmed, ^{68}Ga radiolabelling experiments were carried out. All the attempts to radiolabel NRs at room temperature were unsuccessful, with a yield of around only 5% achieved after 30 min of reaction. However, at 90 °C the kinetics of $^{68}Ga^{3+}$ complexation were found to be very fast, with no free gallium detected after 5 min by radioTLC, see Fig. 5. Surprisingly no significant differences were found between the three kinds of NRs used in this work. The ratio between the DO3A macrocycle available for gallium chelation and the PEGylated arms was varied between 0% (**5**), 50% (**6**) and 100% (**7**) in the initial coating mixture and for all of the different coatings a radiochemical yield (RCY) of 100% was achieved rapidly at 90 °C when using volumes larger than 35 μl of NRs (10 mM) in 0.2 M ammonium

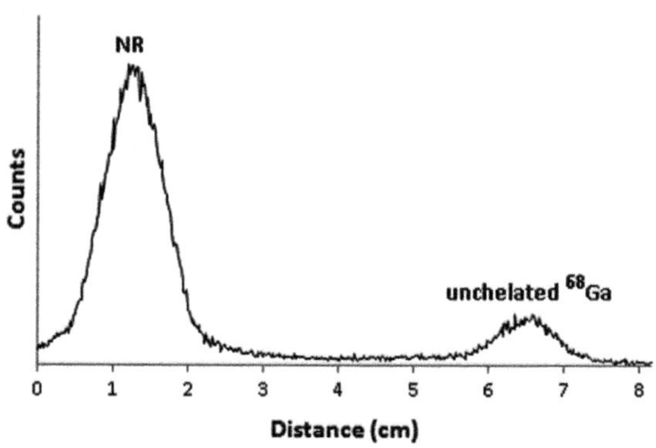

Fig. 5 RadioTLC chromatogram profile for ^{68}Ga radiolabelled NRs ($R_f = 0$) spiked with $^{68}GaCl_3$ ($R_f = 1$).

acetate buffer (pH 5). Lower volumes were seen to generate poorer yields of radiolabelled NRs or to give reduced stability of the suspension. Thus, the radiolabelling conditions for all of the NRs used in this work were optimised to utilise a volume of not less than 35 μl (10 mM), reacting with dried ^{68}Ga at 90 °C for 15 min before being diluted to give a final solution in 95% PBS.

Although a high level of incorporation of ^{68}Ga into PEG-coated NRs lacking a chelator was surprising, we initially thought that there may be a significant difference in the stability of metal retention by a weaker surface interaction compared with macrocyclic coordination. In order to obtain more information about how the presence of the DO3A macrocycle influences the stability of the NRs, some furthers assays were developed to study their behaviour. However, the analysis of a 95% PBS (pH 7.4) solution showed no unchelated ^{68}Ga after 3 h at room temperature for all NRs and so a more rigorous assay was attempted. EDTA is a hexadentate ligand widely known to complex rapidly transition metal ions to form stable complexes.[3] The stability of the gallium labelling was tested by a competition assay by mixing radiolabelled NRs in solution with an excess of EDTA, see Fig. 6. Under these harsh conditions it was found that, after 1 h, approximately 80% of the NRs still remained radiolabelled with no difference (within error) between the three samples. The observed levels of suspension stability were unexpected, with the stability of the ^{68}Ga-PEG-NR comparable to the macrocycle-containing samples in the EDTA competition experiments. Finally, the stability in serum gives a preliminary indication of how well a radiotracer will behave *in vivo*. It is important to be certain that, while in circulation, ^{68}Ga remains coordinated to the NR and is not released into the bloodstream where it will be biodistributed in its free form or associated with transferrin.[54] The stability in serum study showed that after 3 h at room temperature, over 90% of the radiolabel was retained by all NRs, see Fig. 6. Thus, we can conclude that the different coatings we tested did not result in any significant changes in the radioisotope labelling and serum stability behaviour, and are independent of the inclusion of a chelating ligand for $^{68}Ga^{3+}$.

The issue of a fairly rapid (within an hour) and visible precipitation of the radiolabelled NRs from suspension needs to be solved and further studies will be required to quantify the effective amount of NRs sufficient for contrast in MRI

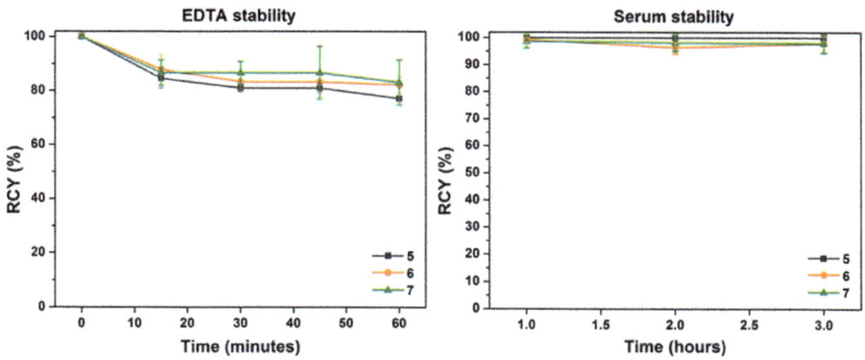

Fig. 6 Percentage of the remaining radiolabelled NRs **5–7** during EDTA and serum stability assays.

experiments. It is important to note the fact that the DO3A unit was not necessary for radiolabelling with ^{68}Ga in our system, which makes the authors query whether macrocyclic units are required for final step radiolabelling of various constructs in other studies using silica or similar type coatings (particularly where the controls could have been more relevant to this aspect).[15,55–57]

The feasibility of successful clinical applications of PET/MRI dual imaging probes is an area currently under debate,[58] the authors believe one of the strongest arguments of those presented is to develop an understanding of the use of MRI contrast agents. The inherent sensitivity limitations of MRI makes quantitative detection of contrast agents in low uptake areas very challenging; however, this problem can be addressed with PET, which is capable of showing where to direct a localised high-resolution MRI scan. The sensitivity of PET also allows more accurate quantification of the biodistribution of MRI contrast agents which is currently not well characterised. Dual modality constructs could allow a better understanding of the effects of different parameters (size, shape, surface coating *etc.*) on the biodistribution and clearance effects, to facilitate the development of the next generation of SPIONs.

An important characteristic for *in vivo* use is the stability of the constructs in suspension preventing aggregation and precipitation. Qualitative observations were carried out to assess the effect of PEG introduction under two sets of conditions: variable pH and increasing salt concentrations. There is little difference between **6** and **7** in the pH stability measurements, with both showing stable suspensions under acidic and neutral conditions. Coating with pure siloxane PEG confers stability at all pH levels; however, the improvement of stability under basic conditions has limited application *in vivo*. The advantage of the PEG incorporation into the coating is much more noticeable in salt stability measurements, given that the NRs lacking a PEG component in the coating precipitate at very low salt concentrations (2 mOsm L^{-1}). Increased the PEG content causes a linear increase in the suspension stability, with 100% siloxane PEG in the coating being stable up to at least 200 mOsm L^{-1}. The size and size distribution of the nanoparticles is important for application as *in vivo* T_2 MRI contrast agents since their size is directly related to their metabolism. If they are too small (<5.5 nm) they can be excreted through the kidneys; if they are too large (>200 nm) they can be easily detected by the immune system and excreted *via* the liver and spleen, resulting for both cases in short blood circulation times.[9,59,60] NTA measurements of the different preparations showed similar modal values (*ca.* 40 nm) and mean values between 50 and 100 nm which are ideal for *in vivo* use. However, it was expected that the radiolabelling procedure would have an effect on the hydrodynamic size due to the temperature and high concentration buffer. Indeed, after radiolabelling, NTA shows a large increase of PEG-containing nanorods to a non-ideal size, with little change noted when PEG was not present, making the use of PEG in the coating detrimental from this perspective.

The magnitude of the zeta potential gives an indication of the potential stability of the water-based dispersed system. The large value of negative or positive charge suggests high electrostatic repulsion between the particles which results in a low tendency for aggregation. Among several parameters strongly affecting stability of the NRs is the pH of the dispersing solution.[61,62] It can be observed that under acidic conditions the particles will acquire more positive charge. When the environment of the particles is changed to neutral or basic, negative values of zeta potential are expected. The type of coating had a significant influence on the behaviour of the nanorods in buffered medium at different pH values. Due to the high importance of the suspension stability for this system *in vivo*, it is a requirement to have nanoparticles that are stable at physiologically relevant pH values. Our results show that an increase in the PEG content of the coating reaction mixture leads to increased negative values and is therefore expected increase the stability in solution.[63] NRs coated with 100% DO3A (**7**) precipitated very quickly in simulated physiological conditions, which is consistent with the recorded low zeta potential value. Additionally, this sample was found to be stable only in acidic conditions which may be due to macrocycle protonation. The addition of 50% PEG (**6**) to the coating mixture significantly improved the stability of the nanorods in neutral buffer.

4. Conclusions

A range of nanorods have been coated with various ratios of PEG and DO3A siloxane compounds and successfully radiolabelled with gallium-68 to form highly stable constructs, even in the absence of the macrocyclic chelating unit. Although the use of DO3A turned out not to be required for radiochemical incorporation it may have some advantages, such as conferring stability towards temperature-induced aggregation and precipitation. Further studies are ongoing towards the development of constructs for *in vivo* studies.

Acknowledgements

The authors would like to thank Ann Lowry for carrying out TEM experiments. N.B. would like to thank the Saudi Arabian Cultural Bureau for a Saudi Government Scholarship (S2042). The research leading to these results has received funding from the People Programme (Marie Curie Actions) of the European Union's Seventh Framework Programme FP7/2007-2013/ under REA grant agreement No. 607868 (iTERM). Some of this work was carried out as STSMs under MPNS EU COST Action TD1007 "Bimodal PET-MRI molecular imaging technologies and applications for *in vivo* monitoring of disease and biological processes".

References

1 P. W. Miller, N. J. Long, R. Vilar and A. D. Gee, *Angew. Chem., Int. Ed.*, 2008, **47**, 8998.

2 E. W. Price and C. Orvig, *Chem. Soc. Rev.*, 2014, **43**, 260.

3 T. J. Wadas, E. H. Wong, G. R. Weisman and C. J. Anderson, *Chem. Rev.*, 2010, **110**, 2858.

4 P. Blower, *Dalton Trans.*, 2006, 1705.

5 F. Gerstl, C. Windischberger, M. Mitterhauser, W. Wadsak, A. Holik, K. Kletter, E. Moser, S. Kasper and R. Lanzenberger, *NeuroImage*, 2008, **41**, 204.

6 R. R. Raylman, S. Majewski, S. Sendhil Velan, S. Lemieux, B. Kross, V. Popov, M. F. Smith and A. G. Weisenberger, *J. Magn. Reson.*, 2007, **186**, 305.

7 R. R. Raylman, S. Majewski, S. K. Lemieux, S. S. Velan, B. Kross, V. Popov, M. F. Smith, A. G. Weisenberger, C. Zorn and G. D. Marano, *Phys. Med. Biol.*, 2006, **51**, 6371.

8 R. R. Raylman, S. Majewski, S. Lemieux, S. S. Velan, B. Kross, V. Popov, M. F. Smith, A. G. Weisenberger and R. Wojcik, *Nucl. Instrum. Methods Phys. Res., Sect. A*, 2006, **569**, 306.
9 M. Colombo, S. Carregal-Romero, M. F. Casula, L. Gutierrez, M. P. Morales, I. B. Bohm, J. T. Heverhagen, D. Prosperi and W. J. Parak, *Chem. Soc. Rev.*, 2012, **41**, 4306.
10 E. C. Cho, C. Glaus, J. Y. Chen, M. J. Welch and Y. N. Xia, *Trends Mol. Med.*, 2010, **16**, 561.
11 A. K. Gupta and M. Gupta, *Biomaterials*, 2005, **26**, 3995.
12 N. Lewinski, V. Colvin and R. Drezek, *Small*, 2008, **4**, 26.
13 Y. X. J. Wang, S. M. Hussain and G. P. Krestin, *Eur. Radiol.*, 2001, **11**, 2319.
14 A. K. Gupta, R. R. Naregalkar, V. D. Vaidya and M. Gupta, *Nanomedicine*, 2007, **2**, 23.
15 C. Glaus, R. Rossin, M. J. Welch and G. Bao, *Bioconjugate Chem.*, 2010, **21**, 715.
16 X. Q. Yang, H. Hong, J. J. Grailer, I. J. Rowland, A. Javadi, S. A. Hurley, Y. L. Xiao, Y. A. Yang, Y. Zhang, R. Nickles, W. B. Cai, D. A. Steeber and S. Q. Gong, *Biomaterials*, 2011, **32**, 4151.
17 R. Qiao, C. Yang and M. Gao, *J. Mater. Chem.*, 2009, **19**, 6274.
18 W.-W. Wang and J.-L. Yao, *Mater. Lett.*, 2010, **64**, 840.
19 A. F. Rebolledo, S. Laurent, M. Calero, A. Villanueva, M. Knobel, J. F. Marco and P. Tartaj, *ACS Nano*, 2010, **4**, 2095.
20 S. Nath, C. Kaittanis, V. Ramachandran, N. S. Dalal and J. M. Perez, *Chem. Mater.*, 2009, **21**, 1761.
21 G. J. Ehrhardt and M. J. Welch, *J. Nucl. Med.*, 1978, **19**, 925.
22 B. P. Burke, G. S. Clemente and S. J. Archibald, *J. Labelled Compd. Radiopharm.*, 2014, **57**, 239.
23 Q. K. T. Ng, T. Segura, A. Ben-Shlomo, T. Krause, T. L. Mindt and M. A. Walter, *J. Nano Res.*, 2012, **20**, 21.
24 M. D. Bartholomä, *Inorg. Chim. Acta*, 2012, **389**, 36.
25 D. J. Berry, Y. Ma, J. R. Ballinger, R. Tavare, A. Koers, K. Sunassee, T. Zhou, S. Nawaz, G. E. D. Mullen, R. C. Hider and P. J. Blower, *Chem. Commun.*, 2011, **47**, 7068.
26 J.-F. Morfin and E. Toth, *Inorg. Chem.*, 2011, **50**, 10371.
27 J. P. L. Cox, A. S. Craig, I. M. Helps, K. J. Jankowski, D. Parker, M. A. W. Eaton, A. T. Millican, K. Millar, N. R. A. Beeley and B. A. Boyce, *J. Chem. Soc., Perkin Trans. 1*, 1990, 2567.
28 E. Wong, P. Caravan, S. Liu, S. J. Rettig and C. Orvig, *Inorg. Chem.*, 1996, **35**, 715.
29 R. E. Mewis and S. J. Archibald, *Coord. Chem. Rev.*, 2010, **254**, 1686.
30 C. F. Ramogida and C. Orvig, *Chem. Commun.*, 2013, **49**, 4720.
31 S. Liu and D. S. Edwards, *Bioconjugate Chem.*, 2001, **12**, 7.
32 G. J. Stasiuk and N. J. Long, *Chem. Commun.*, 2013, **49**, 2732.
33 B. P. Burke and S. J. Archibald, *Annu. Rep. Prog. Chem., Sect. A*, 2013, **109**, 232.
34 J.-s. Choi, J. C. Park, H. Nah, S. Woo, J. Oh, K. M. Kim, G. J. Cheon, Y. Chang, J. Yoo and J. Cheon, *Angew. Chem., Int. Ed.*, 2008, **47**, 6259.
35 L. Frullano, C. Catana, T. Benner, A. D. Sherry and P. Caravan, *Angew. Chem., Int. Ed.*, 2010, **49**, 2382.
36 L. E. Jennings and N. J. Long, *Chem. Commun.*, 2009, 3511.

37 E. Gianolio, L. Maciocco, D. Imperio, G. B. Giovenzana, F. Simonelli, K. Abbas, G. Bisi and S. Aime, *Chem. Commun.*, 2011, **47**, 1539.
38 Y. J. Liu and M. J. Welch, *Bioconjugate Chem.*, 2012, **23**, 671.
39 D. E. Lee, H. Koo, I. C. Sun, J. H. Ryu, K. Kim and I. C. Kwon, *Chem. Soc. Rev.*, 2012, **41**, 2656.
40 M. Swierczewska, S. Lee and X. Y. Chen, *Mol. Imaging*, 2011, **10**, 3.
41 Z. Ali, A. Z. Abbasi, F. Zhang, P. Arosio, A. Lascialfari, M. F. Casula, A. Wenk, W. Kreyling, R. Plapper, M. Seidel, R. Niessner, J. Knoll, A. Seubert and W. J. Parak, *Anal. Chem.*, 2011, **83**, 2877.
42 X. Yang, P. Yu, M. S. Moats and X. Zhang, *Powder Technol.*, 2011, **212**, 439.
43 W. Zhang, S. Jia, Q. Wu, J. Ran, S. Wu and Y. Liu, *Mater. Lett.*, 2011, **65**, 1973.
44 G. J. Stasiuk, H. Smith, M. Wylezinska-Arridge, J. L. Tremoleda, W. Trigg, S. K. Luthra, V. M. Iveson, F. N. E. Gavins and N. J. Long, *Chem. Commun.*, 2013, **49**, 564.
45 Y. Zhang, N. Kohler and M. Q. Zhang, *Biomaterials*, 2002, **23**, 1553.
46 E. K. U. Larsen, T. Nielsen, T. Wittenborn, H. Birkedal, T. Vorup-Jensen, M. H. Jakobsen, L. Ostergaard, M. R. Horsman, F. Besenbacher, K. A. Howard and J. Kjems, *ACS Nano*, 2009, **3**, 1947.
47 M. A. Green and M. J. Welch, *Int. J. Radiat. Appl. Instrum., Part B*, 1989, **16**, 435.
48 J. A. Barreto, M. Matterna, B. Graham, H. Stephan and L. Spiccia, *New J. Chem.*, 2011, **35**, 2705.
49 R. T. M. de Rosales, R. Tavare, R. L. Paul, M. Jauregui-Osoro, A. Protti, A. Glaria, G. Varma, I. Szanda and P. J. Blower, *Angew. Chem., Int. Ed.*, 2011, **50**, 5509.
50 B. R. Jarrett, B. Gustafsson, D. L. Kukis and A. Y. Louie, *Bioconjugate Chem.*, 2008, **19**, 1496.
51 S. T. Selvan, P. K. Patra, C. Y. Ang and J. Y. Ying, *Angew. Chem., Int. Ed.*, 2007, **46**, 2448.
52 J. H. Gao, H. W. Gu and B. Xu, *Acc. Chem. Res.*, 2009, **42**, 1097.
53 M. D. Butterworth, L. Illum and S. S. Davis, *Colloids Surf., A*, 2001, **179**, 93.
54 T. Hara, *Int. J. Nucl. Med. Biol.*, 1974, **1**, 152.
55 D. Patel, A. Kell, B. Simard, B. Xiang, H. Y. Lin and G. H. Tian, *Biomaterials*, 2011, **32**, 1167.
56 H.-Y. Lee, Z. Li, K. Chen, A. R. Hsu, C. Xu, J. Xie, S. Sun and X. Chen, *J. Nucl. Med.*, 2008, **49**, 1371.
57 R. Rossin, D. P. J. Pan, K. Qi, J. L. Turner, X. K. Sun, K. L. Wooley and M. J. Welch, *J. Nucl. Med.*, 2005, **46**, 1210.
58 R. T. M. de Rosales, *J. Labelled Compd. Radiopharm.*, 2014, **57**, 298.
59 H. S. Choi, W. Liu, P. Misra, E. Tanaka, J. P. Zimmer, B. I. Ipe, M. G. Bawendi and J. V. Frangioni, *Nat. Biotechnol.*, 2007, **25**, 1165.
60 P. Bouziotis, D. Psimadas, T. Tsotakos, D. Stamopoulos and C. Tsoukalas, *Curr. Top. Med. Chem.*, 2012, **12**, 2694.
61 M. Ma, Y. Zhan, Y. Shen, X. Xia, S. Zhang and Z. Liu, *J. Nanopart. Res.*, 2011, **13**, 3249.
62 M. Mahmoudi, S. Sant, B. Wang, S. Laurent and T. Sen, *Adv. Drug Delivery Rev.*, 2011, **63**, 24.
63 C. Nazli, T. I. Ergenc, Y. Yar, H. Y. Acar and S. Kizilel, *Int. J. Nanomed.*, 2012, **7**, 1903.

Faraday Discussions

PAPER

Protein A-conjugated iron oxide nanoparticles for separation of *Vibrio cholerae* from water samples

Tran Quang Huy,[a] Pham Van Chung,[a] Nguyen Thanh Thuy,[a] Cristina Blanco-Andujar[bc] and Nguyễn Thị Kim Thanh*[bc]

Received 20th July 2014, Accepted 1st August 2014
DOI: 10.1039/c4fd00152d

Pathogen separation is of great significance for precise detection and prevention of disease outbreaks. For the first time, protein A conjugated with chitosan-coated iron oxide nanoparticles was prepared for pathogen separation at low concentrations from liquid samples. *Vibrio cholerae* O1 (VO1) bacteria were used for testing the effectiveness of this conjugate. Transmission electron microscopy (TEM) was used to confirm the presence of captured VO1. The results showed that, after binding with a specific antibody, the conjugate allows separation of VO1 bacteria from water samples at a concentration as low as 10 cfu mL^{-1}. Moreover, the conjugate can be used in parallel with conventional or modern diagnostic tests for quick and accurate detection of pathogens.

Introduction

Recently, the world has witnessed an increased number of emerging and re-emerging infectious disease outbreaks caused by viruses and bacteria, such as ebola, severe acute respiratory syndrome-associated coronavirus (SARS-CoV), avian influenza (A/H_5N_1, A/H_7N_9), foot–hand–mouth diseases, measles, diarrhea and cholera. These outbreaks not only have public health consequences, but also socio-economic tolls.[1] Rapid and accurate detection of pathogens is of paramount important for the control of these infectious diseases and prevention of their spread to the community.

In particular, cholera outbreak caused by *V. cholerae* is one of the continuous risks to public health and socio-economic development in many developing countries.[2] *V. cholerae* is a Gram-negative, comma-shaped bacterium, which causes abrupt onset of watery diarrhoea, occasionally vomiting, renal failure,

[a]*National Institute of Hygiene and Epidemiology, 1 Yersin Street, Hai Ba Trung District, Hanoi, Vietnam*
[b]*Department of Physics and Astronomy, University College London, London, WC1E 6BT, UK*
[c]*UCL Healthcare Biomagnetic and Nanomaterials Laboratories, 21 Albemarle Street, London W1S 4BS, UK. E-mail: ntk.thanh@ucl.ac.uk; Tel: +44 (0) 207 4916509*

coma and even death. There are more than 200 known *V. cholerae* serogroups, but only the O1 and O139 strains are pathogenic and cause epidemics. *V. cholerae* bacteria can also live for up to one year in an aquatic environment due to a protective bio-film.[2] It is not easy to detect *V. cholerae* using conventional diagnostic methods, because it is non-culturable and generally found in low levels in water.[3] Several techniques have been developed for detection of *V. cholerae* in environmental water including immunomagnetic beads, nucleic acid-based methods (such as PCR and DNA probe hybridization) or fluorescent labeling techniques.[4] Immunomagnetic bead detection offers good sensitivity, but has the possibility of false positives. Nucleic acid-based methods require biological agents such as primers, standardised bio-safe laboratories, and can only detect nucleic acids rather than live bacterial cells. The fluorescent labeling technique is preferred for the detection of *V. cholerae* in environmental water samples, but filtering is necessary to concentrate the bacteria from a large volume of water.

In recent years, many studies have shown the potential applications of magnetic nanoparticles in biomedicine such as imaging, targeted drug delivery and therapy,[5] but only a few studies have used magnetic nanoparticles for the detection of infectious disease pathogens so far.[6] The main challenges in the preparation of magnetic nanoparticles for pathogen detection are surface modification and biofunctionalisation, particularly the immobilisation of biomolecules on the surface of nanoparticles as probes.[7]

In this study, a conjugate of protein A and chitosan-coated magnetic nanoparticles was prepared for separation of *V. cholerae* bacteria at low concentrations from water samples. Chitosan, a polysaccharide obtained from the *N*-deacetylation of chitin, is one of the most abundant polysaccharides in nature. It has been used for a wide range of biological applications, thanks to its biocompatibility and biodegradability.[8] Its use for the functionalisation of nanoparticles not only allows the incorporation of amine groups on the surface of the nanoparticles, but also adds a positive charge to the system, which has been shown to be preferred by some cell lines.[9] On the other hand, protein A is well known as a molecule which binds with high affinity to the Fc region of IgG antibodies and orientates the binding sites of the antibody molecule outwards from the surface of nanoparticles for capturing antigens.[10]

Herein, chitosan-coated iron oxide nanoparticles (IONPs) were first conjugated with protein A and then the conjugate was incubated with the specific IgG antibody against *V. cholerae* strain just before testing. The *V. cholerae* bacteria could be easily magnetically separated and collected by binding to iron oxide nanoparticles.

Experimental procedures

Materials

Protein A (PrA, extracellular from *Staphylococcus aureus*, lyophilized, ≥90% protein basis), glutaraldehyde solution (GA, grade I, 25%), paraformaldehyde (reagent grade), sodium cacodylate trihydrate ($(CH_3)_2AsO_2Na \cdot 3H_2O$, ≥98%), and chitosan $(C_6H_{11}NO_4)_n$, $M_w \sim 60\,000$ g mol^{-1} were obtained from Sigma Aldrich, UK. *V. cholerae* O1 bacteria, anti-*V. cholerae* O1 polyclonal antibody, and *Escherichia coli* (*E. coli*) were provided by Professor Nguyen Binh Minh and Dr Tran Huy Hoang at Department of Bacteriology, National Institute of Hygiene and

Epidemiology (NIHE), Hanoi. Bacterial strains were inoculated at the laboratory of enteric bacteria, certified by ISO 15189: 2012, and all proper and standard biosafety procedure were followed when handling them. All chemicals used were of analytical grade.

Synthesis and characterization of chitosan-coated IONPs

Chitosan-coated IOPNs were synthesized by a chemical method,[11] and then characterized by transmission electron microscope (TEM), dynamic light scattering (DLS), zeta-potential (ζ-potential), and attenuated total reflectance-Fourier transformed infrared spectroscopy (ATR-FTIR).

Bio-conjugation of protein A to IONPs

Protein A was conjugated to the surface of chitosan-coated IONPs *via* a cross-linkage with GA (Fig. 1). The two functional aldehyde groups of GA can easily co-bind with amine groups of the chitosan layer on the surface of IONPs and those of PrA. Firstly, the solution of chitosan-coated IONP (IONP@CHI) was diluted in distilled water to a concentration of 10 $\mu g\ mL^{-1}$, and the solution pH adjusted to pH 6.5. Next, 5% GA was added, incubated for 45 min at room temperature (RT), and washed with phosphate-buffered saline, PBS (0.01 M, pH 7.4) three times to remove residual GA (Fig. 1, step 1). 2 $\mu g\ mL^{-1}$ of protein A was then added to the solution of IONPs and incubated overnight at 4 °C (Fig. 1, step 2). After washing twice with PBS, the non-specific binding sites on the surfaces of the nanoparticles were blocked with 0.5% bovine serum albumin (BSA) in PBS for 20 min at RT. Finally, the conjugate was washed with PBS twice, and then re-suspended in 0.5 mL PBS and kept at 4 °C for testing. Due to the two carbonyl functional groups of GA, as well as multiple amine groups of protein A, cross-linking could occur between IONP@CHI leading to aggregation or precipitation; however, these

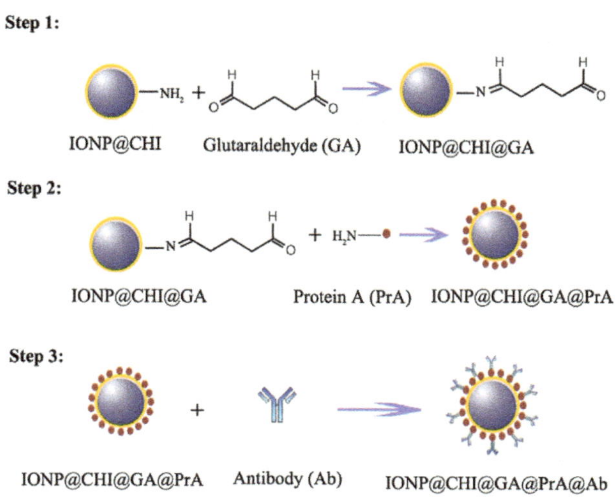

Fig. 1 Step 1: Glutaraldehyde (GA) chitosan was conjugated to chitosan-coated IONP (IONP@CHI) and step 2: reaction of IONP@CHI@GA and protein A (PrA) to form the conjugate of IONP@CHI@GA@PrA. Step 3: preparation of IONP@CHI@GA@PrA linked with specific antibodies against pathogens *via* their bio-affinity.

agglomerates can easily be discarded. This phenomenon can easily be checked by naked eye or using DLS to observe the increase in size of particle aggregates in the solution. Also, round-shaped IONPs may help to minimize the effect of aggregation, because there might be some free spaces remaining on the surface of each IONP@CHI for binding to target molecules, even with several particles stuck together. In our study, the preparation of the conjugate was done in triplicate and no flocculation was detected. The conjugate can then be incubated with specific IgG antibodies against pathogens in advance before separation (Fig. 1, step 3).

To separate *V. cholerae* bacteria from a water sample, 0.5 mL of IONP@CHI@GA@PrA was incubated in advance with polyclonal antibody against *V. cholerae* O1 at a concentration of 2 μg mL^{-1}, for 45 min at RT. In this step, IgG antibody molecules would bind with high affinity to protein A–IONPs (IONP@CHI@GA@PrA). A magnet was used to collect the product of IgG antibody conjugated IONPs (IONP@CHI@GA@PrA@Ab) on the wall of the tube, washed two times with PBS to remove unbound antibodies, and then re-suspended in 0.5 mL of PBS (Fig. 2b). This procedure can be carried out just before separation and detection at RT in the laboratory or prepared in advance and then kept at 4 °C for further detection on-site.

Investigation of bacterial cells before testing

Scanning electron microscopy (SEM) (S-4800, Hitachi) was used to check the presence and morphology of bacterial cells in the samples before testing with the conjugate of IONPs and protein A. At a concentration of 10^6 cfu mL^{-1}, the samples of *V. cholerae* and *E. coli* were fixed in 2% paraformaldehyde/2.5% GA for 1 h, then washed with 0.1 M cacodylate buffer before post-fixing with 2% aqueous OsO_4 in 0.2 M cacodylate buffer for 2 h. After washing with 0.1 M cacodylate buffer again, samples were dehydrated with increasing ethanol solutions (from 30% to 100% with 10% increments), dried using a critical point dryer (Emitech, K850, Quorum Technologies), and coated with Pt–Pd using ion sputter before SEM imaging.

Separation of *V. cholerae* O1 from water samples

V. cholerae O1 bacteria were dispersed in 0.01 M PBS (pH 7.4) with an initial concentration of 10^3 cfu mL^{-1}. Ten-fold serial dilutions in water were then prepared in test tubes to a final volume of 5 mL from the original concentration of bacteria. A 50 μL aliquot of the complex of IONP@CHI@GA@PrA@Ab was added

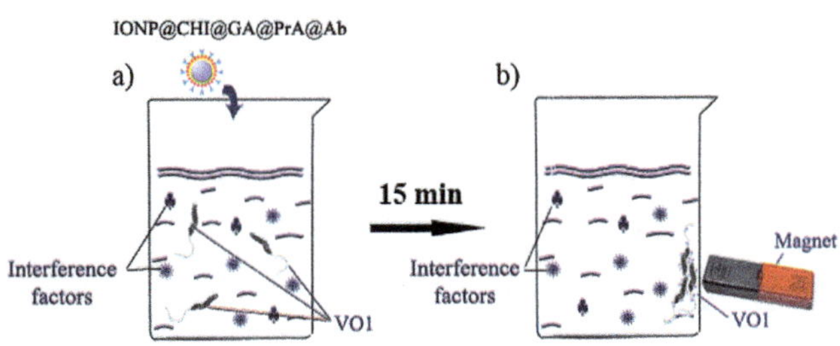

Fig. 2 Separation of VO1 bacterial cells from water samples with interference factors: (a) adding the complex of IONP@CHI@GA@PrA@Ab to the sample, and (b) attracting bacterial cells with a bar magnet.

to the test tube (with the ratio of 1 : 100 v/v), and mixed at RT for about 15 min (Fig. 2a). The IONP-bound VO1 bacteria were collected on the wall of the tube using a bar magnet, and most of the supernatant was removed with a pipette. The conjugate was then re-suspended with the remaining water in the tube (Fig. 2b). A drop of this solution was transferred to a carbon coated copper grid to determine the presence of VO1 bacteria under a TEM (JEM1010, JEOL), operating at 80 kV.

Results and discussion

The synthesised nanoparticles are spheroidal, as observed by TEM (Fig. 3A and B), with an average particle size of 13.4 ± 2.5 nm, as calculated from the measurement of 400 particles (Fig. 3C). The hydrodynamic size of the chitosan-IONPs, as observed by DLS, is 82.9 ± 13 nm. The measured ζ-potential of 45 mV indicated the presence of chitosan around the IONPs (Fig. 3D). From ATR-FTIR spectra (Fig. 3E), we could observe the NH_2 bending bands appearing at 1658 and 1588 cm^{-1} for free chitosan, merging to the 1616 cm^{-1} band after coating. The peak at 1378 cm^{-1} for the ν_{CO} vibration of the hydroxyl group, the C–N stretching at 1153 cm^{-1} and the stretching of the aliphatic CH_2 group around 2900 cm^{-1} were found

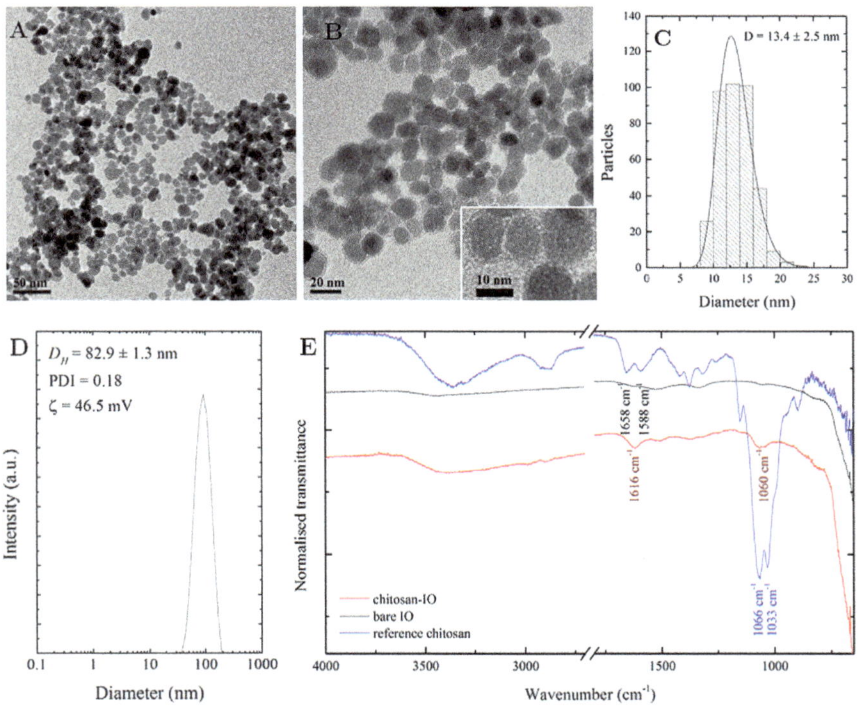

Fig. 3 (A) and (B) TEM images of chitosan-coated IONPs and (C) particle size distribution from image analysis obtained from the measurement of 400 particles. Size distribution was fitted with a log normal function. (D) DLS and ζ-potential of chitosan-IONPs and (E) ATR-FTIR spectra of IONPs before and after chitosan coating. Reference spectrum for chitosan is included for reference. All spectra have been normalized and offset for better comparison. The 1800–2700 cm^{-1} range has been excluded as no significant information was found.

in the coated samples. The coordination of chitosan to the surface of the nanoparticles occurs *via* charge affinity as the reaction occurs at acidic pH where chitosan presents a high positive charge density and thus high affinity for the negatively charged nanoparticle surface. Nonetheless, not all the amine groups are bound to the nanoparticles, as surface charge was 46.5 mV, due to the presence of positively charged groups. As a result, the obtained system is highly stable due to steric repulsion of the nanoparticles provided by the polymer backbone and the surface charge obtained from the amine groups.

Before testing with the complex of IONP@CHI@GA@PrA@Ab, the solution containing bacterial strains was checked by SEM. SEM images revealed the presence of *V. cholerae* and *E. coli* bacterial cells in the samples. *V. cholerae* bacteria were found to be comma-shaped, measuring 0.3 μm in diameter and 1.3 μm in length and have a flagellum at one cell pole (Fig. 4A). *E. coli* bacterial cells are rod-shape, measuring 0.5 μm in diameter and 1.2 μm in length (Fig. 4B).

To test the effectiveness of the conjugate, IONP@CHI@GA@PrA was incubated with anti-*V. cholerae* O1 polyclonal antibody before being dropped into the tubes containing *V. cholerae* bacteria.

The TEM images show that before the test with the complex of IONP@CHI@GA@PrA@Ab, *V. cholerae* bacterial cell could be clearly seen without the presence of IONPs (Fig. 5a). After using IONP@CHI@GA@PrA@Ab to separate the bacterial cells from water, *V. cholerae* cells were found to aggregate on the grid and were bound by IONPs (Fig. 5b). In this study, we used polyclonal IgG antibodies specific to *V. cholerae* O1 bacteria, so IONPs would specifically bind around the bacterial cell wall (Fig. 5c) as well as the bacterial flagellum (Fig. 5d). All *V. cholerae* O1 bacterial cells found on the grids were covered with IONPs, which demonstrates that the capture efficiency of IONP@CHI@GA@PrA@Ab to *V. cholerae* O1 bacterial cells is up to 100%. This may be the result of the diffusion capacity of IONPs in solution, as well as the specificity of the IONP@CHI@GA@PrA@Ab complex to pathogens. By serial dilution of bacterial cultures, the results also showed that VO1 bacteria could be detected at a level as low as 10 cfu mL^{-1}, as indicated by the IONPs presence (repeated three times). Interestingly, most of IONPs were found sticking around bacterial cells, with only a few small clusters of IONPs found on the TEM grids. After separation, the supernatant was also checked by TEM for the presence of bacterial cells remaining there, but none were found on the grids. This confirmed that most of the bacterial cells adhered to IONPs and were attracted by a magnet towards the wall of the tube.

For positive controls, the solutions containing IONP@CHI, and IONP@CHI@GA@PrA were incubated with *V. cholerae* bacteria diluted in water from

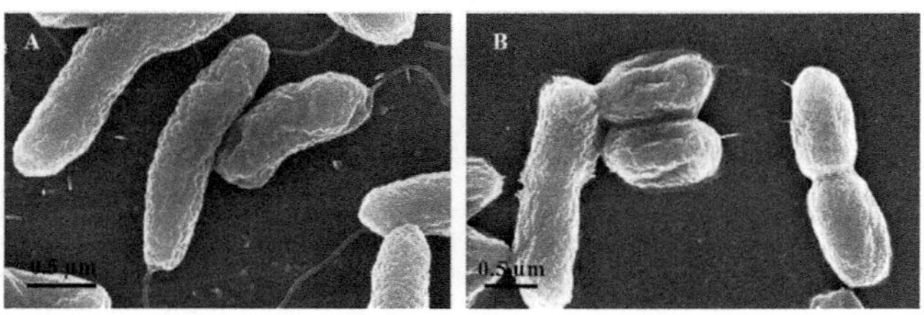

Fig. 4 Scanning electron microscopic images of (A) *V. cholerea* and (B) *E. coli*.

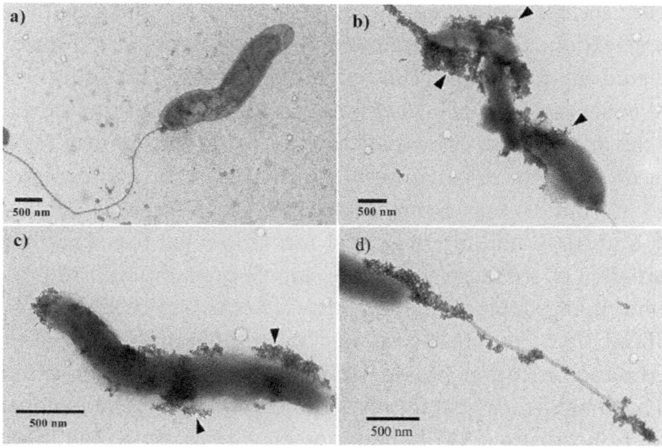

Fig. 5 TEM images of *V. cholerae* bacteria (a) before interaction with the complex of IONP@CHI@GA@PrA@Ab, (b) aggregated and bound by IONPs, (c) IONPs bound to the cell wall, and (d) IONPs bound to the bacterial flagellum.

an initial concentration of 10^3 cfu mL^{-1}. The same procedures used for testing the IONP@CHI@GA@PrA@Ab complex were followed. A significant number of nanoparticles were found clustered on the grid, but no *V. cholerae* bacterial cells were found (data not shown).

To demonstrate the specificity of the complex to the pathogens, as negative controls *E. coli* bacteria were also prepared in 5 mL of PBS with an initial concentration of 10^3 cfu mL^{-1}. Ten-fold serial dilutions in water were also prepared in 5 mL for testing. The complex of IONP@CHI@GA@PrA@Ab (using anti-*V. cholerae* O1 polyclonal IgG antibody) was then added to the tubes and mixed for 15 min at RT. The same procedure used for VO1 was followed; however, TEM images did not show the presence of *E. coli* bacterial cells on the grids.

After incubation of the protein A-conjugated IONPs with anti-*V. cholerae* O1 polyclonal IgG antibody, the separation ability of VO1 bacteria from water samples could be found at concentrations as low as 10 cfu mL^{-1}, as confirmed by TEM. More importantly, the detection level was found to be better than with conventional diagnostic techniques, where bacterial pathogens at concentrations lower than 10^2 cfu mL^{-1} cannot be detected without inoculation, especially non-culturable bacteria that present their own bio-films such as *V. cholerae*.[12] Following isolation and culture, a direct fluorescent antibody (DFA) test is known to be an effective method for direct detection of *V. cholerae*. Hasan *et al.*[13] applied DFA to detect *V. cholerae* O139 in water samples, and the sensitivity of *V. cholerae* O139 detection reached 150 cfu mL^{-1}, which was 15-fold higher than the detection levels reported here. In this work, the complex serves as a convenient and versatile agent for separation of different pathogens on-site or in the laboratory. It can be used for selective enrichment of pathogens for early diagnostic tests and preliminary pathogenic screening for a large number of patients who were infected in a short period of time during outbreaks. Noticeably, before testing, the samples and the complex should be adjusted to pH 7.0–7.4 to ensure their bioactivity, but all procedures for separation and detection can be performed at RT.

There are several reports on the preparation of magnetic nanoparticles for pathogen detection. Gu *et al.*[14] showed that biofunctionalised magnetic nanoparticles could capture vancomycin-resistant *Enterococci* and other Gram-positive bacteria at concentrations of ~10 cfu mL^{-1} within 1 h. In this study, the authors proposed the use of either optical microscopy or SEM to confirm the presence of "magnetized" bacterial cells. However, to the best of our knowledge, it is very difficult to use optical microscopy to identify the differences between bacterial strains, due to the limitation of optical resolution. On the other hand, SEM is time-consuming, as it requires complex sample preparation by dehydration and critical point drying steps before imaging. Gao *et al.*[15] also prepared vancomycin-functionalised FePt nanoparticles, but in combination with a vancomycin-conjugated fluorescent probe to detect bacteria in human blood with a response time of 2 h. Cheng *et al.*[16] combined biofunctional magnetic nanoparticles and adenosine triphosphate (ATP) bioluminescence to detect *E. coli* inoculated into pasteurized milk with a detection limit of 20 cfu mL^{-1} and a detection time of about 1 h. Recently, the development of a magneto-DNA platform based on magnetic nanoparticles that could detect a panel of 13 bacterial species within 2 h was reported.[17] The reported literature demonstrates the use of biofunctionalised magnetic nanoparticles for detection of different pathogens with high sensitivity; however, they all present some intrinsic limitations. Firstly, biofunctionalised nanoparticles need to be prepared in advance against specific pathogens. Moreover, modern complex facilities and equipment are needed, and response time is still longer than 1 h. It is important to consider that for magnetic nanoparticles conjugated directly to antibodies, bio-activity may be reduced due to manipulation or with time.

In this work, the conjugate of protein A and IONPs overcomes some of the aforementioned limitations. Firstly, it serves as a convenient platform for enrichment and separation of various pathogens in different liquid samples, after just a prior incubation with their specific IgG antibodies. Moreover, this conjugate can be used as a smart material to selectively concentrate pathogens from a large volume of water, which may indirectly improve the sensitivity of other diagnostic techniques. Finally, it can also be applied for purifying water infected by pathogens.

Concluding remarks

Our preliminary study showed that the conjugate of protein A and IONPs was successfully prepared and tested for separation of *V. cholerae* bacteria at low concentrations from water samples in the laboratory. The results showed by TEM observation that the conjugate could easily separate *V. cholerae* bacteria from water samples at concentrations as low as 10 cfu mL^{-1}. The selective separation of specific pathogens from liquid samples is achieved by incubation of the conjugate with a specific monoclonal or polyclonal antibody. The pathogens separated from a large volume of liquid samples can then be detected simply by conventional diagnostic methods or immunochromatographic strip tests on-site. More importantly, it has the potential to be used for the separation and detection of even unknown pathogens in clinical samples collected from emerging outbreaks. This conjugate would be incubated in advance with serum collected from patients presenting typical symptoms of the outbreak to perform the immunolabelling reaction. This procedure may allow the preliminary recognition of pathogens by studying their morphology using SEM or TEM. This approach may help other advanced techniques to confirm the causing agent of the outbreak. Further investigation is planned to optimize the bio-activity and stability of the conjugate over time.

Acknowledgements

Nguyễn T. K. Thanh thanks the Royal Society for her Royal Society University Research Fellowship and Foreign Common Wealth Office for a travel grant.

References

1 E. K. Jones, G. N. Patel, A. M. Levy, A. Storeygard, D. Balk, L. J. Gittleman and P. Daszak, *Nature*, 2008, **451**, 990; J. R. Koker, M. B. Hunter, W. J. Rudge, M. Liverani and P. Hanvoravongchai, *Lancet*, 2011, **377**, 599; A. S. Fauci and D. M. Morens, *N. Engl. J. Med.*, 2012, **366**, 454.
2 S. Chaiyanan, A. Huq, T. Maugel and R. R. Colwell, *Syst. Appl. Microbiol.*, 2001, **24**, 331; P. R. Mason, *J. Infect. Dev. Ctries.*, 2009, **3**, 148; N. Sithivong, T. Morita-Ishihara, A. Vongdouangchanh, T. Phouthavane, K. Chomlasak, *et al.*, *Emerging Infect. Dis.*, 2011, **17**, 2060; D. D. Anh, A. L. Lopez, V. D. Thiem, S. L. Grahek, T. N. Duong, *et al.*, *PLoS Neglected Trop. Dis.*, 2011, **5**, e1006; A. M. Devault, G. B. Golding, N. Waglechner, J. M. Enk and M. Kuch, *N. Engl. J. Med.*, 2014, **370**, 334.
3 S. Banoo, D. Bell, P. Bossuyt, A. Herring, D. Mabey, *et al.*, *Nat. Rev. Microbiol.*, 2010, **8**(suppl. 12), S17; F. S. Nolte, *Clin. Infect. Dis.*, 2008, **47**(suppl. 3), S123; P. Yager, G. J. Domingo and J. Gerdes, *Annu. Rev. Biomed. Eng.*, 2008, **10**, 107; A. van Belkum, N. H. Renders, S. Smith, S. E. Overbeek and H. A. Verbrugh, *FEMS Immunol. Med. Microbiol.*, 2000, **27**, 51.
4 K. H. Seo, R. E. Brackett and J. F. Frank, *Int. J. Food Microbiol.*, 1998, **44**, 115; S. Chakraborty, J. Khanam, Y. Takeda and G. B. Nair, *Trans. R. Soc. Trop. Med. Hyg.*, 1999, **93**, 527; E. K. Lipp, I. N. Rivera, A. I. Gil, E. M. Espeland, N. Choopun, V. R. Louis, E. Russek-Cohen, A. Huq and R. R. Colwell, *Appl. Environ. Microbiol.*, 2003, **69**, 3676; D. Wang, X. Xu, X. Deng, C. Cheng, B. Li, *et al.*, *Appl. Environ. Microbiol.*, 2010, **76**, 5520.
5 Q. A. Pankhurst, N. T. K. Thanh, S. K. Jones and J. Dobson, *J. Phys. D: Appl. Phys.*, 2009, **42**, 224001; I. Robinson, L. D. Tung, S. Maenosono, C. Walti and N. T. K. Thanh, *Nanoscale*, 2010, **2**, 2624; O. Veiseh, J. W. Gunn and M. Zhang, *Adv. Drug Delivery Rev.*, 2010, **62**, 284; X. Meng, H. Seton, L. T. Lu, I. Prior, N. T. K. Thanh and B. Song, *Nanoscale*, 2011, **3**, 977; Y. Wang, C. Blanco-Andujar, Z. Zhi, P. W. So, N. T. K. Thanh and J. C. Pickup, *Chem. Commun.*, 2013, **49**, 7255; R. Hachani, M. Lowdell, M. Birchall and N. T. K. Thanh, *Nanoscale*, 2013, **5**, 11362; A. Hervault and N. T. K. Thanh, *Nanoscale*, 2014, DOI: 10.1039/c4nr03482a.
6 N. Tran and T. J. Webster, *J. Mater. Chem.*, 2010, **20**, 8760; P. Tallury, A. Malhotra, L. M. Byrne and S. Santra, *Adv. Drug Delivery Rev.*, 2010, **62**, 424; T. S. Hauck, S. Giri, Y. Gao and W. C. W. Chan, *Adv. Drug Delivery Rev.*, 2010, **62**, 438.

7 N. T. K. Thanh and L. A. W. Green, *Nano Today*, 2010, **5**, 213; T. D. Schladt, K. Schneider, H. Schild and W. Tremel, *Dalton Trans.*, 2011, **40**, 6315; R. A. Sperling and W. J. Parak, *Philos. Trans. R. Soc. London, Ser. A*, 2010, **368**, 1333; H. Xu, Z. P. Aguilar, L. Yang, M. Kuang, H. Duan, Y. Xiong, H. Wei and A. Wang, *Biomaterials*, 2011, **32**, 9758; A. K. Gupta and M. Gupta, *Biomaterials*, 2005, **26**, 3995.

8 M. N. V. Ravi Kumar, R. A. A. Muzzarelli, C. Muzzarelli, H. Sashiwa and A. J. Domb, *Chem. Rev.*, 2004, **104**, 6017; Y. Yi, L. Q. Wu, W. E. Bentley, R. Ghodssi, G. W. Rubloff, J. N. Culver and G. F. Payne, *Biomacromolecules*, 2005, **6**, 2881.

9 E. Frohlich, *Int. J. Nanomed.*, 2012, **7**, 5577; Y. Ge, Y. Zhang, J. Xia, M. Ma, S. He, F. Nie and N. Gu, *Colloids Surf., B*, 2009, **73**, 294; Z. G. Yue, W. Wei, P. P. Lv, H. Yue, L. Y. Wang, Z. G. Su and G. H. Ma, *Biomacromolecules*, 2011, **12**, 2440.

10 I. Takeshi, H. Yumehiro, N. Ken-ichi, I. Yoshiaki, T. Keigo, A. Satoka, H. Ryuichi and K. Akio, *Anal. Biochem.*, 2009, **385**, 132; A. K. Minkstimiene, A. Ramanaviciene, J. Kirlyte and A. Ramanavicius, *Anal. Chem.*, 2010, **82**, 6401; G. Shen, C. Cai, K. Wang and J. Lu, *Anal. Biochem.*, 2011, **409**, 22; T. Q. Huy, N. T. H. Hanh, P. V. Chung, D. D. Anh, P. T. Nga and M. A. Tuan, *Appl. Surf. Sci.*, 2011, **257**, 7090.

11 C. Blanco-Andujar, Sodium carbonate mediated synthesis of iron oxide nanoparticles to improve magnetic hyperthermia efficiency and induce apoptosis, PhD thesis, University College London, London, UK, 2014.

12 M. Alam, M. Sultana, G. B. Nair, A. K. Siddique, N. A. Hasan, R. B. Sack, D. A. Sack, K. U. Ahmed, A. Sadique, H. Watanabe, C. J. Grim, A. Huq and R. R. Colwell, *Proc. Natl. Acad. Sci. U. S. A.*, 2007, **104**, 17801; A. Safa, G. B. Nair and R. Y. C. Kong, *Trends Microbiol.*, 2010, **18**, 46; K. Bharati and S. K. Bhattacharya, *Curr. Top. Microbiol. Immunol.*, 2014, **379**, 87–116.

13 J. A. Hasan, A. Huq, G. B. Nair, S. Garg, A. K. Mukhopadhyay, L. Loomis, D. Bernstei and R. R. Colwell, *J. Clin. Microbiol.*, 1995, **33**, 2935.

14 H. W. Gu, P. L. Ho, K. W. T. Tsang, L. Wang and B. Xu, *J. Am. Chem. Soc.*, 2003, **125**, 15702.

15 J. Gao, L. Li, P.-L. Ho, G. C. Mak, H. Gu and B. Xu, *Adv. Mater.*, 2006, **18**, 3145.

16 Y. Cheng, Y. Liu, J. Huang, K. Li, W. Zhang, Y. Xian and L. Jin, *Talanta*, 2009, **77**, 1332.

17 H. J. Chung, C. M. Castro, H. Im, H. Lee and R. Weissleder, *Nat. Nanotechnol.*, 2013, **8**, 369.

Faraday Discussions

PAPER

A low cost synthesis method for functionalised iron oxide nanoparticles for magnetic hyperthermia from readily available materials†

Joseph C. Bear,*[a] Bin Yu,[a] Cristina Blanco-Andujar,[b] Paul D. McNaughter,[c] Paul Southern,[b] Marc-Krystelle Mafina,[d] Quentin A. Pankhurst[b] and Ivan P. Parkin[a]

Received 9th April 2014, Accepted 15th July 2014
DOI: 10.1039/c4fd00062e

The synthesis of iron oxide nanocrystals from reagents taken from high street sources using thermal decomposition of an iron–fatty acid precursor in a high boiling point solvent in the presence of surfactants is presented. The nanocrystals were characterised using a variety of techniques including: electron microscopy, X-ray dispersive spectroscopy, infrared spectroscopy, X-ray diffraction (XRD), X-ray photoelectron spectroscopy (XPS) and magnetometry. Thermogravimetric analysis (TGA) is also used to compare the decomposition behaviour of iron oleate and iron palmitate, our nanoparticle precursors. The nanoparticles also exhibit shape anisotropy when prepared under optimum conditions. We show that these nanoparticles have potential in magnetic hyperthermia after transfer to aqueous media *via* an amphiphilic polymer.

Introduction

The large scale and inexpensive synthesis of monodisperse nanocrystals from non-hazardous precursors is an area of intense interest in nanoparticulate research.[1–5]

Synthesis methods such as continuous hydrothermal flow synthesis and ball milling generate kilogram-scale volumes of nanoparticles, but often at the

[a]*Materials Chemistry Research Centre, Department of Chemistry, University College London, 20 Gordon Street, London, WC1H 0AJ, UK. E-mail: i.p.parkin@ucl.ac.uk*

[b]*UCL Healthcare Biomagnetics Laboratories, Royal Institution of Great Britain, 21 Albemarle Street, London, W1S 4BS, UK*

[c]*School of Chemistry, University of East Anglia, Norwich Research Park, Norwich, NR4 7TJ, UK*

[d]*School of Engineering and Materials Science, Queen Mary, University of London, Mile End Road, London, E1 4NS, UK*

† Electronic supplementary information (ESI) available: Contains additional TEM, SAED, EDX and FTIR characterisation, differential scanning calorimetry curves of iron oleate and iron palmitate. See DOI: 10.1039/c4fd00062e

expense of product crystallinity and monodispersity.[6–9] On the other hand, the nanoparticle quality control gained using hot-injection and "heat-up" methods is excellent.[3,10,11] However, these methods often utilise expensive reagents and are not applicable for large scale processes due to large temperature gradients placed across synthesis solutions. The vast majority of "heat-up" methods involve the thermal decomposition of iron–fatty acid complexes such as iron oleate. Iron–fatty acid complexes are either isolated and decomposed in the presence of surfactants[12] or formed *in situ*.[13] In their landmark paper, Hyeon and co-workers decomposed a pre-synthesised iron oleate complex in the presence of oleic acid and 1-octadecene to produce highly monodisperse and size-controllable Fe_3O_4 nanocrystals on a batch-level scale.[3] In their paper, they allude to the possibility of a scale-up process and draw attention to the inexpensive nature and low toxicity of the precursors (Fig. 4, ESI†).

In particular, iron oxide has received a great deal of interest for magnetic fluid hyperthermia (MFH); a cancer treatment which uses radiofrequency to heat up superparamagnetic nanoparticles *in vivo* to thermally ablate/treat tumour cells.[14–17] The best nanoparticles for MFH are often a compromise between size, biocompatibility, cost, ease of synthesis, ability to respond to an external stimulus and heating ability.[17–19] Iron oxide has been of particular interest due to its fulfilment of the majority of these desired properties.[20,21]

In this paper, we have developed a synthesis using a similar thermal decomposition method to the one developed by Hyeon *et al.*, but with precursors solely bought from high street shops and supermarkets. We report the batch-synthesis of high-quality iron oxide nanocrystals at a fraction of the cost compared to the Hyeon paper. We also demonstrate their suitability and effectiveness for magnetic hyperthermia, after transfer to aqueous media *via* amphiphilic polymer coating.[22,23]

The nanoparticles demonstrate significant temperature rises at low nanoparticle concentrations, particularly the nanostructures obtained from the decomposition of iron palmitate in shark liver oil. The methods and reagents proposed in this paper therefore represent a viable, facile and cheap route to effective hyperthermia agents.

Experimental details

Characterisation techniques

Transmission electron microscopy (TEM) images were recorded using a JEOL JEM 1200EX with a 4 megapixel Gatan Orius SC200 charge-coupled device (CCD) camera at an acceleration voltage of 120 kV. High resolution transmission electron microscopy (HRTEM) measurements were collected using a Philips CM200 FEG TEM fitted with a Gatan GIF 200 imaging filter and an Oxford Instruments UTW EDX detector running ISIS software. XPS measurements were recorded on a Thermo Scientific K-alpha spectrometer. XRD patterns were obtained on a PanAlytical diffractometer using Co Kα radiation, λ = 1.789010 Å. ATR-FTIR measurements were taken over a range of 450 to 4000 cm^{-1} using a Perkin-Elmer Spectrum-100 equipped with a universal ATR attachment. Magnetisation data was taken using a Quantum Design MPMS SQUID VSM Magnetometer (San Diego, USA) at 300 K using a field range of $\pm$7 T. Mössbauer spectroscopy was carried out using a SEE Co. Model W302 Resonant Gamma Ray Spectrometer, with a $^{57}Co(Rh)$

gamma ray source operating at 300 K. Isomer shifts were measured relative to α-Fe at 300 K. Hyperthermia experiments were carried out using a MACH system (Magnetic Alternating Current Hyperthermia) designed and built by Resonant Circuits Limited.[24] The temperature was monitored using a fluoroptic temperature probe (Luxtron FOT Lab Kit, Lumasense California USA).

Materials

Iron(III) chloride hexahydrate (98%) was purchased from VWR International Ltd. Iron tablets were purchased from Boots Ltd. Olive oil and "Sainsbury basics" soap was purchased form J. Sainsbury Ltd. Shark liver oil (60 × 1000 mg capsules) was purchased from Shark Liver Oil UK.

All other laboratory chemicals used in this investigation were purchased from Sigma-Aldrich Chemical Co; including: 1-octadecene (technical grade, 90%), oleic acid (technical grade, 90%), oleylamine (technical grade, ≤70%), poly(maleic anhydride-*alt*-1-octadecene) (average M_n = 30 000–50 000), tetramethyl ammonium hydroxide pentahydrate (≥97%) and sodium oleate (≤82% fatty acid content). Laboratory solvents were purchased from Fisher Scientific Ltd. and of the highest possible grade.

Nanoparticle synthesis and functionalisation

Iron oxide nanoparticle synthesis. Iron oxide nanoparticles were synthesised by thermal decomposition of iron oleate, according to Park *et al.*[3] The iron oleate was prepared by heating a suspension of iron(III) chloride hexahydrate (10.8 g, 40 mmol) and sodium oleate (36.5 g, 120 mmol) in a solvent mixture of *n*-hexane (140 ml), deionised water (60 ml) and ethanol (80 ml) to 70 °C for 4 hours. The organic layer was separated and washed with 3 × 30 ml portions of distilled water to remove sodium chloride. The dark brown organic layer was separated and dried *in vacuo*, to remove hexane, and isolated as a waxy solid (yield ~70%).

Iron oleate (18.0 g, 20 mmol) and oleic acid (2.35 ml, 10 mmol) were dissolved in 1-octadecene (100 g, 396.1 mmol) and stirred thoroughly at room temperature. The reaction mixture was heated to 320 °C at a rate of 3.3 °C min^{-1} under nitrogen and held at 320 °C for 1 hour. The resulting black solution was allowed to cool to room temperature before addition of ethanol (250 ml) to precipitate the particles. The solution was centrifuged at 600*g* for 10 minutes, giving solid particle precipitates. If the precipitate was not solid, excess oleate-type species were removed by washing with ethanol (2 × 80 ml). The supernatant was discarded and the solid particle precipitates dispersed in chloroform (30 ml total).

Modified Fe_3O_4 nanoparticle synthesis. Iron palmitate was prepared by heating a suspension of homogenised iron tablets (20 tablets each containing 14 mg of iron as iron gluconate) and soap (assumed to be 65% sodium palmitate, 0.321 g, 1.15 mmol) in a solvent mixture of *n*-hexane (140 ml), deionised water (60 ml) and ethanol (80 ml) to 70 °C for 4 hours. The organic layer was separated and washed with 3 × 30 ml portions of distilled water to remove sodium chloride. The dark organic layer was dried *in vacuo* to remove hexane, yielding the iron oleate complex as a waxy solid.

For ~12 nm particles and rods. Iron palmitate (2 g, 2.4 mmol) and olive oil (65% oleic acid, 0.59 ml, 1.2 mmol) were dissolved in 1-octadecene (20 ml, 62.5 mmol) or shark liver oil (20 ml) and stirred thoroughly at room temperature.

A molar ratio of 1 : 1 iron palmitate to the oleic acid content of olive oil was used to grow the rods (Fig. 2(E) and (F)), whilst the solvent was changed to shark liver oil. The reaction mixture was heated to 320 °C at a rate of 3.3 °C min^{-1} under nitrogen and held at 320 °C for 1 hour. The resulting black solution was allowed to cool to room temperature before addition of ethanol (250 ml) to precipitate the particles. The solution was centrifuged at 600*g* for 10 minutes, giving solid particle precipitates. If the precipitate was not solid, excess fatty acid-type species were removed by washing with ethanol (2 × 80 ml). The supernatant was discarded and the solid particle precipitates dispersed in chloroform (30 ml total).

Phase transfer of iron oxide nanoparticles

Oleic acid-coated nanoparticles were transferred from organic solvents to water according to a protocol developed by Lees *et al.*[22] Iron oxide nanoparticles in hexane (5 ml) were precipitated with ethanol (~100 ml), centrifuged and the solid residue re-dispersed in chloroform (10 ml). Poly(maleic anhydride-*alt*-1-octadecene) (10 mg) was dissolved in chloroform (10 ml) and added under vigorous stirring to the nanoparticle solution and left to stir for 1 hour. The chloroform was removed slowly *in vacuo* (750 mbar, 25 °C water bath) and then subsequently placed under a $\sim 10^{-2}$ mbar vacuum for 24 hours. 20 ml of a 0.1 M solution of tetramethyl ammonium hydroxide was added, and left overnight to react. Once the nanoparticles were fully dispersed, the solution was acidified to pH 8 with acetic acid before concentration with centrifuge filters (Centriprep YM-10, 10 kDa MW cutoff for 3 × 20 minute cycles at 3000*g*).

Magnetic hyperthermia testing

A sample of polymer-coated water-transferred iron oxide nanoparticles (1 ml) was placed into a 1.5 ml Eppendorf® tube and situated centrally in a 6 turn room temperature controlled solenoid coil within the MACH system. An alternating field strength of 6.6 kA m^{-1} at a frequency of 945 kHz was used throughout the magnetic hyperthermia experiments. The experiment was run until the sample reached temperature saturation, typically of the order of five minutes. Temperature data was collected using a non-metallic fluoroptic probe in order to prevent eddy current heating within the sample.

Results and discussion

Colloidal iron oxide nanoparticles were prepared by the high temperature thermal decomposition of iron–fatty acid complexes in the presence of surfactants. Reaction solutions turned from brown to black during the reaction, indicating the formation of colloidal nanoparticles. Nanoparticles were isolated by precipitation with ethanol and centrifugation and dispersed readily in hexane. It is noteworthy that the obtained samples had significant organic components, which were difficult to remove without the use of magnetic separation.

Three different samples of iron oxide nanoparticles were prepared herein, as illustrated in Scheme 1. Sample (I) are nanoparticles prepared from laboratory reagents according to the procedure reported by Park.[3] Sample (II) and (III) are both prepared from high street reagents, only differing in the solvent used for the decomposition step.

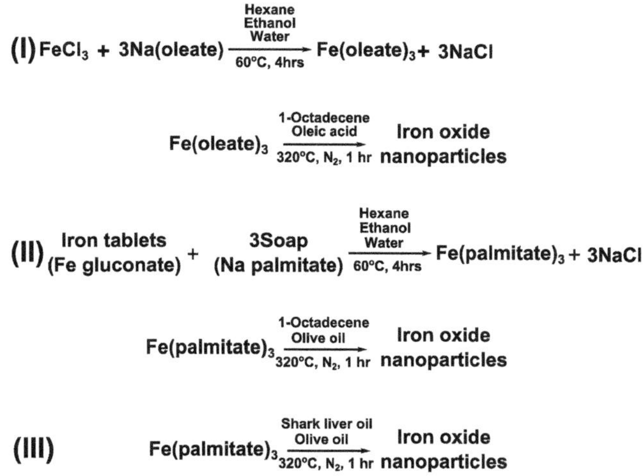

Scheme 1 Sample (I) is the route to iron oxide nanoparticles proposed by Park *et al.*,[3] Sample (II) is the synthesis of iron palmitate from high street sources and its subsequent decomposition in 1-octadecene and Sample (III) the decomposition of iron palmitate in shark liver oil.

The obtained nanoparticles from all reactions were highly crystalline and colloidally stable in organic solvents due to the blanket surface coverage of palmitic acid. This was confirmed by ATR-FTIR analysis as carbonyl stretches are clearly observable on the nanoparticle surfaces at *ca.* 1700 cm^{-1}. Weak stretches from iron oxide (Fe–O) bonds are observed in the fingerprint region (*ca.* 500 cm^{-1}) (Fig. 1).

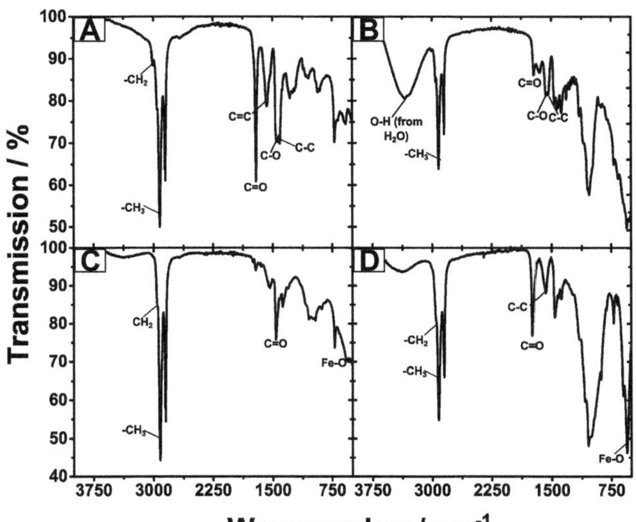

Fig. 1 ATR-FTIR spectra of: iron oleate (A), iron oxide nanoparticles obtained from the decomposition of iron oleate, (Sample (I)) (B), iron palmitate from iron tablets (C) and nanoparticles obtained from iron palmitate decomposition in shark liver oil (Sample (III)) (D).

TEM analysis showed a high degree of monodispersity in all samples of iron oxide nanoparticles. The shape of the NPs was predominantly spherical using iron oleate or iron palmitate as the iron precursor (Fig. 2). It was found that the solvent affected the final shape of the products. When a high ratio of oleic acid to iron palmitate was used for the decomposition in shark liver oil, directional growth of nanorods, as well as iron oxide nanoparticles, was observed (Fig. 2E).

These nanorods were highly crystalline, readily dispersible in organic solvents and exhibited directional growth. The nanorods, however, were not composed of iron oxide, moreover of rhenanite (β-$NaCaPO_4$), determined by EDX and XRD. Ingredients from the initial iron tablets such as calcium phosphate and sodium

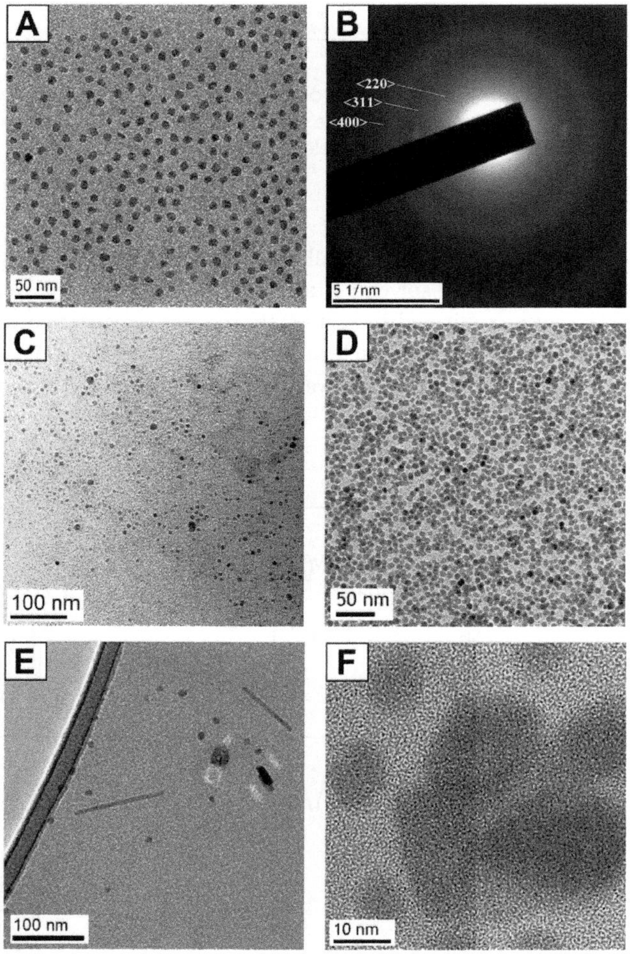

Fig. 2 Iron oxide nanoparticles from the decomposition of iron palmitate in 1-octadecene (Sample (II)) (A). SAED of iron oxide nanoparticles from the decomposition of iron palmitate in 1-octadecene (B). Iron oxide nanoparticles from the decomposition of iron palmitate in shark liver oil (Sample (III)) (C). Iron oxide nanoparticles synthesised according to Park *et al.*[3] (Sample (I)) (D). Iron oxide obtained from a 1 : 1 molar ratio of olive oil (oleic acid) to iron palmitate in shark liver oil (E). HRTEM micrograph of a single iron oxide nanoparticle from a 1 : 1 ratio of iron palmitate shark liver oil (F).

chloride contributed to the nanorod formation. This is possibly due to the organic species present in shark liver oil.

The constituents of shark liver oil itself vary according to the depth at which the shark normally resides.[25–28] Deep-sea sharks can approach neutral buoyancy through storage of low density lipids stored in the liver. A large contributor is squalene (density 0.858 g ml^{-1} at 25 °C, which increases at cold, deep sea temperatures), which has been shown to influence iron oxide nanoparticle shape.[29] Other constituents include: diacyl glyceryl ether, triacylgycerol and wax esters, all of which could contribute to directional growth of nanoparticles.[30–33]

By varying the amount of oleic acid used in the synthesis, the particle size could be tuned, with particle size increasing with smaller amounts of oleic acid (*viz.* olive oil). Average NP sizes were calculated from TEM micrographs as ~12.7, 5.8 and 3.6 nm for 1.2, 6, 24 mmol of oleic acid in olive oil, respectively (when decomposing 2 g of iron palmitate).

EDX analysis detailed the elemental composition of the nanoparticles, confirming the presence of iron and oxygen. EDX and XPS analyses also highlighted additional elementals acquired from the initial iron tablets such as phosphorus and calcium (*vide infra*) (Fig. 3 and S1.7).†

The decomposition of iron palmitate and iron oleate species was followed by TGA. Both iron oleate and iron palmitate samples had very similar decomposition profiles. The primary decomposition temperatures varied slightly between the two samples, but resided between 190 and 210 °C, after which rapid decomposition occurred. There was a significant loss of mass in the iron palmitate sample between 30 and 80 °C, which is associated with the evaporation of residual ethanol from the washing process (Fig. 4).

Particle composition and structure were analysed by XPS and XRD. XRD analysis yielded a pattern consistent with that of an inverse spinel structure

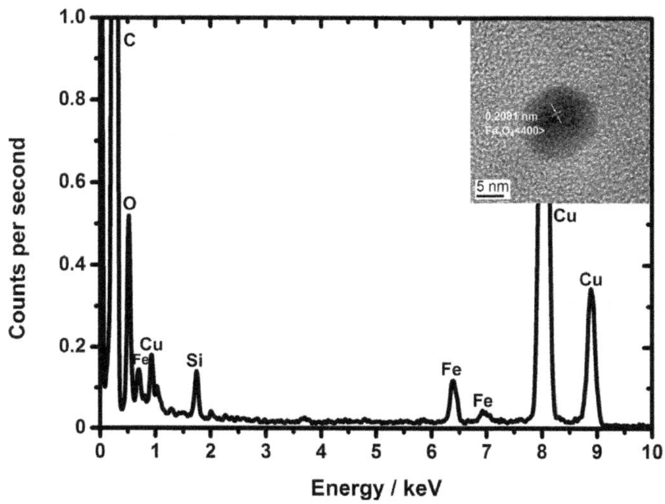

Fig. 3 (Main) EDX spectrum of iron oxide nanoparticles from iron palmitate decomposed in shark liver oil (Sample (III)) and (inset) HRTEM micrograph of a spherical iron oxide nanoparticle, showing the ⟨400⟩ plane of Fe_3O_4. For a zoomed in and labelled EDX spectrum, see the ESI.†

γ-Fe_2O_3/Fe_3O_4; however, it was not possible to discern between the two structures due to the high degree of similarity between their profiles. The nanostructures obtained from a 1 : 1 molar ratio of olive oil to iron palmitate indicated the presence of a multiphase system, as seen from XRD analysis (Fig. 6 ESI†). The sample was found to be mainly constituted by two phases due to the presence of peaks consistent with the diffraction pattern of γ-Fe_2O_3/Fe_3O_4 and also peaks corresponding to the orthorhombic structure of rhenanite (Fig. 5).

XPS analyses revealed that the surface composition of the nanoparticles was influenced heavily by ingredients in the precursor, namely dicalcium phosphate (E341), which is used as a tableting agent. Calcium has been shown to be an effective dopant for iron oxide,[34,35] and iron oxide surfaces have high affinity for phosphorus-containing ligands,[36] hence the high level of surface doping. EDX analysis showed that Ca and P were indeed dopants for the iron, albeit constituting *ca.* 4% of the total nanoparticle mass.

Mössbauer spectroscopy offers insights into the oxidation states and chemical environments of iron atoms present in the iron oxide lattice. All spectra were taken at room temperature. Analysis of the nanoparticles synthesised according to Park gave a very broad single peak, with wings extending out to ± 10 mm s^{-1}. The latter indicates the presence of magnetic hyperfine absorption, as would be expected in a magnetic iron oxide (such as maghemite or magnetite) below its superparamagnetic blocking temperature. The very broad nature of the absorption could be due to temporal or structural factors, or both – *i.e.* temporal as in magnetic relaxation on the nanosecond timescale of the Mössbauer measurement, or structural as in the crystallinity, size or defect structure of the particles. Further experiments would be needed to clarify this.

Mössbauer analysis of Sample (III) shows the superposition of two doublets, one of which has parameters typical of Fe^{3+} (isomer shift $\delta = 0.30 \pm 0.07$ mm s^{-1} and quadrupole splitting $\Delta = 0.70 \pm 0.17$ mm s^{-1}), and another of which has parameters typical of Fe^{2+} ($\delta = 1.18 \pm 0.04$ mm s^{-1} and $\Delta = 2.27 \pm 0.08$ mm s^{-1}).

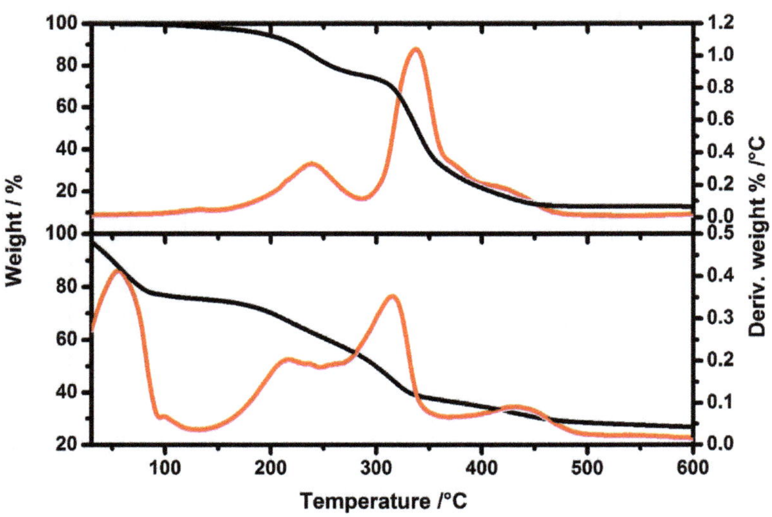

Fig. 4 TGA traces with first derivatives of iron oleate (above) and iron-palmitate (below).

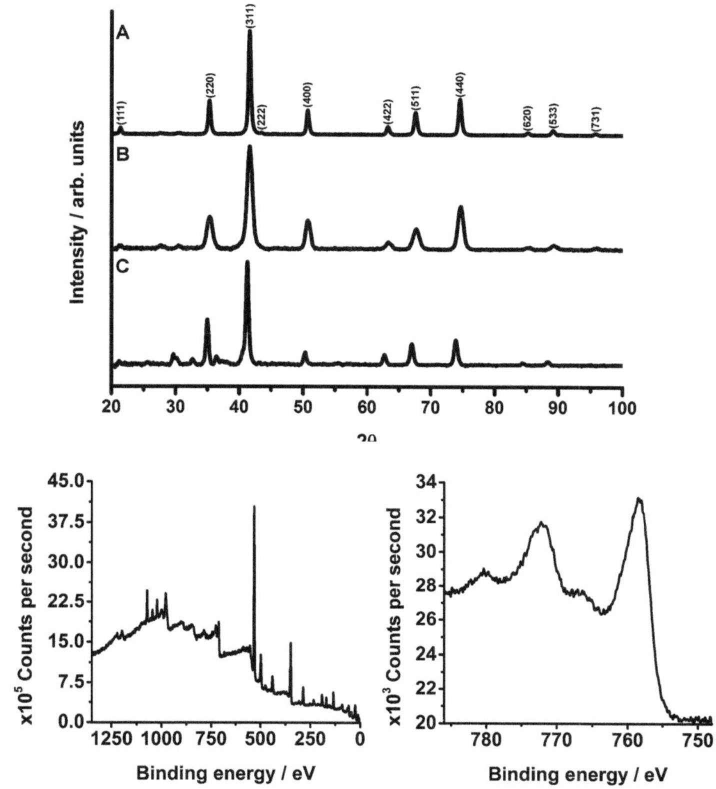

Fig. 5 (Top) XRD patterns of (A) magnetite standard, (B) iron oxide nanoparticles from standard reagents (Sample (I)) and (C) Sample (II). (Bottom) Survey and Fe2p XPS spectra of iron oxide from iron palmitate decomposed in shark liver oil (Sample (III)).

The latter was not expected, and was hypothesised to be due to the presence of a paramagnetic Fe^{2+} species, such as unreacted iron palmitate/iron gluconate from the iron tablets/iron palmitate precursor. The Mössbauer spectrum of a sample of homogenised iron tablets, shown in Fig. 6A, did indeed show such a Fe^{2+} species, with parameters ($\delta = 1.20 \pm 0.01$ mm s^{-1} and $\Delta = 2.23 \pm 0.01$ mm s^{-1}) comparable to those observed in the product. From this analysis, we can say that although there remained some unreacted percursor, an appreciable amount of the precursor has formed magnetite and/or maghemite nanoparticles. The fact that these appear as a Fe^{3+} doublet rather than a magnetic sextet is most likely due to the small particle size, so that the room temperature Mössbauer measurement is above the superparamagnetic blocking temperature of the nanoparticles.

Magnetic hyperthermia measurements were carried out after the nanoparticles were transferred to water. This was achieved using an amphiphilic polymer, poly(maleic anhydride-*alt*-octadecene). The hydrophobic moieties of poly(maleic anhydride-*alt*-octadecene) allow the polymer to form a hydrophobic layer around individual nanoparticles through van der Waals interactions, promoted by a polar solvent, chloroform. Once wrapped around a nanoparticle, the solvent is removed, and maleic anhydride groups in the polymer backbone are opened *via* nucleophilic attack with a base, rendering the nanoparticles dispersible in water.

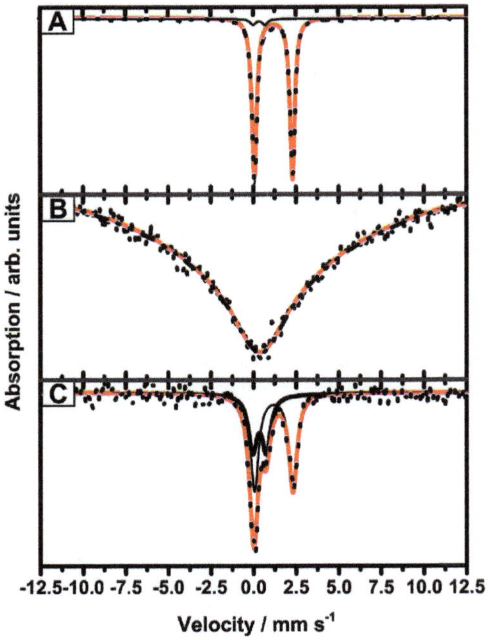

Fig. 6 ^{57}Fe Mössbauer spectra of: homogenised iron tablets (A), iron oxide nanoparticles from standard reagents (Sample (I)) (B), iron oxide nanoparticles from the decomposition of iron palmitate in shark liver oil (Sample (III)) (C). The red trace represents the fit, with the thick black line the Fe^{3+} doublet component and the thin black line the Fe^{2+} doublet component.

Heat loss dissipation measurements of Sample (III) showed a temperature rise of 1.8 °C at a prepared concentration of 2.3 mg ml^{-1}, whilst Sample (I) showed a temperature of 2.1 °C at a prepared concentration of 6.1 mg ml^{-1} with Intrinsic Loss Power (ILP) values of 1.9 and 0.4, respectively, placing them towards the market leaders within the range of commercially available magnetic fluid hyperthermia agents.[37] Sample (II) showed signs of heat dissipation on exposure to the AC magnetic field (Fig. 7D). Whilst this ILP value is not as high as other values quoted in the literature, this method demonstrates a promising outlook for synthesising a low cost functional hyperthermia agent.[37]

SQUID magnetometry shows that Sample (I) and (II) are superparamagnetic at 300 K. The SLO samples exhibited signs of ferromagnetic behaviour with a normalised remanance $M/M_s = 0.2$ and a coercivity of 100 Oe. Samples (I), (II) and (III) have saturation magnetisation values of 52, 2.2 and 0.1 emu per gram, respectively.

Conclusions

In this article, we have successfully synthesised iron oxide nanomaterials from readily available high street sources. The synthesis occurs in two steps, namely the synthesis of the precursor, iron palmitate, and then its decomposition in the presence of surfactants and high boiling point solvents, again shop-purchased. The results from these syntheses were compared to standard techniques, and the

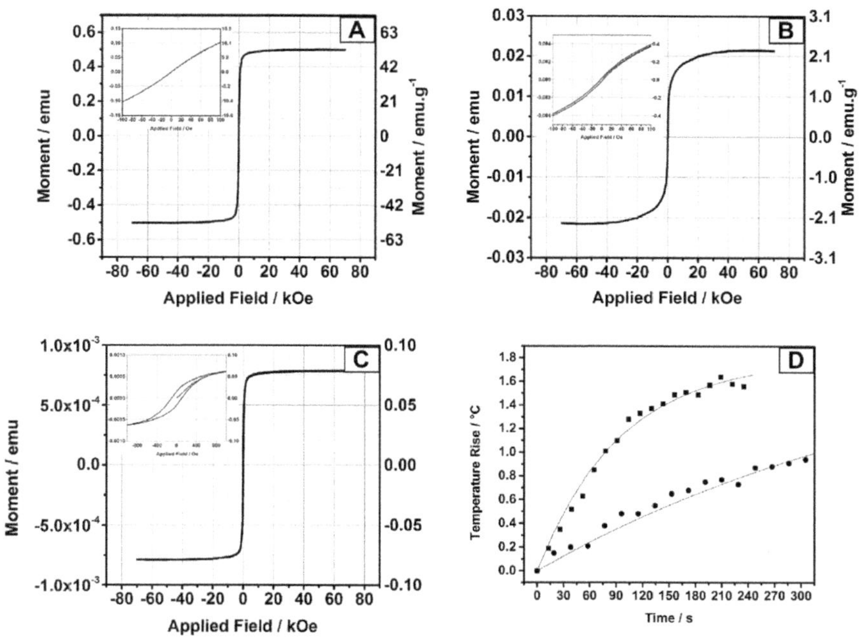

Fig. 7 SQUID magnetic hysteresis loops of: iron oxide nanoparticles from standard reagents (Sample (I)) (A), iron oxide nanoparticles from iron palmitate decomposed in 1-octadecene (Sample (II)) (B) and iron oxide nanoparticles obtained from the decomposition of iron palmitate in shark liver oil (Sample (III)) (C). (D) shows magnetic hyperthermia heating of water-transferred iron oxide nanoparticles from Sample (I) (bottom) and nanoparticles from Sample (III) (top). Insets are magnifications of the hysteresis loops, showing minimal coercivity.

associated costs compared (ESI Table 5†). We have demonstrated that iron oxide nanomaterials can be produced at a far lower cost than standard literature techniques, at the chemist's discretion, due to the availability of the reagents used.

The nanomaterials were characterised using a variety of techniques including: XRD, XPS, EDX spectroscopy, TEM, Mössbauer spectroscopy, TGA/DSC, ATR-FTIR, SQUID, and magnetic hyperthermia measurements. High quality, monodisperse iron oxide nanocrystals were also obtained from the decomposition of iron palmitate in 1-octadecene and shark liver oil, both displaying a degree of heating on exposure to an AC magnetic field.

The inexpensive iron oxide nanomaterials obtained here have tremendous scope for the transfer of this reaction to batch or mass production, for use in numerous applications from medical devices to catalysis. The good performance of the synthesised materials in magnetic hyperthermia tests give scope for the further development of cost effective iron oxide nanomaterials.

Notes and references

1 D. Ling and T. Hyeon, *Small*, 2013, **9**, 1450–1466.

2 N.-N. Song, H.-T. Yang, X. Ren, Z.-A. Li, Y. Luo, J. Shen, W. Dai, X.-Q. Zhang and Z.-H. Cheng, *Nanoscale*, 2013, **5**, 2804–2810.
3 J. Park, K. An, Y. Hwang, J.-G. Park, H.-J. Noh, J.-Y. Kim, J.-H. Park, N.-M. Hwang and T. Hyeon, *Nat. Mater.*, 2004, **3**, 891–895.
4 M. Mahmoudi, S. Sant, B. Wang, S. Laurent and T. Sen, *Adv. Drug Delivery Rev.*, 2011, **63**, 24–46.
5 W. Cai and J. Wan, *J. Colloid Interface Sci.*, 2007, **305**, 366–370.
6 A. Guittoum, A. Layadi, H. Tafat and N. Souami, *J. Magn. Magn. Mater.*, 2010, **322**, 566–571.
7 A. Tavakoli, M. Sohrabi and A. Kargari, *Chem. Pap.*, 2007, **61**, 151–170.
8 A. S. Teja and P.-Y. Koh, *Prog. Cryst. Growth Charact. Mater.*, 2009, **55**, 22–45.
9 S. Laurent, D. Forge, M. Port, A. Roch, C. Robic, L. Vander Elst and R. N. Muller, *Chem. Rev.*, 2008, **108**, 2064–2110.
10 S. Xu, J. Ziegler and T. Nann, *J. Mater. Chem.*, 2008, **18**, 2653–2656.
11 M. Lattuada and T. A. Hatton, *Langmuir*, 2007, **23**, 2158–2168.
12 L. M. Bronstein, X. Huang, J. Retrum, A. Schmucker, M. Pink, B. D. Stein and B. Dragnea, *Chem. Mater.*, 2007, **19**, 3624–3632.
13 S. Sun and H. Zeng, *J. Am. Chem. Soc.*, 2002, **124**, 8204–8205.
14 J. Bear, G. Charron, M. T. Fernandez-Arguelles, S. Massadeh, P. McNaughter and T. Nann, in *BetaSys: Systems Biology of Regulated Exocytosis in Pancreatic β-Cells*, Springer, New York, 2011, vol. 2, pp. 185–220.
15 G. Vallejo-Fernandez, O. Whear, A. G. Roca, S. Hussain, J. Timmis, V. Patel and K. O'Grady, *J. Phys. D: Appl. Phys.*, 2013, **46**, 312001.
16 R. Hergt, R. Hiergeist, I. Hilger, W. A. Kaiser, Y. Lapatnikov, S. Margel and U. Richter, *J. Magn. Magn. Mater.*, 2004, **270**, 345–357.
17 M. A. Gonzalez-Fernandez, T. E. Torres, M. Andrés-Vergés, R. Costo, P. de la Presa, C. J. Serna, M. P. Morales, C. Marquina, M. R. Ibarra and G. F. Goya, *J. Solid State Chem.*, 2009, **182**, 2779–2784.
18 S. E. Barry, *Int. J. Hyperthermia*, 2008, **24**, 451–466.
19 L. A. Thomas, L. Dekker, M. Kallumadil, P. Southern, M. Wilson, S. P. Nair, Q. A. Pankhurst and I. P. Parkin, *J. Mater. Chem.*, 2009, **19**, 6529.
20 I. Baker, Q. Zeng, W. Li and C. R. Sullivan, *J. Appl. Phys.*, 2006, **99**, 08H106.
21 M. Gonzales-Weimuller, M. Zeisberger and K. M. Krishnan, *J. Magn. Magn. Mater.*, 2009, **321**, 1947–1950.
22 E. E. Lees, T.-L. Nguyen, A. H. A. Clayton, B. W. Muir and P. Mulvaney, *ACS Nano*, 2009, **3**, 2049.
23 T. Pellegrino, L. Manna, S. Kudera, T. Liedl, D. Koktysh, A. L. Rogach, S. Keller, J. Rädler, G. Natile and W. J. Parak, *Nano Lett.*, 2004, **4**, 703–707.
24 http://resonantcircuits.com/.
25 B. M. Wetherbee and P. D. Nichols, *Comp. Biochem. Physiol., Part B: Biochem. Mol. Biol.*, 2000, **125**, 511–521.
26 M. J. Bakes and P. D. Nichols, *Comp. Biochem. Physiol., Part B: Biochem. Mol. Biol.*, 1995, **110**, 267–275.
27 J. C. Cain and R. A. Morton, *Biochem. J.*, 1955, **60**, 274–283.
28 I. Batista and M. L. Nunes, *Fish. Res.*, 1992, **14**, 329–334.
29 B. Rodriguez-Gonzalez, M. Spasova, M. Farle, L. Liz-Marzan and A. Shavel, *Adv. Funct. Mater.*, 2007, **17**, 3870–3876.
30 C. Yang, J. Wu and Y. Hou, *Chem. Commun.*, 2011, **47**, 5130–5141.

31 M. V. Kovalenko, M. I. Bodnarchuk, R. T. Lechner, G. Hesser, F. Schäffler and W. Heiss, *J. Am. Chem. Soc.*, 2007, **129**, 6352–6353.
32 J. Cheon, N.-J. Kang, S.-M. Lee, J.-H. Lee, J.-H. Yoon and S. J. Oh, *J. Am. Chem. Soc.*, 2004, **126**, 1950–1951.
33 L. M. Bronstein, J. E. Atkinson, A. G. Malyutin, F. Kidwai, B. D. Stein, D. G. Morgan, J. M. Perry and J. A. Karty, *Langmuir*, 2011, **27**, 3044–3050.
34 L. J. Berchmans, M. Myndyk, K. L. Da Silva, A. Feldhoff, J. Šubrt, P. Heitjans, K. D. Becker and V. Šepelák, *J. Alloys Compd.*, 2010, **500**, 68–73.
35 R. Dom, R. Subasri, K. Radha and P. H. Borse, *Solid State Commun.*, 2011, **151**, 470–473.
36 Y. Sahoo, H. Pizem, T. Fried, D. Golodnitsky, L. Burstein, C. N. Sukenik and G. Markovich, *Langmuir*, 2001, **17**, 7907–7911.
37 M. Kallumadil, M. Tada, T. Nakagawa, M. Abe, P. Southern and Q. A. Pankhurst, *J. Magn. Magn. Mater.*, 2009, **321**, 1509–1513.

Faraday Discussions

PAPER

Design of FeBi nanoparticles for imaging applications

M. Branca,[ab] F. Pelletier,[ab] B. Cottin,[ab] D. Ciuculescu,[ab] C.-C. Lin,[c] R. Serra,[bd] J.-G. Mattei,[bd] M.-J. Casanove,[bd] R. Tan,[be] M. Respaud[be] and C. Amiens*[ab]

Received 10th May 2014, Accepted 7th July 2014
DOI: 10.1039/c4fd00105b

A variety of imaging technologies are now routinely used in the medical field, their use being continuously enlarged through the development of contrast agents. Recently nanoparticles (NPs) proved efficient to improve imaging *in vivo* by increasing contrast and targeting capabilities. The current trend is now focused on the development of dual contrast agents combining two or more functionalities on the same NP. Motivated by this new challenge we developed FeBi NPs as new nanomaterials with potential application as a contrast agent for MRI and CT imaging. In addition to the well-known use of iron in the development MRI contrast agents, we chose Bi as a CT imaging agent rather than the more documented gold, because it possesses a larger X-ray attenuation coefficient and is much less expensive. Two sets of NPs, with sizes around 150 nm and 14 nm, were synthesized using organometallic approaches. In both cases, the NPs are spherical, and contain distinct domains of Fe and Bi, with the surface being enriched with Fe, and a hydrophobic coating. This coating differs from one sample to the other: the surfaces of the 150 nm large NPs are coated by amine ligands, while those of the 14 nm large NPs are coated by a mixture of an amine and its hydrochloride salt. Exchange of the surface ligands to afford water soluble NPs has been attempted. We show that only the larger NPs could be functionalized with water soluble ligands, which is in agreement with the lability of their initial surface coating. Colloidal aqueous solutions of FeBi NPs with glycoPEG ligands have been obtained.

Introduction

Today's capabilities in clinical diagnostic radiology have been made possible by advances not only in diagnostic systems, but also by the contrast media and

[a]*CNRS, LCC (Laboratoire de Chimie de Coordination), 205, route de Narbonne, F-31077 Toulouse, France. E-mail: catherine.amiens@lcc-toulouse.fr*

[b]*Université de Toulouse, UPS, INPT, LCC, F-31077 Toulouse, France*

[c]*Department of Chemistry, National Tsing Hua University, 101, Sec. 2, Kuang-Fu Road, Hsinchu, Taiwan 30013, Taiwan*

[d]*CNRS, CEMES (Centre d'Elaboration des Matériaux et d'Etudes Structurales), France*

[e]*LPCNO, INSA, 135 Avenue de Rangueil, 31077 Toulouse Cedex 4, France*

radiopharmaceuticals that make it possible to see, in great detail, the body's internal structures, organs and tissues that would not otherwise be clearly seen. Among these, nanoparticles (NPs) play an increasingly important role as their larger size compared to molecular contrast agents increases their residence time in the body.[1] Superparamagnetic iron oxide nanoparticles (SPIONs) are already commercially available as contrast agents for magnetic resonance imaging (MRI), fluorescent nanoparticles (quantum dots) are also well-known for optical detection and nanohybrids are being developed to replace the currently used iodine-based contrast agents in computed X-ray tomography (CT), among which bismuth derivatives are reported as being very promising.[2–5] The use of NPs also facilitates the functionalization, and thus targeting, of the contrast agent.[6–9] As no single imaging modality can address all biological questions, the next step to further improve diagnostics lies in the development of dual contrast agents (coupling of T1/T2 MRI contrast agents,[10,11] optical and radioemitting probes,[12] MRI and fluorescent markers[13] or other optical probes,[14] ultrasonography and MRI,[15,16] MRI and CT[17] *etc.*), which will give better resolved images and rule out artifacts, and hence help define the most appropriate treatment.[18] A dual CT/MRI contrast agent, in particular, would help improve the diagnostics of cardiovascular diseases,[19] or the optimization and follow-up of embolization therapies.[20]

In this context, associating iron oxide and bismuth inside a nano-object looks promising as (1) SPIONs are already in use as MRI contrast agents, (2) bismuth NPs have been demonstrated to give good contrast in CT[2–4,21] with low toxicity reported so far, and (3) both materials are of low cost. It is worth noting that research about a nanohybrid associating Fe and Bi was published recently by Andres-Verges *et al.*[22] In this work each NP consisted of an iron oxide core and a bismuth oxide shell. However, optimum arrangement of the Fe and Bi domains in the nanohybrids should comprise a bismuth core (to minimize bismuth interaction with the biological material to be imaged) and an iron shell, thus insulating the bismuth component and maximizing the interaction between the magnetic component and the water protons, which is important to reach a good contrast in MRI as shown by Kim *et al.*[17]

We herein report on the organometallic synthesis of BiFe NPs with a surface enriched with Fe. Depending on the synthetic process, NPs of different average size (150 or 14 nm) and surface chemistry (amine or a mixture of amine and its hydrochloride salt) could be obtained. We show that the initial surface coating is a key-parameter for the further functionalization of the NPs and their transfer into water, the first requisite before any use of these NPs as contrast agents for biomedical imaging. Preliminary results on the X-ray opacity and possible use of these nanohybrids as MRI contrast agents are also reported for the NPs with the optimum size.

Experimental

Materials and methods

All the syntheses were performed under an argon atmosphere by using Fischer–Porter bottle techniques, a glove box and argon/vacuum lines. $[Fe(N(SiMe_3)_2)_2]$ (>99%) was purchased from Nanomeps, HDA (99%), hexamethyldisylazane (HMDS, >99%) and anhydrous anisole from Sigma-Aldrich. Anisole was further purified by distillation under argon (over sodium), degassed by the freeze–pump–thaw technique and dried over a molecular sieve (0.4 nm) before use.

$Bi(N(SiMe_3)_2)_3$,[23] $HDA{\cdot}HCl$[24] and $iPr_2NH{-}BH_3$ (ref. 25) were prepared and purified according to published procedures.

Dihydrogen was purchased from Air Liquide. It contained less than 3 ppm of H_2O and 2 ppm of O_2. To determine the amount of Fe and Bi in the materials, elemental analyses were performed at Antellis[26] by ICP-MS (Inductively Coupled Plasma-Mass Spectrometry), after digestion of the samples in concentrated HNO_3.

Synthetic procedures

Synthesis of 150 nm large BiFe NPs (sample BiFe150). $Fe(N(SiMe_3)_2)_2$ (80 mg, 0.21 mmol), $Bi(N(SiMe_3)_2)_3$ (20 mg, 0.03 mmol), HMDS (343 mg, 2.1 mmol) were dissolved in anisole (18 mL, $H_2O < 3$ ppm) in a Fischer–Porter bottle in a glove box. The homogeneous yellow-green solution was frozen under liquid nitrogen and 1.1 equivalent of diisopropylamine-borane with respect to iron (27 mg, 0.24 mmol) in anisole (4 mL) were canula-transferred into the Fischer–Porter bottle. Slow warm-up of the reaction medium up to room temperature afforded a dark brown solution. This homogeneous solution was then pressurized with 3 bar H_2 and heated for 4 h at 110 °C, and 19 h at 150 °C without any stirring. A black magnetic solid was recovered by filtration, washed with anisole (5 mL) and dried under vacuum (recovered mass: 16 mg; Fe: 11.18%; Bi: 5.52% *i.e.* atomic composition Fe7.6Bi1).

Synthesis of 14 nm large BiFe NPs (sample BiFe14). In a typical synthesis the reaction mixture is prepared in a glove box as follows: in a Fischer–Porter bottle, $Fe[N(Si(CH_3)_3)_2]_2$ (376 mg, 1 mmol) and $Bi[N(Si(CH_3)_3)_2]_3$ (68.9 mg, 0.1 mmol) are dissolved in 20 mL of dry anisole. HDA (483 mg, 2 mmol) and $HDA{\cdot}HCl$ (415 mg, 1.5 mmol) are subsequently added. The color of the solution progressively turns from green-yellow to brown. After this, the reactor is immersed in an oil bath preheated at 150 °C and maintained at this temperature under magnetic stirring for 5 hours. The reaction is then allowed to return to room temperature and the solvent is removed by evaporation under reduced pressure. The residue is washed with 40 mL of toluene (recovered mass: 200 mg; 15% Fe, 4% Bi).

Control experiment. $Bi[N(Si(CH_3)_3)_2]_3$ (68.9 mg, 0.1 mmol) is dissolved in 20 mL of dry anisole. HDA (483 mg, 2 mmol) is subsequently added. After this, the reactor is immersed in an oil bath preheated at 150 °C and maintained at this temperature under magnetic stirring. The color of the solution progressively turns from yellow to brown and black over 1/2 hour. Heating is continued for 5 hours, then the reaction is allowed to return to room temperature and the solvent is removed by evaporation under reduced pressure. The solid recovered (mass: 176 mg) was directly analysed by FT-IR spectroscopy.

Synthesis of modified PEG ligands. LPEG1 and LPEG2 have been prepared according to already published procedures.[27]

Water transfer. The NPs (3 mg) and the selected PEG ligand (5 mg) are suspended in a 1/1 (v/v) water–methanol mixture. After sonication for 20 min, a homogeneous black solution is obtained. The modified NPs are recovered by centrifugation and redispersed in water. Then, two other centrifugation steps are carried out to remove excess PEG ligand. The final powder is dried under vacuum (5 mg). It can be easily redispersed in water but the NPs settle after 6 (LPEG1) to 20 min (LPEG2).

Characterization techniques

FT-IR. FTIR spectra of the Bi nanoparticles, in powder form, were recorded in a glove box using a Bruker ALPHA spectrometer in ATR mode on a diamond crystal.

TEM and EDS experiments. The nano-objects were deposited on carbon-coated copper grids by casting a drop of THF or anisole solution of the nanoparticle powders.

Conventional TEM and HREM images were obtained using JEOL microscopes JEM1011 (100 kV) and JEM2100F (200 kV), respectively, with a spatial resolution of 0.23 nm at the TEMSCAN facility at Université Paul-Sabatier. EDS spectra were recorded in STEM mode using JEOL JEM 2100F TEM-FEG (at TEMSCAN, UPS) or CEM20 FEI (at CEMES) microscopes fitted with a Bruker QUANTAX EDS system.

HREM experiments were carried out at CEMES, using a Cs-corrected FEI TECNAI F20 microscope, with point resolution of 0.12 nm and energy resolution of 0.8 eV. The specimens were analyzed in this microscope before oxidation thanks to the use of a dedicated specimen holder fitted with an environmental cell, and after 2 days of air exposure. HREM images were acquired and processed using the Gatan Digital Micrograph environment.

Titration of galactose. UV-visible spectra were recorded using a Perkin Elmer Lambda 35 spectrophotometer from 2 mL of the solutions placed in 1 cm thick glass cuvets. Galactose, was determined by the anthrone–sulfuric acid method. Briefly, 0.5 mL of D-galactose standard solution (0, 40, 60, 80, 100, 120 $\mu g\ mL^{-1}$), or 0.5 mL of NP solution (50 $mg\ mL^{-1}$) is introduced into a test tube in the presence of 1 mL of anthrone reactant (60 mg of anthrone in 30 mL of concentrated sulfuric acid) and heated at 85 °C for 10 min. 0.5 mL of each of the green solutions obtained is then diluted with 2 mL of concentrated sulfuric acid. The absorbance of each of the standards and samples was measured (626 nm) and the galactose content of the NPs was interpolated from a standard curve prepared in this manner.

The coverage of galactose at the surface of the NPs was thus derived knowing the % of surface atoms in the NP dispersion (3.4% per NP given their size and morphology). LPEG1: 20%; LPEG2: <10%.

Results and discussion

Synthesis and structural characterisation

From a thermodynamic point of view, as Fe and Bi are immiscible, Bi having the lowest surface tension and the largest atomic radius should segregate at the surface of the NP. To circumvent this problem we have relied on kinetically controlled growth processes. We have already demonstrated in the case of an FeRh system that controlling the kinetics of production of each element allowed precise control over the chemical order inside the NP from a Rh core–Fe shell structure to the reverse situation, namely an Fe core and Rh rich shell.[28,29] It is worth noting that this second arrangement corresponds to a metastable chemical distribution. We thus decided to follow this strategy to build Bi@Fe core–shell NPs. Based on previous results, we have designed two different pathways starting in each case from a homogeneous solution of the Bi and Fe homoleptic amido complexes: $[Bi(N(SiMe_3)_2)_3]$ and $[Fe(N(SiMe_3)_2)_2]$.

In the first process (P1) we use amine-borane as a reducing agent to produce Fe atoms from $[Fe(N(SiMe_3)_2)_2]$. These atoms are thus generated in the presence of the oxidizing $[Bi(N(SiMe_3)_2)_3]$ complex. A fast galvanic displacement follows, regenerating the Fe precursor and producing Bi atoms which evolve into Bi NPs, the core of the expected Bi@Fe NPs. Following this first step, pressurization of the reacting medium under 3 bar H_2 ensures reduction of the Fe precursor and deposition of Fe atoms on top of the first formed Bi seeds.

Analysis of the sample evidenced the existence of NPs of average diameter 150 nm with a bismuth enriched core and iron rich surface (Fig. 1), as the result of the very precise design of the kinetics of reaction of each metal complex, where iron acts as a catalyst for the reduction of the poorly reactive bismuth precursor.[30] The sample obtained is called BiFe150 in the following paragraphs.

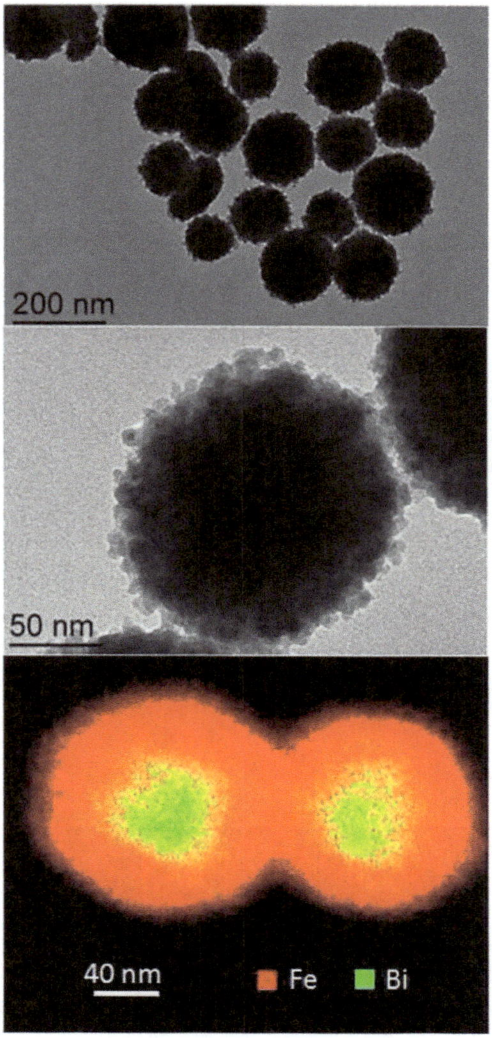

Fig. 1 Top and middle: typical TEM images of NPs from BiFe150; bottom: EDS analysis showing the chemical distribution in the NPs.

A second possibility to kinetically differentiate the reduction of these Fe and Bi precursors relies on the use of long chain amines as reducing agents. To investigate this possibility we chose hexadecylamine which is frequently used as both a reducing and stabilizing agent in nanochemistry. A control experiment followed by FT-IR spectroscopy shows that the Bi precursor starts decomposing at room temperature and is fully decomposed after 5 h of reaction at 150 °C (not shown). On the other hand, literature reports that the Fe precursor requires a temperature above 110 °C to be reduced,[31] and is only fully decomposed after 48 h at 150 °C. This confirms the possibility of controlling the chemical order of the final BiFe NPs based on the kinetics of reactions.

TEM analysis of this material (Fig. 2) evidences an average size of 14 nm, with a standard deviation of 4 nm according to a Gaussian fit of the size distribution. Accordingly the sample will be called BiFe14 hereafter. A light shell and a dark crystalline core are observed suggesting a core–shell distribution of the two elements with bismuth, the heaviest element, located in the core of the NPs. However, a more careful analysis is needed to ascertain this point, as this contrast is also the expected contrast for a spherical object, the thicker region of the core being much more absorbent for the incident electron beam than the surface region. EDS analysis has been carried out on individual NPs. As can be observed

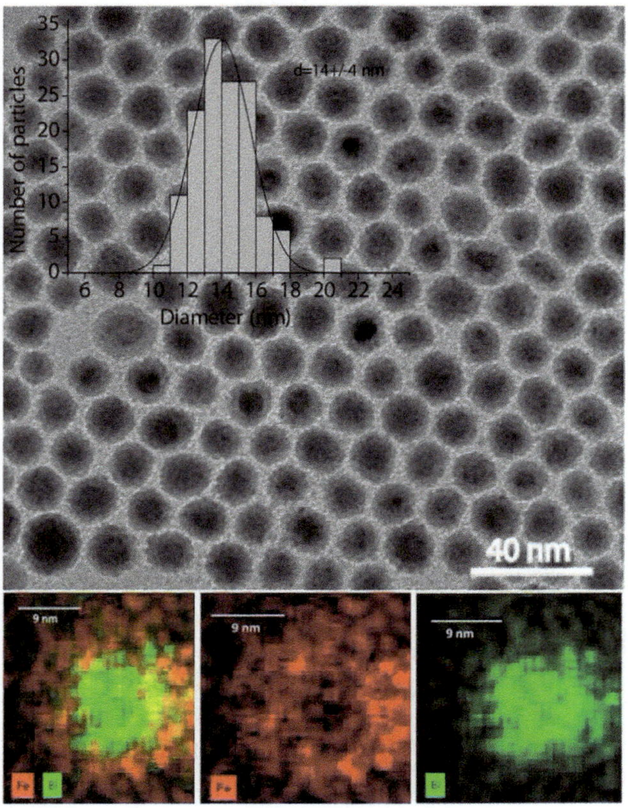

Fig. 2 TEM image with a size histogram of NPs from BiFe14 (top) and EDS analysis on an isolated NP (bottom).

in Fig. 2 (bottom), the signal over noise ratio is low due to the small size of the objects being studied (although EDS analysis was carried out on the largest nanoparticles found in the preparation). However, it also suggests the presence of Bi in the core of the NPs while the surface seems to be mainly constituted of iron. Analysis of the HREM images further allows determination of the nature and structure of the nanodomains. For this purpose, HREM images were recorded using a dedicated TEM specimen holder fitted with an environmental cell in order to protect the specimen from oxidation during transfer from the glove box to the microscope. Investigation into the outer iron layer of the NPs evidences the presence of densely packed iron nanograins with some of them adopting a *bcc* structure. For example, in Fig. 3, a small *bcc* iron grain is observed along the [111] zone axis in the left side of the particle. In the cores of the NPs, lattice planes can be observed, the spacing of which (0.328 nm) is in agreement with the formation of a single crystalline bismuth domain of *rhombohedral* structure.

The observed chemical order is thus in complete agreement with the relative kinetics of reduction of the two metal precursors which should generate Bi atoms in the reacting medium first, and Fe ones later on. From the average thickness of the iron shell (around 1.5 nm) and average size of the bismuth core (around 11 nm) one can estimate the atomic composition of the Bi@Fe nanoparticles to be around $Bi_{20}Fe_{80}$.

Water transfer

The as produced BiFe NPs are hydrophobic and soluble only in organic solvents such as THF, anisole, or toluene. It is worth noting that the most reliable and versatile routes to obtain large quantities of NPs of good size, shape, crystallinity, and surface control, are developed in organic solvents and afford hydrophobic NPs. Therefore many methodologies have been developed to transfer them into water when required by the application (*e.g.* use in green catalysis or in the biomedical field). Among these, three main paths arise: ligand exchange, coating

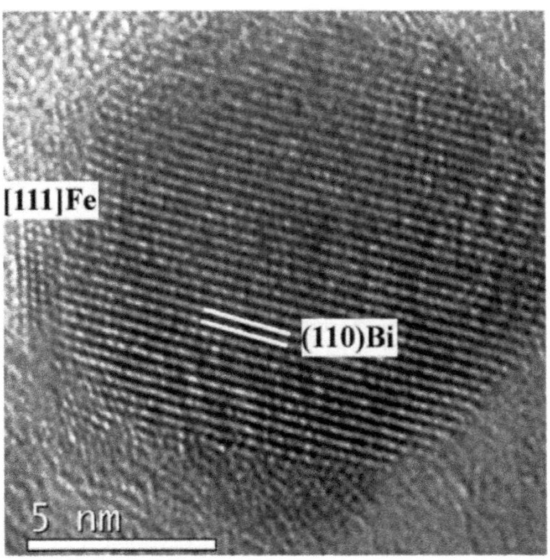

Fig. 3 HREM image of an isolated NP from BiFe14.

with amphiphilic ligands or polymers, or coating with a water soluble oxide shell such as silica. These main paths are depicted in Fig. 4.

As ligand exchange at the surface of Fe NPs is well-known, we attempted this approach to change the hydrophilic balance of the BiFe NPs. Anticipating surface oxidation upon water transfer, we resorted to ligands capable of binding strongly both on iron and iron oxide surfaces such as carboxylic acid moieties.[32] The chosen ligands, hereafter called LPEG1 and LPEG2 are depicted in Fig. 5. They are composed of a PEG chain and a galactose unit to provide water solubility, biocompatibility and future targeting towards hepatocytes and hepatomas. Indeed, it has been reported that galactose-derivatized superparamagnetic iron oxide NPs can be targeted specifically to hepatocytes and hepatomas due to the presence of asialoglycoprotein receptors on their cell surface.[33] Targeting BiFe NPs would allow accurate monitoring of liver cancers by coupling CT and MRI measurements. Furthermore these ligands bear a pendant carboxylic acid function to attach onto iron and iron oxide surfaces.

Incubation of BiFe150 NPs with galactose ligands leads to their anchoring at the surface of the NPs, as schematically represented in Fig. 6 (top). Water transfer could be achieved. This further attests the high lability of amines, especially HMDS, at the NP surface.

Stability of the colloidal solution varies as a function of the ligand used. As expected, longer PEG chains ensure better solubility.

UV-vis estimation of the galactose content was carried out following the Dreywood's analytical method.[34] Surface coverage was found to be around 20% for LPEG1 and below 10% for LPEG2. The lower surface coverage for LPEG2 can be

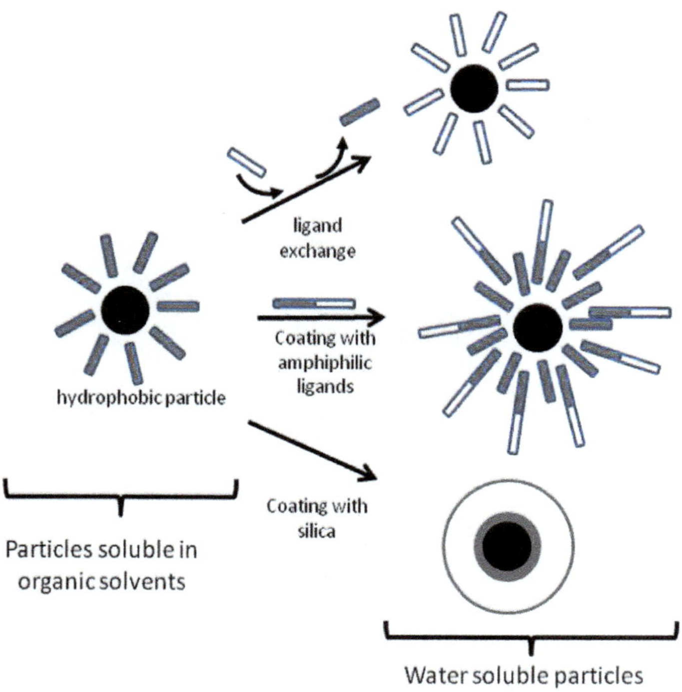

Fig. 4 Main approaches for transferring hydrophobic NPs into water.

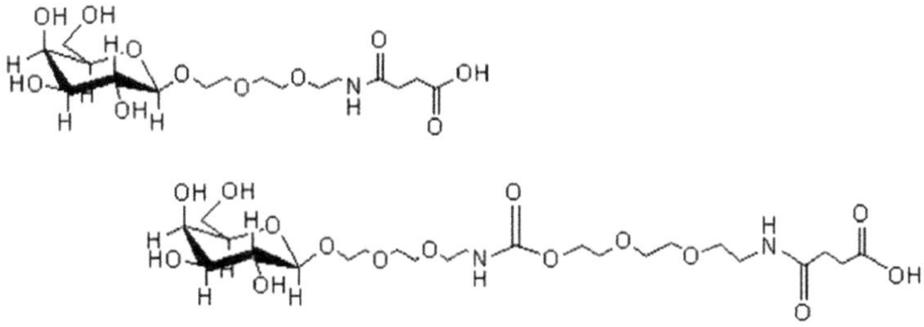

Fig. 5 The structures of LPEG1 (top) and LPEG2 (bottom).

explained by the ability of the PEG chains to coil upon themselves.[35,36] However, this generally occurs for much longer polymer chains and suggests that the galactose moiety has a role to play in the conformation of the ligand at the surface of the NPs. In particular, one cannot discard the possibility of interactions between the –OH groups of galactose and the Fe surface atoms. This phenomenon can occur already for LPEG1 as the surface coverage is quite low compared to that reported for *e.g.* mercapto-PEG ligands of comparable length and could explain the limited stability of the aqueous solutions obtained.[37] However, despite a lower coverage, NPs stabilized with LPEG2 display a better solubility in water than those stabilized by LPEG1. Interestingly the NPs still responded to an applied magnetic field. This observation is important in view of the application as MRI contrast agents as this technique relies on magnetization.

Unfortunately, if the amine coating could be displaced easily from the BiFe150, BiFe14 NPs prepared in the presence of HDA, and its hydrochloride salt could not be functionalized by carboxylic acids, even with the simple oleic acid, as evidenced by IR spectroscopy. The passivation of the Fe rich surface is tentatively attributed to the coordination of chloride ions from HDA·HCl. Indeed, anchoring of chloride ions has been evidenced for cobalt NPs when chloroammonium salts were used for their stabilization,[38] and halide ions are reported to decrease the catalytic activity of iron NPs due to strong grafting on their surface[39] and tune the structure of Fe nanoparticles, once again because of the strong interaction between Fe and these ions.[40] Furthermore, it is noteworthy that these NPs were slightly soluble in water after extensive washing with ethanol. We suggest that the ammonium chloride organizes into a double layer around the NPs thus exposing polar heads towards the solution as shown in Fig. 6 (bottom). However, the colloidal solutions obtained this way were only stable for a few minutes.

From these results it is apparent that different strategies towards surface modification must be envisaged depending on the surface state of the NPs. BiFe14 NPs especially display the most interesting size in view of biomedical application, a proper method for their efficient stabilization in aqueous solutions must be designed.

Spontaneous oxidation

The nanoobjects described above have been prepared under an inert atmosphere and are prone to oxidation as soon as they are exposed to air given the high reactivity of iron, especially at the nanoscale. It is worth noting that this oxidation

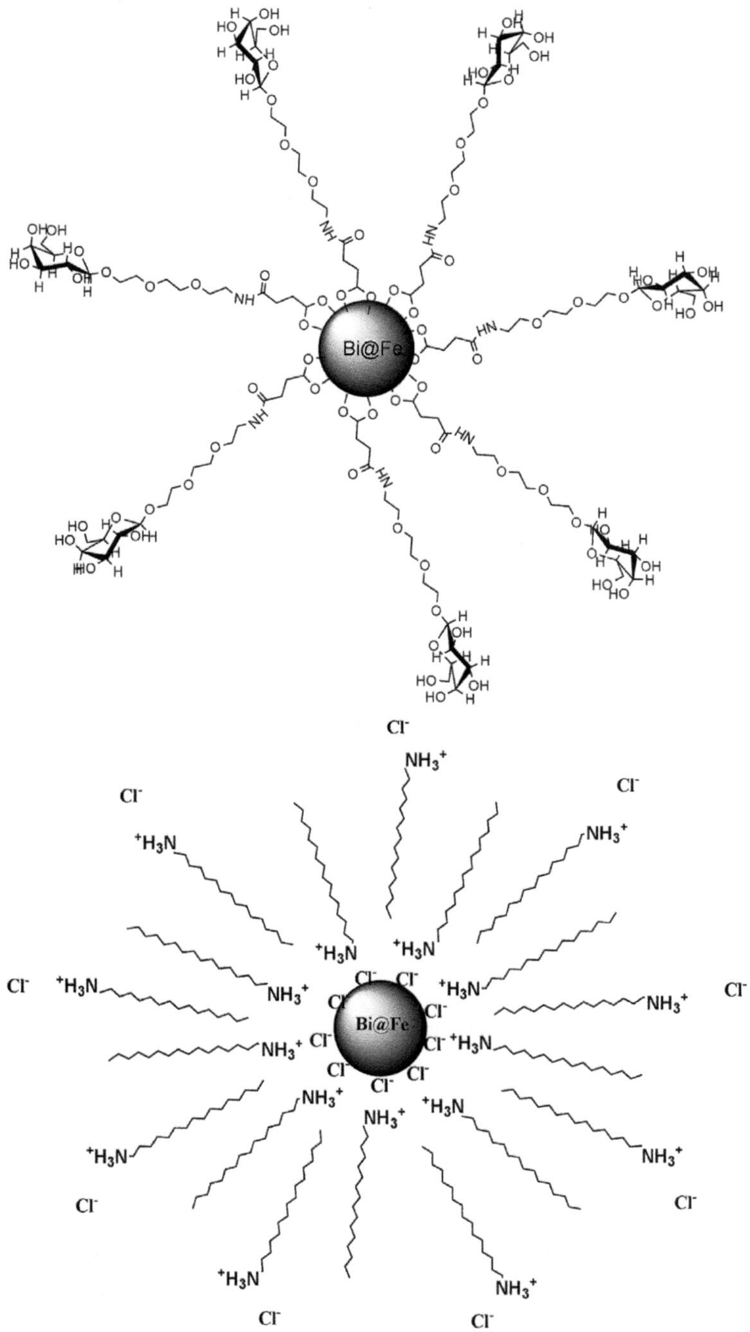

Fig. 6 Tentative schematic presentation of: top: the LPEG1 coating on BiFe150. Bottom: the HDA·HCl double layer at the surface of BiFe14.

is a desirable event in view of biomedical applications as it will lead to nanohybrids displaying a surface state comparable to that of the SPIONs presently developed as MRI contrast agents.

As BiFe14 NPs displayed the most interesting size in view of biomedical applications, spontaneous oxidation of these NPs was thus investigated. After one day of air exposure, observation of the (110) lattice planes of *rhombohedral* Bi (Fig. 7, top) is a clear indication that the Bi core remains unchanged. However, the iron shell around it now seems fully oxidized as attested by the observed lattice plane spacing (*ca.* 0.25 nm). Due to the small size of the iron oxide grains, it is difficult to establish their structural phase, especially as the observed lattice spacing is shared by all major iron oxides. Also, given the spontaneous character of the oxidation, different kinds of oxides and/or hydroxides are likely to form simultaneously. It is worth noting that Mossbauer spectra present a weak signal over noise ratio probably due to the high absorption of the bismuth core preventing any investigation into the iron oxide nature by this technique.

After longer exposure times, the crystallinity of the Bi core is lost (Fig. 7, bottom) and full oxidation of the NPs is achieved. It is important to note that

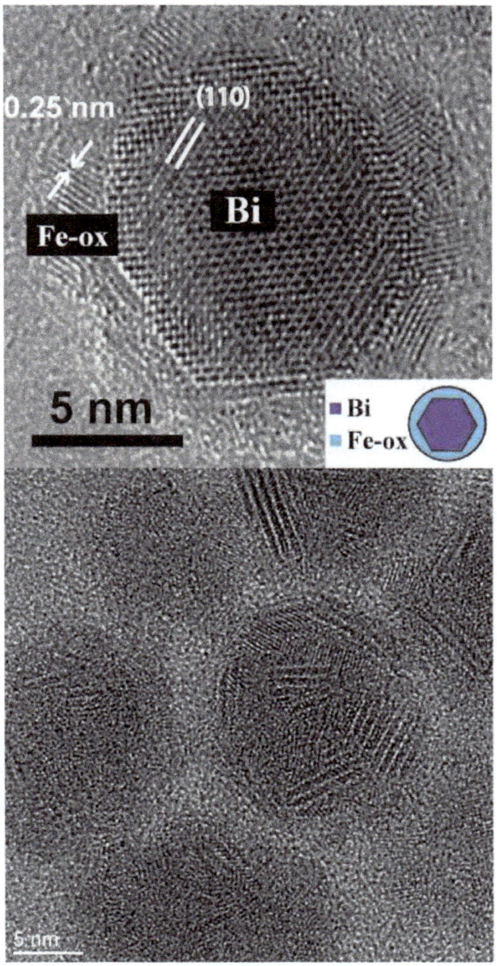

Fig. 7 TEM images of BiFe14 after air exposure. Top: after 2 days; bottom: after one year.

oxidation of the Bi core is not a limitation for the application envisaged as the X-ray absorption does not depend on the oxidation state of an element.

X-ray opacity and first MRI imaging results

The possible activity of BiFe14 NPs as a CT or MRI contrast agent was then investigated as the suspensions of NPs in water are magnetically responsive, evidencing the formation of a magnetic iron oxide. However, given the configuration of the external shell (similar to that of a hollow sphere), closure of stray fields should lead to a very weak overall magnetization. As a consequence of the small thickness of the shell (1 to 2.5 nm, with an inner diameter around 10 nm), strong magnetic disorder and non-collinear spin structures may occur, which could also lead to small magnetic moments carried by the nanoparticles. These features could lead to low activity as an MRI contrast agent (to get high MRI sensitivity, the magnetic moment should be as high as possible) if any at all. However, at the nanoscale many surface defects are present and deviation from a perfect spherical symmetry is expected which may overcome the limitations discussed above. Preliminary tests were carried out to check if the local field produced by the oxide shell is strong enough to induce contrast. As the stability of the NPs in water is limited, homogeneity of the aqueous medium has been ascertained by gel formation with agar. Spin–spin relaxation time (T2) weighted spin echo MR images of the Bi–Fe nanohybrids are shown in Fig. 8. Contrast enhancement varies as a function of the increase in the concentration of the nanomaterial, which suggests that it can be used as a contrast agent for MRI.

Compared to classical iodine based CT contrast agents, bismuth presents the advantage of a higher attenuation coefficient of X-rays at 100 keV due to the position of its *K* edge (90.8 keV). (The X-ray absorption coefficient of bismuth is 5.74 $cm^2\ g^{-1}$ and that of iodine is 1.94 $cm^2\ g^{-1}$, for comparison water absorption is only 0.171 $cm^2\ g^{-1}$). Bismuth based NPs are thus perfectly adapted to color CT imaging, the newest advancement in CT imaging technology.[3] However, classical CT scanners operated at 100 kV, emit an X-ray spectrum from 30 keV to about 100 keV, with a maximum emission intensity around 60 keV. Therefore, the attenuation properties of the elements over the range 30–100 keV need to be considered as iodine actually has better attenuation than bismuth at voltages below bismuth's *K*-edge. We thus investigated the properties of BiFe14 and compared them to those of a commercial CT contrast agent: Omnipaque. This study, carried out on a CT scanner used for small animals operated at 120 kV, evidences an absorption coefficient twice that of Omnipaque, as observed for Bi_2O_3 NPs of a

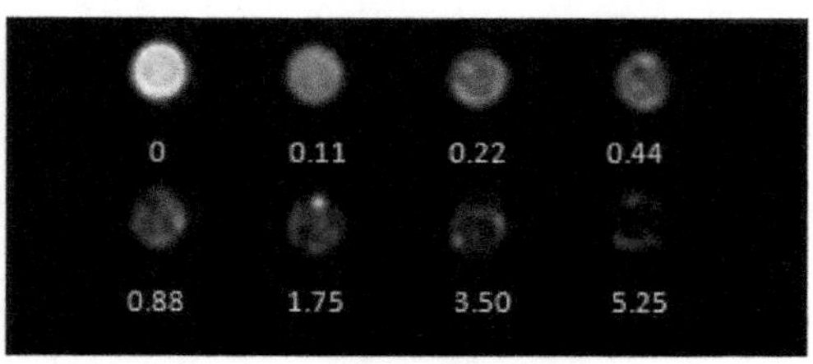

Fig. 8 T2-weighted MRI images of the Bi–Fe hybrid nanoparticles in agar gel at 3T as a function of Fe concentration (mmol L^{-1}).

similar size.[4] BiFe14 is thus a promising nano-platform for dual CT and MR imaging.

Conclusions and perspectives

In conclusion we have designed an efficient synthesis route to well crystallized core–shell Bi–Fe NPs that spontaneously oxidize into an intermediate Bi@Fe oxide nanohybrid. As few data are available for Bi in the NP form,[41,5] the presence of Fe in the shell of the NPs, providing a protecting barrier and isolating the Bi domain from biological media, is important for safety reasons. These NPs give an enhanced contrast in CT imaging and significant contrast in MRI *in vitro*. They are thus promising cheaper alternatives to the FeAu dual contrast agents reported so far. Beyond their possible use in biomedical imaging, such nanohybrids allow the prospect of better tuning the radiation dose in a precisely localized area of the body during X-ray radiotherapy.[42] On-going work focuses on the optimization of their surface functionalization, and on the investigation into their magnetic behavior to correlate with their effect on T2 weighted MRI images.

Acknowledgements

This work has been supported by the ANR, the French National Research Agency (Mag@M project ANR-09-BLAN-0002-01). M. Branca thanks the Region Midi-Pyrénées and University Paul-Sabatier for a grant. We are also indebted to Dr Puerto Morales for fruitful discussions, to Voxcan for CT measurements, Pr Marc Verelst, Hélène Gros-Dagnac and M. Semyou-Ayele Osseni for their help in recording MRI images.

Notes and references

1 A. Jakhmola, N. Anton and T. F. Vandamme, *Adv. Healthcare Mater.*, 2012, **1**, 413–431.
2 K. Ai, Y. Liu, J. Liu, Q. Yuan, Y. He and L. Lu, *Adv. Mater.*, 2011, **23**, 4886.
3 D. Pan, E. Roessl, J.-P. Schlomka, S. D. Caruthers, A. Senpan, M. J. Scott, J. S. Allen, H. Zhang, G. Hu, P. J. Gaffney, E. T. Choi, V. Rasche, S. A. Wickline, R. Proksa and G. M. Lanza, *Angew. Chem., Int. Ed.*, 2010, **49**, 9635.
4 H. Aviv, S. Bartling, I. Grinberg and S. Margel, *J. Biomed. Mater. Res., Part B*, 2013, **101**, 131.
5 A. L. Brown, P. C. Naha, V. Benavides-Montes, H. I. Litt, A. M. Goforth and D. P. Cormode, *Chem. Mater.*, 2014, **26**, 2266.
6 D. Danila, E. Johnson and P. Kee, *Nanomedicine: Nanotechnology, Biology and Medicine*, 2013, **9**(7), 1067–1076.
7 T. Zhou, B. Wu and D. Xing, *J. Mater. Chem.*, 2012, **22**, 470.
8 P. Debbage and W. Jaschke, *Histochem. Cell Biol.*, 2008, **130**, 845.
9 M. Liong, J. Lu, M. Kovochich, T. Xia, S. G. Ruehm, A. E. Nel, F. Tamanoi and J. I. Zink, *ACS Nano*, 2008, **2**, 889–896.
10 T. Courant, V. G. Roullin, C. Cadiou, M. Callewaert, M. C. Andry, C. Portefaix, C. Hoeffel, M. C. de Goltstein, M. Port, S. Laurent, L. V. Elst, R. Muller, M. Molinari and F. Chuburu, *Angew. Chem., Int. Ed.*, 2012, **51**, 9119.

11 D. Choi, A. Han, J. P. Park, J. K. Kim, J. H. Lee, T. H. Kim and S.-W. Kim, *Small*, 2009, **5**, 571.
12 X. Shao, H. Zhang, J. R. Rajian, D. L. Chamberland, P. S. Sherman, C. A. Quesada, A. E. Koch, N. A. Kotov and X. Wang, *ACS Nano*, 2011, **5**, 8967.
13 R. Ta, M. Suchy, J. H. K. Tam, A. X. Li, F. S. Martinez-Santiesteban, T. J. Scholl, R. H. E. Hudson, R. Bartha and S. H. Pasternak, *Contrast Media Mol. Imaging*, 2013, **8**, 127.
14 G. Pourroy, S. Begin-Colin and D. Felder-Flesch, French patent no. 1053139 (CNRS), France, 2010.
15 E. Pisani, N. Tsapis, B. Galaz, M. Santin, R. Berti, N. Taulier, E. Kurtisovski, O. Lucidarme, M. Ourevitch, B. T. Doan, J. C. Beloeil, B. Gillet, W. Urbach, S. L. Bridal and E. Fattal, *Adv. Funct. Mater.*, 2008, **18**, 2963.
16 F. Yang, Y. Li, Z. Chen, Y. Zhang, J. Wu and N. Gu, *Biomaterials*, 2009, **30**, 3882.
17 D. Kim, M. K. Yu, T. S. Lee, J. J. Park, Y. Y. Jeong and S. Jon, *Nanotechnology*, 2011, **22**, 155101.
18 A. Louie, *Chem. Rev.*, 2010, **110**, 3146.
19 K. Nikolaou, M. Poon, M. Sirol, C. R. Becker and Z. A. Fayad, *Cardiol. Clin.*, 2003, **21**(4), 639–655.
20 A. Hagit, B. Soenke, B. Johannes and M. Shlomo, *Biomacromolecules*, 2014, **11**, 1600.
21 O. Rabin, J. M. Perez, J. Grimm, G. Wojtkiewicz and R. Weissleder, *Nat. Mater.*, 2006, **5**, 118.
22 M. Andreés-Vergeés, M. d. P. Morales, S. Veintemillas-Verdaguer, F. J. Palomares and C. J. Serna, *Chem. Mater.*, 2012, **24**, 319–324.
23 M. Vehkamaki, T. Hatanpaa, M. Ritala and M. Leskela, *J. Mater. Chem.*, 2004, **14**, 3191–3197.
24 K. Soulantika, A. Maisonnat, M.-C. Fromen, M.-J. Casanove and B. Chaudret, *Angew. Chem., Int. Ed.*, 2003, **42**, 1945.
25 I. G. Green and B. P. Roberts, *J. Chem. Soc., Perkin Trans. 2*, 1986, 1597.
26 http://www.antellis.com/.
27 C.-H. Lai, C.-Y. Lin, H.-T. Wu, H.-S. Chan, Y.-J. Chuang, C.-T. Chen and C.-C. Lin, *Adv. Funct. Mater.*, 2010, **20**, 3948.
28 D. Ciuculescu, C. Amiens, M. Respaud, A. Falqui, P. Lecante, R. E. Benfield and B. Chaudret, *Chem. Mater.*, 2007, **19**(19), 4624.
29 N. Atamena, D. Ciuculescu, G. Alcaraz, A. Smekhova, F. Wilhelm, A. Rogalev, B. Chaudret, P. Lecante, R. E. Benfield and C. Amiens, *Chem. Commun.*, 2010, **46**, 2453–2455.
30 J. G. Mattei, F. Pelletier, D. Ciuculescu, P. Lecante, J. C. Dupin, N. Yaacoub, J. Allouche, J. M. Greneche, D. Gonbeau, C. Amiens and M. J. Casanove, *J. Phys. Chem. C*, 2013, **117**, 1477.
31 A. Meffre, S. Lachaize, C. Gatel, M. Respaud and B. Chaudret, *J. Mater. Chem.*, 2011, **21**, 13464.
32 X. Wang, R. D. Tilley and J. J. Watkins, *Langmuir*, 2014, **30**, 1514.
33 G. Huang, J. Diakur, Z. Xu and L. I. Wiebe, *Int. J. Pharm.*, 2008, **360**, 197.
34 D. L. Morris, *Science*, 1948, **107**, 254.
35 K. Rahme, L. Chen, R. G. Hobbs, M. A. Morris, C. O'Driscoll and J. D. Holmes, *RSC Adv.*, 2013, **3**, 6085.
36 A. G. Kanaras, F. S. Kamounah, K. Schaumburg, C. J. Kiely and M. Brust, *Chem. Commun.*, 2002, 2294.

37 E. E. Foos, A. W. Snow, M. E. Twigg and M. G. Ancona, *Chem. Mater.*, 2002, **14**, 2401.
38 H. Modrow, S. Bucher, J. Hormes, R. Brinkmann and H. Bönnemann, *J. Phys. Chem. B*, 2003, **107**, 3684.
39 K. Moore, B. Forsberg, D. R. Baer, W. A. Arnold and R. L. Penn, *J. Environ. Eng.*, 2011, **137**(10), 889–896.
40 S. Zhang, G. Jiang, G. T. Filsinger, L. Wu, H. Zhu, J. Lee, Z. Wu and S. Sun, *Nanoscale*, 2014, **6**, 4852.
41 Y. Luo, C. Wang, Y. Qiao, M. Hossain, L. Ma and M. Su, *J. Mater. Sci.: Mater. Med.*, 2012, **23**, 2563–2573.
42 J. Zheng, G. Perkins, A. Kirilova, C. Allen and D. A. Jaffray, *Invest. Radiol.*, 2006, **41**, 339.

DISCUSSIONS

Magnetic Nanoparticles: general discussion

Peter Dobson, Lanry L. Yung, Liane Rossi, Sara Carreira, Paresh Ray, Catherine Amiens, Katherine Brown, Dejian Zhou, Asterios Gavriilidis, Nguyen T. K. Thanh, Sandhya Moise, Lucio Litti, Hedi Mattoussi, Matthew Todd, Scott Mitchell, Stefan Borsley, Amelie Heuer-Jungemann, Oliver Reiser, Edman Tsang, Maya Thanou, Tom Berkleman, Dalibor Soukup, Kerry Chester, Ivan Parkin, Yuri Antonio Diaz Fernandez and Benjamin P. Burke

DOI: 10.1039/C4FD90078B

Oliver Reiser opened the discussion of the paper by Peter Dobson: Is a size of below 20 nm critical to achieve cell penetration by nanoparticles?

Peter Dobson responded: No, I chose this as a typical size. I believe that the most important point is the surface chemistry and that particle size could be a secondary factor. On the other hand, smaller particles of <20 nm will induce a distortion of the outer cell membrane enabling the particle to penetrate by endocytosis, but this requires a strong affinity brought about by the nanoparticle–cell surface interaction.

Ivan Parkin remarked: The antimicrobial properties of nanodiamond may have been misrepresented in the literature as the commercial solutions contain a variety of antimicrobial agents. Studies must make sure that these are not present if antimicrobial analyses are to be performed.

Peter Dobson replied: I fully agree, but on the other hand nanodiamond has got unique energy levels with respect to water and related systems.[1]

1 C. E. Nebel, *Nature Materials*, 2013, **12**, 780.

Dejian Zhou remarked: What is your favourite inorganic drug nanocarrier and why?

Peter Dobson responded: I think this has to be porous nanoparticles of silica. They are relatively easy to make and to load. The problems come in allowing them to circulate and target the right cells and then release their payload. I select this choice also because there do not appear to be any adverse reactions to small quantities of silica in the body. Silica is also close to ideal for the functionalisation protocols that will need to be adopted. So, to summarise: porous silica is my favourite, but I am less sure about rendering them with the right functionality for circulation, attachment and drug release.

Dejian Zhou commented: How do you achieve controlled release of drugs from mesoporous silica nanoparticles to only release the drug load when it reaches the target, without release occurring during drug transport? We have recently developed a pH-responsive DNA-based drug carrier that displays reversible pH-triggered conformational changes between a four-stranded I-motif and single-stranded structure.[1] Do you think that this DNA system could be used for controlled release of mesoporous silica nanoparticles?

1 L. Song, V. H. B. Ho, C. Chen, Z. Yang, D. Liu, R. Chen and D. Zhou, *Adv. Healthcare Mater.*, 2013, **2**, 275–280.

Peter Dobson answered: I have little to add because triggering release is not going to be easy. If the proposed pH-triggered method works under physiological pH conditions then it is worth a try. My first understanding of the work published by Song *et al.* is that a large pH swing from 7.6 to 5.1 is needed.[1] I am not sure, but believe the local pH at a tumour site is closer to 7. Another factor of concern is the increased complexity that is being introduced. Is it because the DNA-based idea does not carry a sufficient amount of drug?

1 L. Song, V. H. B. Ho, C. Chen, Z. Yang, D. Liu, R. Chen and D. Zhou, *Adv. Healthcare Mater.*, 2013, **2**, 275–280.

Nguyen T. K. Thanh opened the discussion of the paper by Edman Tsang: How did you inject the magnetic nanoparticles into the rat brain?

Edman Tsang responded: Before the micro-surgery, the SD (Sprague Dawley) rats were first anesthetized with sodium pentobarbital (60 mg kg^{-1}; i.p.; Saggittal). A small 1 cm midline sagittal skin incision was cut approximately on the scalp to expose the skull. Two holes (Bregma: +0.05 cm, Medline: $\pm$0.1 cm) with a diameter of 0.2 cm were stereotaxically drilled in the skull for the injection of particles into the left hand side of the sub-ventricular zone (SVZ). The particles (5 µl, 2000 µg ml^{-1}) were then stereotaxically administered into the target sites (Dura: $-$0.5 cm; 1 µl min^{-1}) and were allowed to incubate for different amounts of time (0 hour [$n = 5$], 1 hour [$n = 5$], 3 hours [$n = 5$], 6 hours [$n = 5$] and 24 hours [$n = 5$]).

Nguyen T. K. Thanh remarked: Could you please comment on the translational aspect of this technology? How close is it to the clinical trial? What else is needed to be done before neural stem cell therapies can be a reality? What type of rats did you use; you mentioned Institute of Cancer Research rats?

Edman Tsang answered: Magnetic separation of biological species, magnetic hyperthermia and magnetic controlled drug release using modified magnetic nanoparticles have already found clinical applications. The use of iron oxide based magnetic nanoparticles in particular is deemed acceptable with FDA approval. However, the technique involving the use of such particles to extract neuron stem cells from a live brain *in vivo* has only recently been disclosed by our group. We are at the stage of refining and optimizing important parameters such as minimization of unnecessary damage to underneath brain tissue and reduction in the retention of magnetic particles in the living brain, *etc.* These are important parameters that we need to address before any further clinical trial is undertaken. We are therefore unable to put forward a timeline but so far the results look encouraging. The type of rat: SD (Sprague Dawley) rats (male and female), 150–180 g. The animal experiments have been carried out with our research partners: one based at Hong Kong Baptist University and the other at the Taiwan Mouse Clinic.

Dejian Zhou said: Where are the magnetic nanoparticles located? On the inside or outside of the cells?

Edman Tsang replied: Initially the magnetic nanoparticles are attached to the exterior of the cells due to the surface recognition groups ($CD133^+$). However, confocal microscopy showed that some magnetic particles are internalised into the cells depending on various factors such as incubation time and particle concentration.

Dejian Zhou said: Have you determined whether the magnetic nanoparticles are located inside or outside of the cell surface? This can be done by cryo-TEM imaging by which we have recently shown that nanoparticles are mostly trapped in endosome-like intracellular compartments after cell uptake.[1]

1 L. Song, V. H. B. Ho, C. Chen, Z. Yang, D. Liu, R. Chen and D. Zhou, *Adv. Healthcare Mat.*, 2013, **2**, 275–280.

Edman Tsang responded: Thank you for the suggestion of using cryo-TEM imaging. Although the confocal microscopy shows evidence for the internalization of these particles into the cells the image resolution is not satisfactory. The mentioned technique could be useful for further study in this area.

Dejian Zhou remarked: When cells differentiate, where do the nanoparticles end up?

Edman Tsang answered: This is an interesting question. The simple answer is that we do not know as we have not yet monitored the differentiation process of cells in the presence of the magnetic nanoparticles. Their presence and their location could be important regarding cell differentiation.

Asterios Gavriilidis asked: How do the microemulsion properties affect nanocomposite shell thickness?

Edman Tsang responded: Basically, the thickness of the silica shell increases as the amount of TEOS increases; a situation in which core-free silica particles appear when the concentration of TEOS is too high should be avoided. Also, there are other minor factors that can affect the shell thickness *e.g.* the ratio of base to amphiphilic surfactant, reaction time, the size of aqueous domains and the number of hydrophobic NPs added into this system, *etc.* For detailed information please refer to the paper titled "Fe_3O_4@SiO_2 Core/Shell Nanoparticles: The Silica Coating Regulations with a Single Core for Different Core Sizes and Shell Thicknesses".[1]

1 H. L. Ding *et al.*, *Chem. Mater.*, 2012, **24**, 4572–4580.

Asterios Gavriilidis remarked: What is the reason for some nanoparticles having two iron cores?

Edman Tsang answered: There are many possible reasons leading to two iron cores in the same shell, which has been commonly observed when core–shell particles are synthesized *via* a reverse micro-emulsion technique. However, we believe that the main reason in our case is that the number of magnetic nanoparticles added (excess) did not match with the number of aqueous domains (micelles) in the reverse micro-emulsion system, hence two iron cores shared the same shell. It should be noted that it is very difficult experimentally to match their numbers.[1] However, by optimizing our system, we managed to produce 90% of our Fe_3O_4@SiO_2 particles with a single core (please refer to the low magnification TEM image Fig. S2 in the supporting information of our paper (*Faraday Discuss.*, 2014, DOI:10.1039/C4FD00132J)).

1 H. L. Ding *et al.*, *Chem. Mater.*, 2012, **24**, 4572–4580.

Sara Carreira said: How many of the magnetic NPs attach to the surface of the neural stem cells *in vivo*? Can you at least estimate it, maybe from the number of CD133 receptors present on the cell surface?

Edman Tsang responded: We are at present unable to give the number of magnetic NPs that attach to each neural stem cell (NSC). It is evident that the process of attachment of ependymal cells by magnetic NPs is rather dynamic in a living subject. They continue to attach and detach in a flowing fluid and can also be taken up by inner cells. As we mentioned in the article, several processes have to be considered: (1) the diffusion and adsorption of particles on NSCs, (2) the removal of particles by cerebrospinal fluid (CSF), (3) dynamically binding/(4) unbinding to NSCs and (5) internalization of bound particles by NSCs. Accordingly, the initial decrease in the T_2 CNR signal can be attributed to the higher rate of particle removal (processes 2 and 4) compared to the counter processes (1), (3) and (5). In order to effectively isolate the NSCs magnetically we injected excess magnetic NPs for maximal attachment, however, we are not yet able to estimate the number of magnetic NPs on a single NSC.

Katherine Brown said: Are you able to determine or quantify the amount of active antibody on your nanoparticles?

Edman Tsang replied: We incubated excess antibodies to the given weight of nanoparticles after chemical functionalization (see the Experimental section of our paper (*Faraday Discuss.*, 2014, DOI: 10.1039/C4FD00132J)) in order to maximize the carrying capacity of the nanoparticles and reduce multiple attachment to two or more nanoparticles. However, we did not determine the exact amount of grafted antibody per nanoparticle. No doubt that different morphological forms of bound antibodies of different activities arise. We are currently working to quantify their attachment, morphology and their activity.

Hedi Mattoussi asked: Is there a minimum required number of antibodies attached to a nanoparticle to promote strong binding onto the target cell membranes?

Edman Tsang answered: In our experiment, we aimed to attach as many antibodies as possible to a single nanoparticle. This not only enhances its chance of binding to the CD133 receptors of a neuron stem cell (NSC) but it also strengthens the interactions between the composite particle carrying the antibodies and NSCs through multiple antibody–antigen bonds (surface recognition groups). As a result, we have not yet studied the minimum number of antibodies per particle for the binding of NSCs. It is noted that the binding force between such particles and NSCs could depend on their relative numbers and morphologies, which are rather difficult to assess.

Matthew Todd asked: If the aim of using a nanoparticle appended to an antibody is to extract cells, don't you ideally want one antibody per particle? That way you will have multiple nanoparticles attached per cell, increasing the chance of extraction. On the other hand, if you make particles each appended to multiple antibodies, you will more likely achieve aggregation behaviour, since one particle could (cross-)link to multiple cells.

Edman Tsang answered: Thank you for your interesting suggestion. However, we feel that it is rather difficult to engineer one single molecule on a single particle surface since chemical modification is normally carried out on a generic surface with many related energetic surface sites. Although there are reported methods to immobilize a single guest molecule on a particle surface, the process is always time consuming and complicated.[1] This is achieved by limiting the chemical functionalities on a single particle; the stability of this antibody–particle ensemble through limited surface bonding is still open to question. Thus for us, a more practical way is to graft more than one guest molecule onto a single particle. We understand the concern of aggregation, however, the relative size of the magnetic nanoparticle is about 50 nm and the size of the $CD133^+$ cell (antigen) is tens of microns. Thus, there will not be many cells that can be spatially attached to the magnetic nanoparticle–antibody composite. Besides, the antibody–antigen composite is highly specific and complimentary recognition groups must be present between them before they can be attached in a specific orientation. We feel the chance of the same bound antibody becoming fixed with one antigen to bind with another bound antibody and leading to extensive aggregation is not high.

1 M.-L. Ho *et al.*, *J. Phys. Chem. C*, 2009, **113**, 1686–1693.

Nguyen T. K. Thanh said: Your work is so fascinating, so I hope you do not mind another question.

You said that you would like to isolate a single cell, but that would be very hard with magnetic separation, how will you tackle this?

Edman Tsang answered: We are pleased that this new work is appreciated by our fellow researchers. Yes, it will be a very challenging goal. Firstly, according to our results, the neuron stem cells (NSCs) appear to be located in specific areas of a living rat brain rather than being evenly distributed on all brain tissue. Specifically, they are located by MRI in the epithelial regions of ventricles and SVZ areas. However, there is no information on how these NSCs are distributed depth-wise. On the other hand, many brain diseases are thought to relate to their deprival in some functional areas. It will therefore be exciting to move the NSCs from rich areas to deficient areas in future nano-surgery. At this stage, we are unfortunately unable to image them with sufficient resolution in order to differentiate them on a single cell basis. To move this research forward, we believe the first stage is to achieve single cell imaging of our optimized particles (t1/t2 optimization). Then we would be able to relocate each of them by magnetic means in a highly precise way (computer aided technology). We agree that the magnetic separation would depend on the overall magnetic susceptibility of the particles (the total volume of the magnetic phase). In order to achieve this goal, we would need to design the particles with a high response as well as a small but strong magnetic probe for computer operated relocation for the required precision.

Peter Dobson opened the discussion of the paper by Oliver Reiser: I believe that the particles you refer to in your work are ferromagnetic so the magnetic moments will be much larger than the superparamagnetic iron oxide particles that have been referred to in earlier papers in the meeting.

Oliver Reiser responded: Yes, this is correct. The "naked" carbon coated nanoparticles have a magnetic moment of approx. 140–150 emu g^{-1}, which will decrease once you attach molecules, for example polymers. We feel that this can be an advantage since you are able to generate highly functionalized materials that still have a considerable high magnetic moment, but the disadvantage is that agglomeration, especially with the unfunctionalized particles, is more severe.

Peter Dobson commented: Because the particles are ferromagnetic with a fairly high magnetic moment, they will be difficult to disperse I believe?

Oliver Reiser replied: The unfunctionalized particles are not dispersible, *e.g.* in water. With the appropriate coating, *i.e.* the PEI coating shown here (and described in our paper (*Faraday Discuss.*, 2014, DOI: 10.1039/C4FD00108G) the particles are perfectly dispersible in water.

Edman Tsang asked: In the encapsulation of a metal phase by carbon nanotubes and related structures, there is a defective region in the carbon structure which may lead to leaching of the metal phase into corrosive solution. Has this been observed in your experiments? Are any measures to reduce such metal leaching in place?

Oliver Reiser responded: Thank you for this important comment. We constantly discuss this with our collaborators at the ETHZ who manufacture the particles. In fact, we have noticed differences in quality in various batches that we have received from them (you can easily check by adding EDTA to see if you extract cobalt). TURBOBEEDS who sells the particles is therefore performing quality checks to make sure that they not distribute particles which have such effects that would cause metal leaching. All in all, credit to the procedure developed at the ETHZ as the coating is generally uniform to prevent metal leaching.[1]

1 R. N. Grass, E. K. Athanassiou, W. J. Stark, *Angew. Chem.*, 2007, **46**, 4909.

Edman Tsang commented: Is there any issue in the material preparation that means that encapsulated metal particles are not totally separated from each other. Would this create a problem in magnetic separation if they were actually aggregated during graphitization?

Oliver Reiser replied: Thank you for the question, all credit goes to our collaborators at ETHZ who manufacture the particles.[1] Based on TEM analysis of these particles, single entities and non-aggregated ones are formed during graphitization, so we did not run into this problem. Magnetic separation works very well with these particles.

1 R. N. Grass, E. K. Athanassiou, W. J. Stark, *Angew. Chem.*, 2007, **46**, 4909.

Dejian Zhou said: Are these magnetic nanoparticles cytotoxic? The stability of the amine grafted magnetic nanoparticles can be further stabilised by PEGylation. We have found that PEGylated nanoparticles can sustain freeze-drying without undergoing aggregation.

Oliver Reiser responded: Thanks for the comment and suggestion. We do not yet know the cytotoxicity of the particles, but it is currently under investigation by our colleagues in bioanalytics. Thank you also for pointing out the PEGylation – we need to look into this more. We have some preliminary results though that PEGylated PEI nanoparticles are less stable than the PEI only particles.

Liane Rossi commented: Regarding the stabilization of the cobalt nanoparticles against oxidation: is it related to the carbon-coating or a matter of particle size?

Oliver Reiser responded: The stabilization is due to the thin (3–5 layers) graphene-type layer around the particles.

Liane Rossi commented: What characterization techniques were used to show that the cobalt nanoparticles are metallic and do not oxidize after exposure to air?

Oliver Reiser replied: The following work was not done by us but by our collaborators (Prof. W. Stark, ETH Zurich; Turbobeeds Inc.) who manufacture the nanoparticles:

A combination of three methods was used:

a) X-ray diffraction to validate the identity of the metallic crystalline phase. No oxidic phases could be identified.

b) Magnetic measurements; cobalt oxide is not ferromagnetic, so any saturation magnetization has to be allocated to the metallic phase. The saturation magnetization of the carbon coated cobalt nanomaterials is very close to the saturation magnetization of bulk cobalt.

c) Thermogravimetry: Due to the weight gain upon oxidation of the metallic material at elevated temperatures (at 400 °C) to the oxides, the initial metal content can be calculated.

Nguyen T. K. Thanh commented: Did they determine that it is a graphene layer on the surface of nanoparticles, and not any other form of carbon?

Oliver Reiser replied: Yes, our collaborators at the ETHZ did Solid State NMR, Raman, IR and also measured electronic transport, everything – as well as the chemical reactivity we see – is consistent with a graphene-like surface.[1]

1 W. Stark *et al.*, *Acc. Chem. Res.*, 2013, **47**, 2297.

Matthew Todd commented: Covalent attachment to a nanoparticle means that the attached compound is fixed to the surface. π-π attachment is interesting because the appended species could move over the surface. This could, for example, permit initial binding interaction followed by the slower development of polyvalency in binding. Do you see evidence of such movement on the surface?

Oliver Reiser answered: We have deposited Pd-nanoparticles on the carbon surface of the magnetic nanoparticles, and indeed, we see movement here in ref. 1. We also see that the attachment through π-stacking (pyrene) units is reversible with temperature, so indeed I would expect to have movement on the surface.

1 Q. M. Kainz *et al.*, *Adv. Funct. Mat.*, 2014, **24**, 2020.

Nguyen T. K. Thanh asked: In Fig. 2 (in *Faraday Discuss.*, 2014, DOI: 10.1039/C4FD00108G) you mentioned: "The dispersibility of 12 was by far superior in dichloromethane than in water."

There are some ligand-ligand interactions, and 12 might not form a well-packed passivating layer due to side chain packing/bonding properties.

We have studied the effect of 58 different peptide sequences on the electrolyte-induced aggregation of the nanoparticles in our paper.[1]

1 R. Levy, N. T. K. Thanh, R. C. Doty, I. Hussain, R. J. Nichols, D. J. Schiffrin, M. Brust, D. G. Fernig, *J. Am. Chem. Soc.*, 2004, **126**, 10076–10084.

The stabilities conferred by the peptide ligands depended on their length, hydrophobicity, and charge.

Oliver Reiser replied: Yes, very nice work, and I completely agree with your statement that the stability of the dispersions is dependent on length, hydrophobicity, and the charge of the outer layer. What we wanted to show with Fig. 2 in our paper (*Faraday Discuss.*, 2014, DOI: 10.1039/C4FD00108G) is that (1) typical coatings like MeOPEG 2000 that have been shown to stabilize metaloxide nanoparticles are not sufficient for the highly magnetic metal particles we have studied and (2) that seemingly very similar coatings (*e.g.* compare 12 and 16) can still confer quite different stabilities to the nanoparticles.

Nguyen T. K. Thanh asked: If the nanoparticles are polydisperse, different sized particles would not only have different physical properties, but they would also have different surface areas. Therefore, the coating might not be so effective, as different curvatures require different coatings (*e.g.* molecular structures). Moreover, they would have different numbers of active molecules, such as antibodies. So, the bioactivities are also different; this would cause non-reproducibility of subsequent biological assays.

Oliver Reiser replied: Thank you, this is a very good comment, and we will need to address this when we move to biological assays. The nanoparticles have by and large a diameter of 50 nm, but indeed, smaller particles are also present.

Hedi Mattoussi remarked: We know that cobalt metal is easily oxidized by simple exposure to air and/or water.

Does the graphene protective layer used in your Co nanoparticles provide enough shielding to avoid the issue of cobalt oxidation?

Oliver Reiser answered: Yes, while the oxidation of metallic cobalt surfaces even occurs at room temperature the carbon layer shields the surface from this oxidation. This can be evidenced by differential calorimetry combined with thermogravimetry, where no material oxidation (= weight gain and energy release) is observed at temperatures <180 °C. Similarly, the oxidative properties in water over a wide pH range have been quantified by leaching studies over the course of several weeks[1] and support the excellent shielding properties of the carbon layers. Our collborators at ETH Zurich and Turbobeeds who manufacture the particles have been using the materials for more than seven years, and still have old samples – they can find no indication of cobalt oxidation during storage at ambient conditions over this time frame.

1 A. Schaetz *et al.*, *Chem. Eur. J.*, 2011, **17**, 10566–10573.

Catherine Amiens said: When you functionalised the graphene layer around the cobalt nanoparticles, you either use or create defects. Does it alter the protection that this graphene layer affords against oxidation of the cobalt core?

Oliver Reiser answered: The functionalization takes place only on the first carbon layer exposed to the surface. Since there are multiple carbon layers around the nanoparticle (3–5), the stabilization is not affected.

Catherine Amiens commented: To estimate the mobility of the ligands physisorbed on the graphitic surface, you could work with ligands bearing radicals at the end of their chains (such as TEMPO).

This method has been followed by V. Chechik to demonstrate the mobility of thiols on nanoparticles.[1]

1 P. Ionita, A. Volkov, G. Jeschke, V. Chechik, *Anal. Chem.*, 2008, **80**(1), 95–106.

Oliver Reiser replied: Thank you for this excellent suggestion, we will definitely follow up on this, especially, since we had have already experience with (covalently bound) TEMPO to these nanoparticles.[1]

1 A. Schaetz *et al.*, *Chem. Eur. J.*, 2008, **14**, 8262.

Edman Tsang asked: The total encapsulation of graphitic layers on Co particles is rather challenging to achieve. The Co particle may have different shapes and crystallographic facets. The graphitic layers composed of sp^2 carbon atoms are also spatially rigid, and hence will not be able to join up seamlessly around the particle to offer the perfect encapsulation. My main question is are there substantial defect regions between the junctions of the graphitic layers, particularly at the interface between the metal facets that cause metal leaching into the solution?

Oliver Reiser answered: Thank you, this is a point well taken, and we constantly discuss this with our collaborators at the ETHZ who manufacture the particles. In fact, we have noticed differences in quality in various batches we have received from them (you can easily check by adding EDTA to see if you extract cobalt). TURBOBEEDS, who sells the particles, is therefore performing quality checks to make sure that they not distribute particles that would cause metal leaching. All in all – credit to the procedure developed at the ETHZ the coating is generally uniform to prevent metal leaching.[1]

1 R. N. Grass, E. K. Athanassiou, W. J. Stark, *Angew. Chem.*, 2007, **46**, 4909.

Maya Thanou opened the discussion of the paper by Kerry Chester: Are there any potential risks in the use of dextran sulfate as a plasma expander?

Kerry Chester answered: To the best of our knowledge, unsulfated dextrans are the agents clinically used as plasma substituents.[1,2] Dextran sulfate 500 has been used clinically as an anti-coagulant (as it is a synthetic analogue to heparin)[3] and an antilipemic, as well as an antiviral against human immunodeficiency virus (HIV);[4] it was shown that dextran sulfate inhibits the binding of viruses to target cells and was used as an anti-HIV agent in patients with AIDS. However, in the doses applied it showed limited anti-viral efficacy as well as some toxicities (like reversible thrombocytopenia and alopecia) and was not taken further as an antiviral agent.[5]

1 C. Y. Quon, Clinical pharmacokinetics and pharmacodynamics of colloidal plasma volume expanders, *Journal of Cardiothoracic Anesthesia*, 1988, **2**, 13–23.
2 R. McCahon and J. Hardman, Pharmacology of plasma expanders, *Anaesthesia and Intensive Care Medicine*, 2010, **11**, 75–77.
3 K. Walton, Experiments with dextran sulphate as an anticoagulant, *Proc. R. Soc. Med.*, 1951, **44**, 563–4.
4 H. Mitsuya, D. J. Looney, S. Kuno, R. Ueno, F. Wong-Staal and S. Broder, Dextran sulfate suppression of viruses in the HIV family: inhibition of virion binding to CD4+ cells, *Science*, 1988, **240**, 646.
5 C. Flexner, P. A. Barditch-Crovo, D. M. Kornhauser, H. Farzadegan, L. J. Nerhood, R. E. Chaisson, K. M. Bell, K. J. Lorentsen, C. W. Hendrix and B. G. Petty, Pharmacokinetics, toxicity, and activity of intravenous dextran sulfate in human immunodeficiency virus infection.[1]
1 C. Flexner *et al.*, *Antimicrob. Agents Chemother.*, 1991, **35**, 2544.

Maya Thanou remarked: Near IR fluorescence may be affected (quenched) depending on the distance from the core NP and or the density of grafting. Do you observe such a phenomenon?

Kerry Chester replied: This is an interesting point, but we have not attempted to measure quenching.

Nguyen T. K. Thanh remarked: In your paper you wrote: "We therefore developed methods to functionalize SPIONs with near-infrared (NIR) dyes in order to trace their biodistribution" In Fig 6 in your paper, it says "NIR signal of SPIONs in blood measured on Odyssey scanner". Could you please clarify how you trace their "biodistibution"?

Kerry Chester replied: We only measured the blood but *in vivo* imaging can be done using a mouse pod attached to the LiCor Odyssey scanner: http://www.licor.com/bio/products/accessories/odyssey/mousepod/. Examples of some papers that used the scanner to trace and quantify dye labelled conjugates *in vivo* can be found in refs. 1–4.

1 J. L. Kovar, M. A. Simpson, A. Schutz-Geschwender, D. M. Olive, A systematic approach to the development of fluorescent contrast agents for optical imaging of mouse cancer models, Biochemistry, Faculty Publications, 2007, **9**.
2 J. L. Kovar, W. Volcheck, E. Sevick-Muraca, M. A. Simpson and D. M. Olive, Characterization and performance of a near-infrared 2-deoxyglucose optical imaging agent for mouse cancer models, *Anal. Biochem.*, 2009, **384**, 254–262.
3 J. L. Kovar, W. Volcheck, J. Chen and M. A. Simpson, Purification method directly influences effectiveness of an epidermal growth factor-coupled targeting agent for noninvasive tumor detection in mice, *Anal. Biochem.*, 2007, **361**, 47–54.
4 L. Sampath, S. Kwon, S. Ke, W. Wang, R. Schiff, M. E. Mawad and E. M. Sevick-Muraca, Dual-labeled trastuzumab-based imaging agent for the detection of human epidermal growth factor receptor 2 overexpression in breast cancer, *J. Nucl. Med.*, 2007, **48**, 1501–1510.

Maya Thanou asked: At what time post-injection can you observe NIRF in the blood samples?

Kerry Chester replied: We have only investigated the NIR signal 1 hour post injection but it will be interesting to further investigate the fate of the NIR labelled SPIONs at different time points.

Hedi Mattoussi questioned: (1) Does the sulfate agent used in your experiment block binding to cells by interacting with the receptor(s) or by interacting with the nanoparticle itself?

(2) Have you considered using polyethyelne glycol (PEG) to coat the nanoparticle surfaces as a means of reducing nonspecific interactions?

Kerry Chester answered: (1) We have not carried out a formal experiment but our working hypothesis is that dextran sulfate 500 interacts with the receptors and not the SPIONs themselves. Ferucarbotran is rapidly cleared from circulation mainly (but not exclusively) by the scavenger receptors present on the liver Kupffer cells. Dextran sulfate 500 is a known blocker of this type of receptor and we achieve blocking more effectively when we give dextran sulfate 500 at 2 and

24 hours prior to Ferucarbotran than when we give both agents simultaneously. Furthermore, both Ferucarbotran and dextran sulfate 500 are negatively charged, suggesting that direct interactions with each other are less likely. Please refer to the first and third paragraph of the paper discussion for references.

(2) Yes, we have considered PEG but have not yet explored this option for clinical use for a number of reasons, including the complexity, heterogeneity and biodegradability of PEGylated products and our experience that, in some instances, PEG modification of SPIONs can increase unspecific or untargeted internalisation and uptake (see reference 37 in the paper).

Maya Thanou enquired: For NP injected directly into the brain, is a stealth coating required?

Kerry Chester responded: Within the brain tumour microenvironment there are a range of cell types which include: tumour core, microglial cells (brain macrophages), astrocytes, oligodendrocytes and neurons. All these cell types could take up the SPIONs, specifically the macrophages. Therefore, despite the possible evasion of the RES by direct intratumoural injection of SPIONs, stealth is still advisable.

Dejian Zhou opened the discussion of the paper by Benjamin P. Burke: Where are the gallium ions loaded on the magnetic nanorods?

Benjamin P. Burke answered: This is a question that we are very interested in answering but have not yet identified an experiment to satisfy our curiosity. The issue lies in the potential difference between the larger amounts of gallium(III) used in a standard characterisation experiment and the picomolar amounts of radioactive gallium ions. It has proved challenging to assign data in some spectroscopic methods used to analyse the non-radioactive gallium(III) species, due to overlap of Fe–Si and Fe–O stretches. Even if an assignment could be made it may not be an analogous mechanism for the much smaller amounts of radioactive gallium.

It is interesting to note that the formation of stable multinuclear gallium oxide cations has been observed in zeolite pores.[1] This could be of relevance to the observed high stability interaction with the silica surface.

1 E. A. Pidko *et al.*, *Phys. Chem. Chem. Phys.*, 2009, **11**, 2893–2902.

Tom Berkleman asked: Could you be seeing deposition of gallium oxide, perhaps templated or promoted by your silica layer?

Benjamin P. Burke replied: This is one possible mechanism. However, it seems unlikely that it would be a coating, as the radiosynthesis is carried out at acidic pH and the subsequent stability is high. We think that it is more likely to be a gallium species (which could be oxides or hydroxides) formed in pores on the silica surface. This hypothesis requires further investigation (see also the answer to the previous question).

Sara Carreira asked: What was your motivation for using rod-shaped particles rather than spherical ones? Are there any advantages of using that shape?

Maya Thanou asked: Rods interfere with the cell membrane in a different way to spherical nanoparticles. What is the rationale of using rods for this application?

Benjamin P. Burke answered both questions: Rod shaped particles were selected in part for reasons of curiosity. Iron oxide based PET/MR imaging agents have been developed a handful of times previously, but always using nanospheres. We wanted to see if we could not only develop a novel method of radiolabelling for PET, but also using a previously untested core material shape. We anticipate that similar procedures will be equally valid with spherical iron oxide nanoparticles. The literature is less well developed for iron oxide nanorods than it is for nanospheres, however preliminary studies offer possible advantages in magnetic susceptibility and cell penetration.

Nguyen T. K. Thanh asked: What is the crystal phase of your iron oxide nanorods?

Benjamin P. Burke answered: The preparation of iron oxide nanorods in this publication utilises literature methods (references 42 and 43 in our paper) which are known to form magnetite.

Nguyen T. K. Thanh remarked: What do you think about the regulatory issues of such a novel ligand for clinical use?

Benjamin P. Burke responded: Novel MRI contrast agents have to meet the standard regulations and approval processes for the use of new agents in humans. The key advantage of this work is its potential applicability in the modification of currently licensed agents to a form of PET/MRI multimodal imaging constructs.

Nguyen T. K. Thanh asked: How do you quantify the radioactive dose if your particles are very different sizes?

Benjamin P. Burke replied: The radioactive dose is quantified as activity (MBq)/Fe weight (g) and is therefore not size dependent.

Edman Tsang remarked: Are there any advantages or disadvantages of your approach compared to other labelling techniques?

Benjamin P. Burke answered: The advantages come from the simplicity of the approach: the preparation is facile as no chelator is attached to the surface, only short heating times are needed and it is a final step radiolabelling process. The tracer can therefore be produced rapidly (which is an important issue with short lived radioisotopes). The biggest disadvantage is the temperature required for the radiolabelling reaction, which could lead to stability issues with some surface coatings.

Edman Tsang asked: Would any radioactive species leach out to the external environment when applied ?

Benjamin P. Burke replied: During *in vitro* experiments using human serum we see no evidence of radioactivity being transchelated to transferrin at up to 3 hours, which is the maximum applicable time for acquiring image data using the short lived ^{68}Ga isotope.

Matthew Todd remarked: The outlier result from the Ga labelling in Figure 3 is the increase in size for 5. Does this imply that the metal is loosely chelated by something other than the macrocycle, and in a way that leads to aggregation of several rods through chelation of a metal by more than one rod?

Benjamin P. Burke replied: Although the proposed mechanism is possible, we believe it to be temperature related. Reactions with ^{68}Ga at room temperature do not seem to cause aggregation. In our opinion the most likely explanation is temperature based aggregation of PEG chains, which is documented.

Amelie Heuer-Jungemann opened the discussion of the paper by Nguyen TK Thanh: Have you encountered any problems during the coupling, *e.g.* aggregation due to one glutaraldehyde molecule coupling to two different particles?

Lanry L. Yung asked: Can you give any information about the experimental design, with regard to minimizing particle crosslinking when functionalizing the surface of the particles with glutaraldehyde?

Edman Tsang asked: How can you avoid the problem of condensation between amine and aldehyde onto the same particle?

Nguyen T. K. Thanh responded to all three questions: In our experiment we used huge a excess of glutaraldehyde (GA). Density of GA per nm^2: 4.10×10^8. therefore each particle is immediately coated with GA, so we did not observe any cross linking.

Paresh Ray queried: How many antibodies are attached to each particle?

Nguyen T. K. Thanh answered: We do not normally calculate the number of antibodies attached to each particle, for further information please see:

B. Kozissnik, L. A. W. Green, K. Chester and N. T. K. Thanh, Strategy for functionalisation of magnetic nanoparicles for biological targets, in Magnetic nanoparticles: from fabrication and clinical applications, ed. N. T. K. Thanh, CRC Press, Taylor and Francis, Boca Raton, London & New York, 2012, pp. 129–150.

Tom Berkleman said: I would like to point out that with the high concentration of glutaraldehyde used, it would be in orders of magnitude excess over the amine functionality on the surface of the nanoparticle. You actually wouldn't expect much crosslinking.

Nguyen T. K. Thanh answered: We agree.

Scott Mitchell asked: In terms of control experiments for the interactions with *V. cholerae* bacteria, I was wondering what the TEM images of the IONPs@CHI look like. Do the IONPs@CHI interact in the same way as the complex of ION-P@CHI@GA@PrA@Ab? Surely this is an essential control that must be shown to test the hypothesis of the presence and importance of the Ab conjugated to the polymer matrix? It is an important study to show: Abs are expensive, so in this case are they really necessary?

Furthermore, what about the toxicity of the IONPs@CHI and IONP@CHI@GA@PrA@Ab? The growth of the bacteria appears to be affected in the TEM images. Did the authors carry out Resazurin cell viability assays (or similar) to quantify the toxicity of the particles, or lack thereof?

Nguyen T. K. Thanh responded: It is clearly stated in the paper: "For positive controls, the solutions containing IONP@CHI and ION- P@CHI@GA@PrA were incubated with *V. cholerae* bacteria diluted in water from an initial concentration of 103 cfu mL^{-1}. The same procedures used for testing the IONP@CHI@GA@PrA@Ab complex were followed. A significant number of nanoparticles were found clustered on the grid, but no *V. cholerae* bacterial cells were found (data not shown)." I do not see the relevance of cell viability assays in this work.

Dejian Zhou said: How stable is your magnetic nanoparticle-protein A conjugate? Are the results repeatable after being stored for an extended period? This is important in terms of practical applications. For example, we have found that our covalently conjugated magnetic nanoparticle-DNA conjugate is stable for about 6 months,[1] which is often considered the minimum stability requirement for commerical reagents.

1 Y. Zhang, C. Pilapong, Y. Guo, Z. Ling, P. Quirke and D. Zhou, *Anal. Chem.*, 2013, **85**, 9238.

Nguyen T. K. Thanh replied: To retain the bioactivity of protein A molecules, their stability can be maintained for years if kept at −20 °C. In our experiments, the conjugate was tested every month; during the first 3 months in storage at 4 °C we did not see any difference in the tested results between different time points. We expect it to remain stable for even longer.

Dejian Zhou remarked: Detection of bacteria in your method takes place by SEM or TEM – can you quantify the number of bacteria? What is the dynamic range of detection?

Nguyen T. K. Thanh responded: In our method, the number of bacteria can be quantified by TEM or SEM. We reported that the conjugate could easily separate *V. cholerae* bacteria from water samples at concentrations as low as 10 cfu mL^{-1} by TEM observation.

Liane Rossi asked: Did you reduce the imine group obtained by the condensation of the aldehyde group of glutaraldehyde and the terminal amino group on the nanoparticle surface? Please add this information in the experimental section.

Nguyen T. K. Thanh answered: No, we did not.

Lucio Litti commented: You used imine formation to cross-link the nanoparticle and protein A. As you know, Schiff bases are pH-sensitive and can undergo hydrolysis. Did you verify the stability of your systems in terms of protein A antibody release in aqueous media? On the other hand, you could use the complete hydrolysis of your Schiff bases for a quantitative estimation of the total number of antibodies loaded onto your nanoparticles.

Nguyen T. K. Thanh replied: We used an excess of glutaraldehyde in the conjugation step. The imine hydrolysis is assisted through the use of an acid catalyst. Under our conditions, the final products IONP@CHI@GA@PrA@antibody were formed, and evidently they were selectively bound to the bacteria. We tested our conjugates after three months and they were still stable.

Kerry Chester opened the discussion of the paper by Ivan Parkin: In Figure 2 in the paper there appear to be particles that are rod-shaped and others that are spherical, why is this?

Ivan Parkin replied: This is because they are of different types of material – one is due to the iron oxide (spherical) and the other to rhenanite (rods).

Kerry Chester asked: Is there a way to control the synthesis process to obtain uniform particles?

Ivan Parkin answered: We are looking into this but do not have a synthesis for mono-dispersed particles. We will be exploring a microfluidics approach outlined in Scheme 1 in our paper "Sample (I) is the route to iron oxide nanoparticles proposed by Park *et al.*[3]"

Catherine Amiens commented: In your paper you mentioned that nanorods are epitaxially grown. What do you mean exactly? Do the iron oxide rods grow on top of the $NaCaPO_4$ ones? What is the epitaxial relationship in this case?

Ivan Parkin replied: The paper has since been rectified, as the rods were not epitaxially grown, rather the particular blend of surfactants afforded by the shark liver oil/olive oil mix was enough to promote directional growth of $NaCaPO_4$ (β-rhenanite) rods. As for iron oxide, there were spherical nanoparticles observed in the sample, and a little iron content was observed for the rods by EDX spectroscopy, so the possibility of the growth of the rods from iron oxide nanoparticles remains a distinct possibility.

Catherine Amiens asked: Do you mean that in the preparation you observe $NaCaPO_4$ rods and iron oxide rods at the same time?

Ivan Parkin responded: Just the one type of rod was observed, with a portion of iron oxide nanoparticles. The composition of said rods was difficult to determine, but is probably $NaCaPO_4$ (β-rhenanite). However the possibility that the rods grew from a seed iron oxide nanoparticle is possible.

Catherine Amiens commented: Can you please give the mean size of the nanoparticles used during hyperthermia experiments. It would be useful for comparison purposes.

Ivan Parkin replied: From this and previous work, the optimum size for nanoparticles for hyperthermia are single magnetic domain particles of around 25 nm in size. However this is related to the frequency used in the hyperthermia system. In theory it should be possible to sweep the frequency of the hyperthermia apparatus and find an optimum resonance value that could be used for a wide range of nanoparticle sizes. However this does not seem to be easy to achieve and hence we have focussed our efforts on making near monodisperse particles around the best heating values. In this paper, nanoparticle sizes for magnetic hyperthermia were recorded as follows: Standard synthesis ~8.5 nm, iron-palmitate decomposed in 1-octadecene ~10 nm and iron-palmitate decomposed in 1-octadecene ~12 nm nanoparticles with rods in the order of ~150 nm in length.

Matthew Todd commented: You are applying a field and generating motion. Presumably you can run this the other way – move a liquid in which the same nanoparticles are suspended and generate a field? *I.e.* conversion of mechanical power to electrical?

Ivan Parkin responded: The nanoparticles used in this paper are super-paramagnetic so do not have a net magnetic moment under normal conditions. This may be possible with ferrofluids of greater particle size.

Nguyen T. K. Thanh asked: In Scheme 1 you state "Sample (I) is the route to iron oxide nanoparticles proposed by Park *et al.*,[3] Sample (II) is the synthesis of iron palmitate from high street sources and its subsequent decomposition in 1-octadecene and Sample (III) the decomposition of iron palmitate in shark liver oil." In the reprint, SQUID magnetometry shows:

"Samples (I), (II) and (III) have saturation magnetisation values of 52, 2.2 and 0.1 emu per gram, respectively". Could you please explain why Sample (III) has lost its saturation magnetisation by ~900 fold?

Ivan Parkin responded: We are not sure why this occurred – probably due to different sized crystallites.

Sandhya Moise asked: What is the dominant mechanism of heat generation for your nanoparticles during magnetic hyperthermia? What are the field conditions used?

Ivan Parkin answered: There are thought to be three mechanisms in operation – changes in Brownian motion, reversal by thermal activation and suppression of the anisotropic barrier by a magnetic field for single domain nanoparticles. For multidomain particles domain nucleation and domain wall switching can be important. I am not a physicist but I normally think about this heating effect in terms of a friction induced by the fact that the external field can switch faster than the particles, and the lag generates a heating effect. The field conditions are constant and set by the MACH instrument and are as follows: frequency = 945 kHz, field strength = 6.6 kA m^{-1} with a mass of 1 mg ml^{-1} of iron in the nano-particles. The coil itself was 6 turns, internally cooled with water, 36 mm in length with 18 mm internal diameter.

Peter Dobson remarked: Regarding the mechanisms for hyperthermia, there are broadly two possibilities. For the larger single domain ferromagnetic particles the heating effects will be because of magnetic domain reversal; for the smaller superparamagnetic particles the heating will be associated with the movement of spins within the particle and this will manifest itself in high losses at certain frequencies. This requires a more detailed knowledge of the real and imaginary parts of the magnetic susceptibility of the materials.

Dejian Zhou opened the discussion of the paper by Catherine Amiens: Did you use any base to deprotonate your carboxylate ligands before ligand exchange?

Catherine Amiens answered: We tried to directly use sodium oleate but the exchange was still not effective.

Dejian Zhou asked: Both ligands contain just a single carboxylate to coordinate to the magnetic nanoparticle. The binding strength here may not be strong enough completely displace the original hyrdophobic ligands. It would be good to use chelating ligands to increase the binding strength, especially those with a dendritic shape, which can match the surface curvature of the nanoparticle much better and hence provide much more stable capping of the nanoparticle surface.[1,2]

1 Y. A. Wang, J. J. Li, H. Chen and X. Peng, *J. Am. Chem. Soc.*, 2002, **124**, 2293.
2 D. Zhou, Y. Li, E. A. H. Hall, C. Abell and D. Klenerman, *Nanoscale*, 2011, **3**, 201.

Catherine Amiens answered: This is a very good suggestion. We also plan to try ligands with stronger binding abilities towards the iron surface, such as phosphonate end groups.

Edman Tsang asked: Could you rationalize the chemical reactivity of FeBi? In one instance, Fe is preferentially oxidized but in another instance the reverse is taking place.

Catherine Amiens answered: As iron is located mainly at the surface of the nanoparticles it is not surprising that it should oxidize first. Then in time one can observe that the core of the nanoparticles is also modified. The long distances measured on the HREM images are indicative of oxidation, but we could not determine the type of oxide formed.

Edman Tsang asked: How much scrambling of the atoms within FeBi is taking place in your system?

Catherine Amiens responded: This is a crucial question. It is indeed very important to know if there is any redistribution of the Fe and Bi atoms, especially upon oxidation or after prolonged exposure to biological media. We are beginning work on this study, which is highly challenging due to the small size of the objects.

Edman Tsang said: Did you try EXAFS to characterize your samples?

Catherine Amiens replied: This has been done in the case of the BiFe150 sample, and was reported in the paper by G. Mattei *et al.*[1] We plan to study the smaller nanoparticles during our next beamtime.

1 G. Mattei *et al.*, *J. Phys. Chem. C*, 2013, **117**, 1477.

Ivan Parkin remarked: Bismuth is very dense and often it is hard to prove if a composite material is formed or if it is a mixture of the elements. We have previously made BiP and it took over 12 separate techniques to provide proof that an actual compound, and not an intimate mixture of the elements, had been formed.

Edman Tsang asked: Have you seen any phase segregation upon oxidation?

Catherine Amiens responded: So far oxidation experiments have been carried out on already segregated nanoparticles only, *i.e.* on nanoparticles consisting of a bismuth core and iron shell.

In this case, HRTEM images clearly show that when working in the solid state, oxidation can be stopped at the very first stage when only the iron shell is oxidized. Upon prolonged oxidation (over weeks), full oxidation of the core is evidenced, however at this stage we cannot tell if the initial segregation between Fe and Bi is retained or if some bismuth atoms migrate towards the surface.

This is an important question that we want to investigate further, especially in the case of oxidation in aqueous solutions, as bismuth leaching would probably prevent the use of these nano-objects *in vivo*.

Lanry L. Yung said: Both Prof. Nguyen Thanh and Peter Dobson have mentioned the potential of applying nanotechnology for applications in infectious disease, such as pathogen separation and diagnostics (Prof. Thanh) and antimicrobial agents (Prof. Dobson). What are the key issues we should address from the materials and chemistry point of view?

Peter Dobson replied: There are several requirements for sensing bacteria and fall broadly into the following categories:

(1) To sense and quantify the presence of bacteria, and to some extent this is now being done in the food industry but it needs to be more widely adopted. The methods in current use measure oxygen consumption as an indicator of the presence of microbes, or, in some cases, the emission of other gaseous products.

(2) There needs to be methods of identifying the species of bacteria present and quantifying them. This should be very rapid and applied at "point of use".

(3) It is desirable to identify the strain of a bacterium and this will require a rapid miniature DNA sequencing technique.

Nguyen T. K. Thanh replied: For magnetic detection or separation of pathogens you need to have stable superparamagnetic nanoparticles with high magnetic moments. Also, the nanoparticles should have functional groups that are specific to the pathogens of interests and have sensitivities that allow early detection.

Ivan Parkin observed: Surfaces can also be made antimicrobial by incorporation of dyes and gold nanoparticles.

Peter Dobson remarked: I was asked to comment on the scientific challenges. Firstly I suggest that nanoparticles could transform the way we try to design new antimicrobials; secondly there is still a fundamental question about the nature of the electron spins at the surface of magnetic particles, and if these differ from the spins in the bulk of the nanoparticle. These altered spins could be further modified by the binding of ligands onto the surface. This could have implications for the design of particles for use in hyperthermia or MRI.

Dalibor Soukup remarked: Currently, interparticle dipole–dipole interactions seem to be a hot topic in the field of magnetic hyperthermia. Some papers show that they can enhance the heating efficiency of magnetic nanoparticles, some show the opposite. They can also reduce the magnetisation saturation, thus decreasing the heating efficiency in turn. What is your opinion on these dipole–dipole interactions with respect to magnetic hyperthermia? Routinely, magnetic nanoparticles are tested in solutions in which they are homogeneously distributed, however, once internalised in cells, they are usually stored in endosomes/lysosomes where the interparticle distance is much smaller than that in solutions, which favours dipole–dipole interactions. Do you think there is a need for better methods of testing that would show biologically relevant heating efficiencies?

Peter Dobson responded: This is a difficult question! Frankly I do not know. I have tried to find an answer, and a recent paper by M. E. Sadat *et al.* seems to give part of the answer to your question and they imply that there are very significant differences between “free” and “confined” nanoparticles.[1]

1 M. E. Sadat *et al.*, *Mat. Sci. Eng.*, 2014, **42**, 52.

Ivan Parkin remarked: In magnetic hyperthermia it is critical to have particles of exactly the correct size and shape for maximum effect.

Nguyen T. K. Thanh responded: I agree.

Catherine Amiens commented: Dispersibility is not always well described in the papers.

Most of the time one simply discusses the stability of the colloidal solutions on the macroscopic observation of the precipitation of nanoparticles with time in different conditions (variation of pH, ionic strength *etc.*). No microscopic (or even nanoscopic) description is made of the organisation of the nanoparticles in solution (real dispersion of small aggregates already forming), although this point is very important when discussing certain properties *e.g.* relaxivities in MRI. We need standard conditions to be able to compare results from one paper to the other and an agreement on the level of detail (macroscopic or otherwise).

Nguyen T. K. Thanh responded: We need cryo-TEM for the small aggregates you mention.

Edman Tsang asked: Is there any subtle change in physiochemical properties with respect to the change in particle size?

Nguyen T. K. Thanh said: Reproducibility in nanoparticle synthesis is one of the biggest challenges. The physical properties of nanoparticles depend on their size, but also the number of ligands coated on the surface. The interaction between ligands is also dependent on the nanoparticle surface area. Therefore, it is very important to produce monodisperse nanoparticles. Currently most of the synthesis of nanoparticles is carried out in batch, subjects it to many factors that could affect the synthesis such as temperature, even at ambient temperature, mixing, and diffusion. Therefore the reproducibility is limited. On a fundamental level, each material will nucleate and grow depending on its reaction conditions and even a small change in conditions, such as pH, can lead to a completely different mechanism.[1]

1 N. T. K Thanh, N. Maclean and S. Mahiddine, *Chem. Rev.*, 2014, **114**, 7610–7630.

Lanry L. Yung commented: Since research knowledge often goes missing when a research student/postdoc leaves the group, I would be interested if anyone has any suggestions on how to keep knowledge within the research group.

Matthew Todd remarked: The development of new antimicrobial agents is a continual battle against resistance. It is easier to develop resistance to some drugs. Is there something unusual about the mechanism of bacterial killing employed by these surfaces (or indeed nanoparticles more generally) that might mean the bacteria would find it challenging to develop resistance?

Ivan Parkin responded: Yes. In general our work on light activated coatings as antimicrobials functions *via* the formation of reactive oxygen species and singlet oxygen. These typically function in a non-specific way to destroy the bacteria through thousands of possible pathways simultaneously. This is different from an antibiotic which typically has a very small number of pathways to attack the bacteria. Hence in the radical or singlet oxygen based approach it is very unlikely the bacteria could develop resistance.

Peter Dobson asked the delegates: How can one measure the temperature changes in hyperthermia? In fact, one might also call into question the very meaning of "temperature" at this nano-level.

Yuri Antonio Diaz Fernandez responded: I appreciate this topic being raised, since it is extremely relevant. The applicability of the definition of macroscopic temperature on the nanoscale is a fascinating and not always sufficiently addressed problem. To my understanding, this question can only be answered in the context of nonequilibrium thermodynamics and fluctuation theory, which, even after several years of development, are still works in progress, obscure to those from other disciplines. Those in the nanomaterial scientific community who have a particular interest in hyperthermia research, will benefit from establishing systematic discussion opportunities with mathematicians and experimental and theoretical physicists, bridging the gaps among these fields. Probably a Faraday Discussion on "Physical Chemistry Experimental Tools and

Theoretical Concepts Applied at the Nanoscale" could be the best starting point for this cross-disciplinary exchange. Regarding available methods to characterize hyperthermia, Pallavicini, Chirico *et al.* have recently demonstrated an indirect spectroscopic method able to probe the local temperature on the surface of different nanomaterials used for hyperthermia.[1]

1 S. Freddi *et al.*, *Nano Lett.*, 2013, **13**(5), 2004–2010.

Scott Mitchell also responded: Our group recently published a paper using DNA conjugated to iron oxide NPs as an indirect method for probing local temperature on the surface of magnetic nanoparticles during magnetic hyperthermia.[1]

1 J. T. Dias, M. Moros, P. del Pino, S. Rivera, V. Grazú and J. M. de la Fuente, *Angew. Chem.*, 2013, **125**, 11740–11743.

Maya Thanou raised the question: How would these magnetic nanoparticles (MNP) compare with already advanced or available MNPs, with respect to their ability to induce hyperthermia?

Ivan Parkin answered: The particles show good hyperthermia response but they are not as efficient as the best commercial materials.

Maya Thanou said: Hyperthermia (mild temperature increase) does not kill tumour cells, however it sensitizes them to other treatments. What other treatments would these be?

Ivan Parkin responded: We have not investigated this. Potential ideas we are developing include having a low melting point wax containing an anticancer drug.

Nguyen T. K. Thanh commented: The thermal energy needed for induction of cell death has been found to be close to the energy needed for protein denaturation, leading to the conclusion that the main cytotoxic effect of hyperthermia is based on the denaturation of membrane and cytoplasmic proteins. For more on this please see ref. 1.

1 A. Hervault and N. T. K. Thanh, *Nanoscale*, 2014, **6**, 11553–11573.

Maya Thanou asked: What is the required time of exposure of tumours to hyperthermic temperatures for sensitisation?

Nguyen T. K. Thanh answered: Please see a paper by M. Johannsen *et al.*

1 M. Johannsen, U. Gneveckow, L. Eckelt *et al.*, *Int. J. Hyperthermia*, 2005, **21**(7), 637–647.

Sandhya Moise addressed Ivan Parkin and Kerry Chester: During magnetic hyperthermia, the temperature rise is very local and falls with distance from the nanoparticle surface. Is bulk temperature measurement of nanoparticle suspensions accurate and representative of temperature changes encountered by the cells?

Kerry Chester replied: Yes it is true that bulk measurements may not be the best comparison to *in vivo* or *in vitro* measurements, however depending upon the local concentration of nanoparticles within a cellular structure we have observed similar experimental bulk temperature rises *in vitro* and *in vivo*. In these cases all we can realistically say is that the temperature local to the cells has reached its equilibrium point. Constructing an experiment to measure nanoparticle temperatures would be very difficult and at best would have to be modelled/simulated.

Ivan Parkin responded: The local heating is one of the key reasons that hyperthermia works. The bulk temperature measurement will greatly underestimate the actual temperature at the surface of the nanoparticle. However it is a guide to indicate heating and a measure of relative effectiveness.

Stefan Borsley asked Nguyen TK Thanh: You are using imine formation with a short, flexible di-aldehyde (gluteraldehyde, GA) to bind to amine coated nanoparticles. This equilibrium process generally favours the carbonyl/amine side in water, as imine formation results in liberation of water. Imine formation is acid catalysed, so the pH of the solution will also influence the rate of the reaction. Bidentate binding of the GA to two amine ligands on the same nanoparticle also must be considered, this might be expected to be a favourable process due to the flexibility and proximity of the unbound end of the GA, essentially resulting in a higher effective molarity.

Given these concerns, do you have any evidence for determining the number of proteins bound per nanoparticle?

Stefan Borsley communicated: Further to this, the reversible nature of imine formation means that, once bound, proteins may be released as the environmental conditions change, favouring the equilibrium towards the aldehyde/amine side, a particular concern when diluting the samples. Have you verified that you do not get release of proteins from your protein-nanoparticle conjugates?

Nguyen T. K. Thanh communicated in reply: We used an excess of glutaraldehyde in the conjugation step. The imine hydrolysis is assisted through the use of an acid catalyst. Under our conditions, the final products IONP@CHI@GA@PrA@antibody were formed, so evidently they were bound to the bacteria selectively. We tested our conjugates after three months and they were still stable.

Faraday Discussions

PAPER

A multifunctional amphiphilic polymer as a platform for surface-functionalizing metallic and other inorganic nanostructures

Wentao Wang, Fadi Aldeek, Xin Ji, Birong Zeng and Hedi Mattoussi*

Received 21st July 2014, Accepted 21st July 2014
DOI: 10.1039/c4fd00154k

We designed a new set of polymer ligands that combine multiple metal-coordinating groups and short polyethylene glycol (PEG) moieties in the same structure. The ligand design relies on the controlled grafting of a large number of amine-terminated histamines and PEG short chains onto a poly(isobutylene-*alt*-maleic anhydride) backbone, *via* a one-step nucleophilic addition reaction. This addition reaction is highly efficient, can be carried out in organic media and does not require additional reagents. We show that when imidazole groups are used the resulting polymer ligand can strongly ligate onto metal nanostructures such as nanoparticles (NPs) and nanorods (NRs) made of gold cores. The resulting polymer-coated NPs and NRs exhibit good colloidal stability to pH changes and added electrolytes. This constitutes a departure from the use of thiol-based ligands to coordinate on Au surfaces. The present chemical approach also opens up additional opportunities for designing hydrophilic and reactive platforms where the polymer coating can be adjusted to various metal and metal oxide surfaces by simply modifying or combining the addition reaction with other metal coordinating groups. These could include iron oxide NPs and semiconductor QDs. These polymer-capped NPs and NRs can be used to develop biologically-active platforms with potential use for drug delivery and sensing.

Introduction

An effective integration of inorganic nanoparticles (NPs), such as those made of metal and metal oxide cores, within biological systems has the potential to provide novel hybrid platforms with unique photo-physical properties.[1,2] Such platforms can advance our understanding of various biologically challenging problems.[3–5] This, however, requires access to hydrophilic NPs that are homogenously dispersed and reactive. It has been shown that the most successful routes to prepare high quality nanocrystals of semiconductors, magnetic

Department of Chemistry and Biochemistry, Florida State University, Tallahassee, FL 32306, USA. E-mail: mattoussi@chem.fsu.edu

materials and even gold, with controllable size and high crystallinity, often rely on the thermal decomposition of metal precursors in hot surfactant solutions.[6–11] However, nanocrystals synthesized *via* pyrolysis of organometallic precursors are capped with hydrophobic ligands, which make them dispersible only in organic solvents. Consequently, integration within bio-inspired assemblies requires additional chemical manipulation and post growth surface modification to render the nanocrystals colloidally stable in buffer media and biocompatible.[3,12–15]

Use of amphiphilic polymers (*e.g.*, block copolymers) and phospholipids to encapsulate the as-prepared hydrophobic NPs within micelle-like structures has been employed by several groups to prepare various water-soluble NPs.[16–23] This route relies on the entropy-driven interdigitation between the hydrophobic segments of the polymers (or phospholipids) and the native cap. As a strategy, encapsulation is believed to better preserve the physical properties of the nanocrystals (*e.g.*, optical or magnetic), because it maintains the native organic cap on the NP surfaces. It does, however, tend to substantially increase the hydrodynamic size of the NPs and may yield more than one nanocrystal per micelle,[16,23] which eventually limits their use in applications requiring small size probes. An alternative and more advantageous strategy relies on the removal of the hydrophobic cap, and the replacement of it with bifunctional ligands: ligand exchange. These ligands present anchoring groups that coordinate onto the metal-rich surface of the nanocrystals along with hydrophilic moieties for promoting affinity interactions with the surrounding medium (*i.e.*, buffer). These anchors interact with the metal surface *via* Lewis-base type coordination. This route can provide compact NPs and better colloidal stability in physiological conditions. When this strategy is applied using ligands with one anchoring group (*i.e.*, monodentate ligands), the binding affinity to the NP surface is rather weak. ligand desorption from the surfaces, which negatively affects the NP stability in biological media at low concentrations. Furthermore, ligands with weak affinity can be easily displaced by biomolecules bearing amine and carboxylic functional groups, eventually promoting NP aggregation.[15] Multidentate ligands, in comparison, bind onto the inorganic nanocrystal surface, *via* multisite coordination, and can provide much stronger binding affinity. This drastically reduces the rate of ligand desorption compared to mono-dentate ligands, which improves the NP colloidal stability in biological media.[24–30]

In the present study, we design a new set of amphiphilic polymers that combine imidazole-to-metal coordination and hydrophilic polyethylene glycol (PEG), as ligands that can promote the dispersion of Au NPs and nanorods (NRs, as well as other inorganic nanocrystals) in buffer media. The polymer platform is synthesized *via* nucleophilic addition of several histamine groups and biocompatible hydrophilic moieties onto a poly(isobutylene-*alt*-maleic anhydride), PIMA, precursor.[30] This ligand design greatly benefits from the high efficiency of the addition reaction, while allowing simultaneous coupling of all desired groups in one step. We would also like to note that the histidine-coordination onto Au surfaces may be weaker than other groups (*e.g.*, thiol-to-Au). However, this mode of ligand anchoring can be beneficial when competition to the ligands by other small molecules for binding to the Au surfaces are used for sensing schemes or drug delivery vehicles.[31]

Experimental section

Materials

Poly(isobutylene-*alt*-maleic anhydride) (PIMA) (average M_w: ~6000 Da, 85%), poly(ethylene glycol) (average M_w: ~600 Da), poly(ethylene glycol) methyl ether (average M_w: ~750 Da), histamine (>97.0%), hexadecyltrimethylammonium bromide (CTAB, >98.0%), gold(III) chloride trihydrate ($HAuCl_4 \cdot 3H_2O$, >99.9%), L-ascorbic acid (BioUltra, ≥99.5%), silver nitrate ($AgNO_3$, >99%), sodium borohydride ($NaBH_4$, 99%), and hydrochloric acid (HCl, 37 wt% in water), dimethylformamide, along with most of the chemicals used were purchased from Sigma Aldrich (St Louis, MO). Sodium oleate (purity > 97.0%) was purchased from Tokyo Chemical Industry Co, Ltd. Solvents were purchased from Sigma Aldrich (St Louis, MO). Column purification chromatography was performed using silica gel (60 Å, 230–400 mesh from Bodman Industries, Aston, PA). Deuterated solvents used for H NMR experiments were purchased from Cambridge Isotope Laboratories (Andover, MA). Ultrapure water obtained from a Milli-Q Integral 5 system was used in all experiments. The precursors H_2N–PEG750–OCH_3 was synthesized in our laboratory following the procedures detailed in previous work.[32]

The syntheses were carried out under N_2 passed through an O_2 scrubbing tower unless otherwise stated. Air sensitive materials were handled in a glovebox, and standard Schlenk techniques were used to handle air-sensitive materials and reactions.

Instrumentation

^{1}H NMR spectra were recorded using a 600 MHz spectrometer (Bruker SpectroSpin, Billerica, MA). Optical absorption data of various AuNP and AuNR dispersions were collected using a UV-Vis absorption spectrophotometer (UV 2450 model, Shimadzu, Columbia, MD) and UV-Vis-NIR spectrophotometer (Perkin Elmer Lambda 950). Fourier Transform Infrared (FT-IR) spectra were collected using an FT-IR spectrometer (Perkin Elmer Spectrum 100). Transmission electron microscopy (TEM) images were acquired using either a 200 kV JEOL-2010 instrument or a Philips FEI CM-120 microscope. Samples for TEM were prepared by drop casting the NP dispersion onto the holey carbon film on a fine mesh Cu grid (400 mesh). Dynamic light scattering measurements were carried out using an ALV/CGS-3 Compact Goniometer System (ALV-GmbH, Langen, Germany). This system is equipped with a HeNe laser (illumination at 632.8 nm), an ALV photon correlator and an avalanche photodiode for signal detection. Solvent evaporation was carried out using a rotary evaporator R-215 (Buchi, New Castle, DE).

Synthesis of the His–PIMA–PEG ligands

We briefly describe the synthesis of the histamine- and PEG-modified PIMA polymer where 50% of the anhydride rings were targeted for reaction with histamines, while the other 50% were reacted with amine–PEG–OCH_3 moieties. In a typical reaction, 0.385 g of PIMA ($M_w \sim 6000$ g mol^{-1}, 2.5 mmol monomer units) was dissolved in 10 mL of DMF and added into a 50 mL three-neck round-bottom flask equipped with a magnetic stirring bar and an addition funnel. The solution was purged with nitrogen and then heated up to 45 °C. Separately, histamine (0.139 g, 1.25 mmol) and H_2N–PEG–OMe (0.995 g, 1.25 mmol) were dissolved in 4 mL of DMF. This solution mixture was added dropwise (*via* the addition funnel) to the PIMA solution above and the reaction mixture was further stirred at 45 °C for 8–10 hours (*i.e.*, overnight). After removing the DMF under vacuum, 3 mL of chloroform was added, and the compound was purified on a silica column with chloroform as the eluent. The final product was a gel-like compound, with a reaction yield of ~84%.

Growth of gold NPs and gold NRs

Three sets of Au nanostructures were prepared.

(1) The first set is made of hydrophobic oleylamine-capped AuNPs (~5 nm in radius). The NPs were synthesized by rapidly injecting $HAuCl_4$ precursors into a pre-heated surfactant solution, as reported by Swihart and coworkers.[33] Briefly, 5 mL of oleylamine (OA) was refluxed at 150 °C in a 100 mL flask under argon. A mixture of 0.3 mmol of $HAuCl_4 \cdot 3H_2O$ in 1 mL of oleylamine was rapidly injected into the above flask, and the mixture was further heated and annealed for 1.5 h. One round of centrifugation/purification to remove unreacted precursor was applied, then the solution was diluted with hexane and stored until further use.

(2) The second set is made of citrate-stabilized 13 nm diameter AuNPs, grown following a procedure reported in the literature.[34,35] Briefly, 50 mL of 1 mM $HAuCl_4 \cdot 3H_2O$ aqueous solution were added to 100 mL Erlenmeyer flask equipped with a magnetic stirring bar. The solution was heated to 100 °C (boiling conditions). Then, 50 mg of sodium citrate tribasic dihydrate dissolved in 2 mL DI water were rapidly added. The reaction mixture was stirred for an additional 15 min, and heating was stopped when a deep red color was obtained. The solution was then purified by applying one round of concentration/dilution using a membrane filtration device.

(3) The third set is made of CTAB-coated AuNRs (with dimensions of 110 nm in length and 20 nm in diameter: 110 nm × 20 nm). These NRs were prepared following the scheme initially developed by Murray and co-workers to grow gold NRs using a seed solution and a binary surfactant mixture (CTAB and sodium oleate, NaOL).[36] The seed solution for the NR growth was prepared as follows: 5 mL of 0.5 mM $HAuCl_4$ was mixed with 5 mL of 0.2 M CTAB solution in a 20 mL scintillation vial. 0.6 mL of fresh 0.01 M $NaBH_4$ was diluted into 1 mL of water, and the mixture was then injected into the Au(III)–CTAB solution under vigorous stirring (1200 rpm). The solution color changed from yellow to brownish yellow and the stirring was stopped after 2 min. This seed solution was aged at room temperature for 30 min before use.

To prepare the growth solution required for the NRs growth, 7.0 g (0.037 M in the final growth solution) of CTAB and 1.234 g of NaOL were dissolved in 250 mL of warm water (~50 °C) using a 1 L Erlenmeyer flask. The solution was allowed to cool down to 30 °C and 24 mL of $AgNO_3$ (4 mM) solution was added. The mixture was kept undisturbed at 30 °C for 15 min after which 250 mL of 1 mM $HAuCl_4$ solution was added. The solution became colorless after 90 min of stirring (700 rpm). 3.6 mL of HCl (37 wt% in water, 12.1 M) was then introduced to acidify the solution. After 15 min of slow stirring at 400 rpm, 1.25 mL of 0.064 M ascorbic acid (AA) was added and the solution was stirred vigorously for 30 s. Finally, 0.4 mL of seed solution was injected into the flask. The resulting mixture was stirred for 30 s and left undisturbed at 30 °C for 12 h (overnight) to allow for growth and homogenization of the NRs. The materials were isolated by centrifugation at 7000 rpm for 30 min followed by removal of the supernatant.

Ligand exchange with His–PIMA–PEG

Ligation with the above His–PIMA–PEG polymer ligand was applied to oleylamine-coated AuNPs, citrate-stabilized AuNPs and to CTAB/NaOL-coated AuNRs (referred to as CTAB-coated AuNRs).

Cap exchange of oleylamine-capped AuNPs. 400 μL of stock solution (~0.9 μM) was precipitated using ethanol and redispersed in 200 μL of chloroform. Then 15 mg of His–PIMA–PEG dissolved in 200 μL of chloroform was mixed with the AuNP solution. The vial was sealed with a rubber septum and the atmosphere was switched to nitrogen by applying 2 to 3 rounds of mild vacuum, followed by purging with nitrogen. The solution was then left stirring overnight while maintaining the temperature at 50 °C. The AuNPs were then precipitated by adding excess hexane and centrifuged for 5 min at 3700 RPM to provide a dark pellet. The clear supernatant was discarded, and the pellet redissolved in 200 μL of chloroform, followed by precipitation using excess hexane. The turbid dispersion was centrifuged, the supernatant was discarded and the residual precipitate was dried under vacuum. This yielded a pellet which was easily dispersed by adding 500 μL of DI water. After sonication, the aqueous dispersion was filtered using a 0.45 μm disposable syringe filter, then excess free ligands were removed by applying 3 to 4 rounds of concentration/dilution using a centrifugal filtration device (Millipore, M_w cutoff = 50 kDa); 37 000 rpm and 7 min were used.

Cap exchange of citrate-stabilized AuNPs. In a typical reaction, 4 mL of citrate-capped AuNPs (at a concentration of 14 nM) were added in a glass vial equipped with a magnetic stirring bar. In a separate vial, 2 mg of His–PIMA–PEG ligands were dispersed in 2 mL of 10 mM phosphate buffer (pH = 8.5). The ligand solution was added dropwise to the AuNP dispersion and stirred for 15 h at room temperature. The cap exchanged AuNPs were then purified from free ligands by applying 3 rounds of concentration/dilution using a membrane filtration device as done above. The AuNPs were finally dispersed in DI water or buffers (if needed) and stored at 4 °C for later use.

Ligation of CTAB/NaOL-coated AuNRs. Ligation of the CTAB-coated NRs was carried out using one phase ligand exchange in DI water. 120 μL of stock solution (2.82 nM, $\varepsilon = 7.4 \times 10^9\ M^{-1}\ cm^{-1}$) was purified by centrifugation (7000 rpm, 10 min) to remove excess CTAB and sodium oleate. The concentrated AuNRs were then redispersed in 200 μL of DI water. 15 mg of ligands were dissolved in 200 μL of DI water and mixed with AuNR dispersion. The mixture was stirred at 50 °C overnight, then further diluted with 1 mL of DI water and sonicated for 2 min. The aqueous dispersion was filtered with a 0.45 μm disposable syringe filter, then excess free ligands were removed by applying 2 rounds of concentration/dilution using a centrifugal filtration device (M_w cutoff = 50 kDa) as described above.

Dynamic light scattering (DLS) measurements

These measurements were applied to AuNPs only. Dynamic light scattering profiles were collected from dispersions of oleylamine-capped AuNPs in hexane, or citrate-stabilized AuNPs in DI water, and from the same NPs following ligand-exchanged with His–PIMA–PEG and dispersed in buffer media (pH = 7.5). Data were collected using an ALV/CGS-3 Compact Goniometer System equipped with a He–Ne laser source ($\lambda = 633$ nm) and a variable delay time (multi-tau) ALV-7004 correlator (ALV-GmbH, Langen, Germany). All the dispersions were filtered through a 0.22 μm disposable filter (Millipore) before the scattered signal was collected. The scattered signal which consists of the average result of 3 acquisition periods of 10 seconds each was collected at angles varying between 60 and 120 degrees. The intensity count rates were maintained at ~150–300 Hz, achieved through appropriate control over the NP concentrations and density filters used. The resulting auto-correlation function, generated from the above scattered intensity profiles using the ALV correlator, was fitted to a cumulant series using ALV software. For every sample we verified that the measured hydrodynamic size (extracted from the Laplace transform of the auto-correlation function) is independent of the scattering angle, as anticipated.[37] The data on the hydrodynamic size reported here were collected at 90 degrees scattering angle. To track possible changes in the NP size for a particular dispersion with time, the signal was collected and analyzed after different storage time periods.

NMR characterization of the hydrophilic AuNPs

We used pulsed field gradient water suppression ^{1}H-NMR to collect our spectra. This allowed collection of high signal-to-noise spectra using aqueous dispersions with relatively low concentration. The sample concentrations used for our measurements were about 0.10 μM. Prior to spectra collection the solvent was switched from DI water to deuterium oxide (D_2O) by applying two rounds of concentration/dilution with 2 mL × 2. The final volume of AuNPs in D_2O was adjusted to 500 μL before NMR measurements. For instance, the ^{1}H-NMR spectra shown in Fig. 4 were collected using His–PIMA–PEG-ligated-AuNPs at a concentration of 0.113 μM (and a volume of 500 μL) and averaged over 500 scans. The sample used for the surface ligand counting experiments was prepared following the same protocol, but 0.8 μL pyridine (9.93 μmol) was added to provide a standard to which the signature of the NPs were compared (*i.e.*, the AuNP dispersion had a total volume of 500 μL and a concentration of 0.113 μM).

Results and discussion

Ligand synthesis

The polymer ligands used for the surface modification of AuNPs and AuNRs were prepared *via* a nucleophilic addition reaction between PIMA and amine-containing molecules (here histamine and amine–PEG–methoxy). The synthetic design offers a few unique and advantageous features. The nucleophilic addition between anhydride rings and amine groups is highly efficient and can be carried out without the need for additional reagents, which simplifies the purification and characterization of the final products. In addition, the availability of several maleic anhydride rings per polymer chain permits a straightforward and controlled insertion of a large number of lateral groups with distinct chemical functionalities. This provides polymer ligands that simultaneously present several imidazole and PEG moieties, which increases the ligand coordination onto the NP surface, while promoting high affinity to water and reducing non-specific interactions. This synthetic design was successfully applied to prepare multi-dopamine-based ligands to functionalize magnetic NPs.[30]

Fig. 1 shows a schematic representation of the synthetic approach we used to prepare the His–PIMA–PEG ligand. Typically, we used a molar amount of aminated molecules (histamine and amine-PEG moieties) that was equal to the molar concentration of maleic anhydrides presented on the PIMA chains. In this study we introduced a 50 : 50 molar mixture of histamine and H_2N–PEG–OMe, but the ratios can be easily varied to tune the final composition of the polymer ligand.

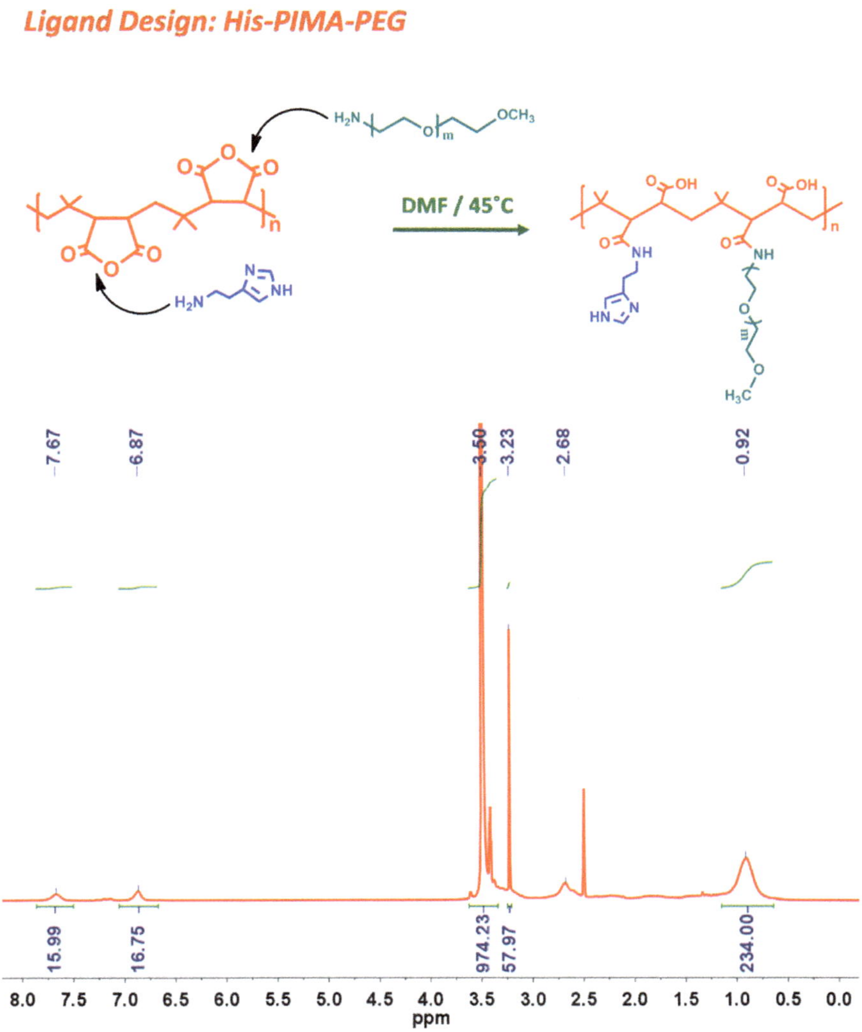

Fig. 1 (Top) schematic representation of the nucleophilic addition reaction used to introduce both histamine groups and amine-modified PEG moieties along the same PIMA polymer chain. (Bottom) ^{1}H NMR spectrum collected from a solution of His–PIMA–PEG ligand (made with 50 : 50 mixture of imidazole and PEG along the PIMA backbone) dissolved in DMSO-d_6.

Overall, these conditions are expected to introduce a total of ~20 imidazoles and ~20 PEG moieties; it also frees ~40 carboxyl groups along the polymer chain.

The His- and PEG-modified PIMA polymer was characterized using ^{1}H NMR spectroscopy. The spectrum in Fig. 1 shows two clearly defined peaks at 6.8 and 7.6 ppm, characteristic of the protons in the imidazole ring. Additional signatures include a strong peak at ~3.5 ppm and a sharp peak at ~3.23 ppm, which can be attributed to the protons in the PEG segments and the terminal methoxy protons, respectively, while a broad peak at ~0.9 ppm is attributed to the protons of the methyl groups in the polymer chain. The degree of grafting is estimated by comparing the relative signal integrations of the 2 protons from the imidazole ring ($\delta \sim 6.8$ and 7.6 ppm, corresponding to 16.7 H and 15.9 H, respectively), the 3 protons from the methoxy group of the PEG moieties ($\delta \sim 3.23$, 57.9 H), and the protons of the two methyl repeat units of the PIMA chain ($\delta \sim 0.9$, 234 H). For the polymer compound above we measured ~16–17 imidazoles and ~19 PEG moieties per PIMA chain; such values are in agreement with the anticipated ones based on the nominal number of maleic anhydride rings per polymer chain, and the fact that the histamine-to-maleic anhydride and PEG-to-maleic anhydride molar ratios used in the addition reaction were 0.5.

Ligand exchange

The hydrophobic NPs capped with oleylamine, with average radius R_0 of 5 nm (extracted from TEM, as shown in Fig. 2), were synthesized using thermolysis of

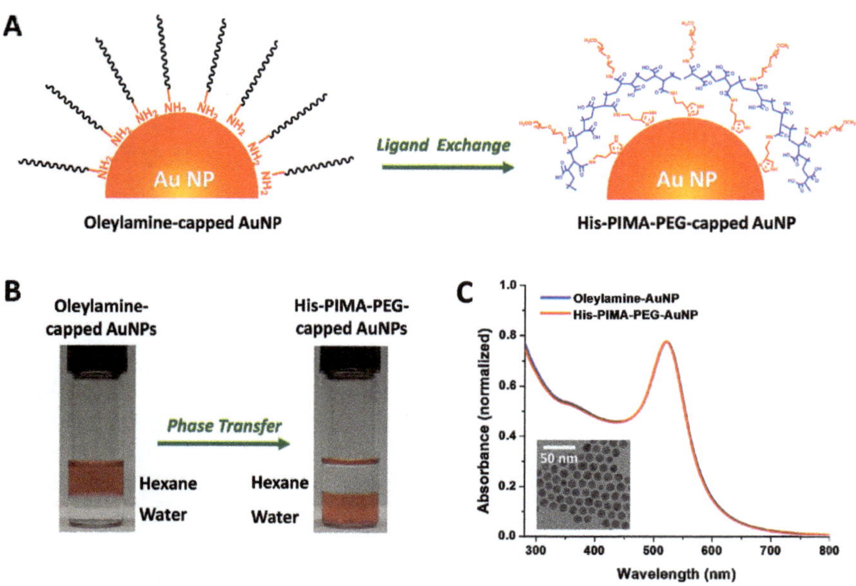

Fig. 2 (A) Schematic depiction of an oleylamine-capped AuNP (left) and the same NP following ligation with His–PIMA–PEG (right). (B) White light images of a two-phase dispersion showing oleylamine–AuNPs dispersed in the hexane (top) phase and after ligand exchange with His–PIMA–PEG where the NPs are homogeneously dispersed in the water bottom phase. (C) UV-Vis absorption spectra of dispersion of hydrophobic oleylamine–AuNPs in hexane and following phase transfer to water using His–PIMA–PEG polymer. Inset shows the TEM image of oleylamine-capped AuNPs (5 nm in radius).

gold precursors, as reported previously.[33] The TEM images indicate that these NPs are spherical with uniform cores and reduced size distribution. The NPs were ligated with His–PIMA–PEG by mixing the polymer with the oleylamine capped-AuNPs in chloroform and stirring overnight at 50 °C, followed by purification. The NPs capped with the new ligands were readily dispersible in water. An additional purification step to remove excess free ligands and solubilized oleic acid was carried out using a membrane filtration device (see Experimental section). Fig. 2A shows schematically the ligation and phase transfer of an oleylamine–AuNP using His–PIMA–PEG ligand. The white light images of AuNPs dispersed in immiscible phases made of hexane and water shows that oleylamine-capped AuNPs are limited to the hexane phase, but following ligand exchange with His–PIMA–PEG the NPs are homogeneously transferred to water (Fig. 2B).

Characterization of the hydrophilic AuNPs

Optical characterization. Fig. 2C shows the absorption spectra of the AuNPs before and after ligand exchange with His–PIMA–PEG. The spectrum of AuNPs capped with His–PIMA–PEG-AuNPs dispersed in DI water is nearly identical to that collected from NPs capped with oleylamine dispersed in hexane. This indicates that the presence of the imidazoles in the polymer ligand indeed promote ligation of the NPs with no measurable deterioration or etching of the inorganic cores.

Dynamic light scattering measurements. Fig. 3A shows representative plots of the autocorrelation function collected from oleylamine–AuNPs diluted in hexane (blue symbols) and His–PIMA–PEG-capped AuNPs in water (red symbols). Data show that ligand exchange with the polymer ligands slightly increases the hydrodynamic size of the NPs, as indicated by the slower decay of the correlation function (slope in Fig. 3B) measured for His–PIMA–PEG–AuNPs compared with oleylamine–AuNPs.[37] Plots of the intensity *vs.* hydrodynamic size histograms extracted from the Laplace transform of the autocorrelation function collected for both sets of NP dispersions are shown in Fig. 3C. Overall, cap exchange with the polymer ligand yields dispersions of AuNPs with narrow size distribution, similar to what is measured for the hydrophobic dispersions (polydispersity index value, PDI, ≤0.1). The hydrodynamic radius (R_H) measured for oleylamine–AuNPs in

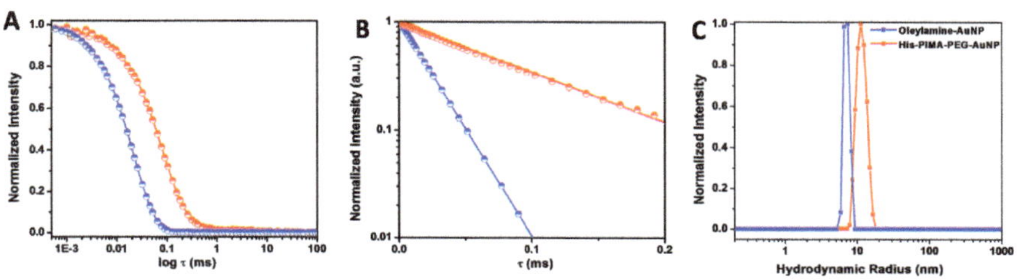

Fig. 3 Representative plots of the autocorrelation function, $g^{(2)}$, collected from dispersions of 5 nm AuNPs before and after cap exchange with the His–PIMA–PEG; shown are: (A) ($g^{(2)}$) *vs.* log(lag time, τ), and (B) log($g^{(2)}$) *vs.* lag time (τ). (C) Histograms showing the distribution of the intensity *vs.* hydrodynamic radius, extracted from the Laplace transform of the intensity auto-correlation function from the above dispersions of oleylamine- and His–PIMA–PEG-capped AuNPs.

toluene is 7.5 nm and increases to 11.2 nm for the AuNPs capped with His–PIMA–PEG dispersed in water (Fig. 3B), a relatively small increase (~3–4 nm) compared to those measured for NP encapsulated with amphiphilic polymers reported in the literature.[38,39]

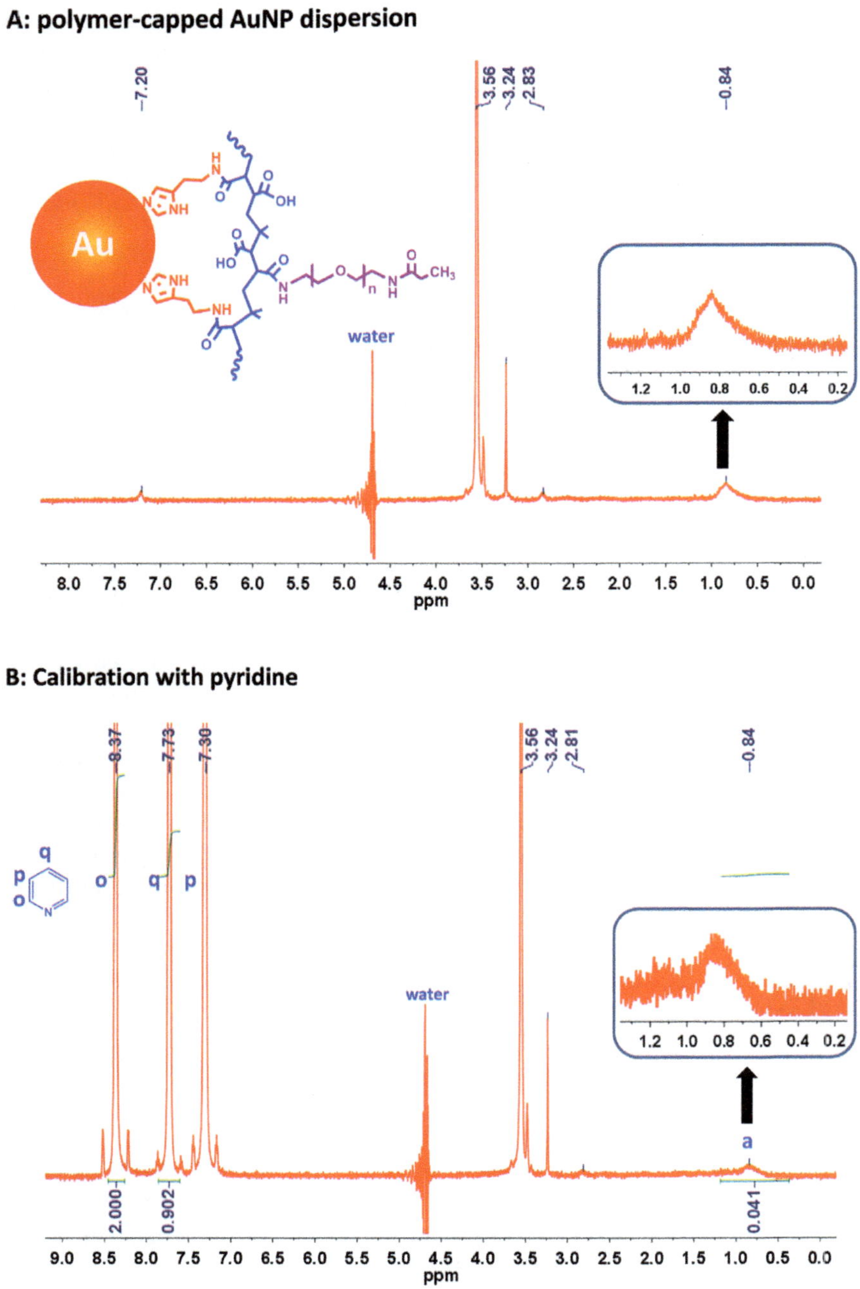

Fig. 4 (A) ^{1}H-NMR spectrum of the hydrophilic His–PIMA–PEG-capped AuNPs. (B) ^{1}H-NMR spectrum of the His–PIMA–PEG polymer calibrated with a fixed amount of pyridine (used as a control).

^{1}H NMR characterization. The dispersion of AuNPs ligated with His–PIMA–PEG (in D_2O) was characterized using pulsed field gradient-based water suppression (of the peak at 4.7 ppm) ^{1}H-NMR spectroscopy. Fig. 4A shows the ^{1}H-NMR spectrum of a dispersion of hydrophilic AuNPs ligated with His–PIMA–PEG, which is overall similar to the spectrum of ligands alone. The spectra show pronounced resonances at 3.24 ppm and 3.56 ppm corresponding to the methoxy groups and PEG moieties, while the broad peak at 0.84 ppm is attributed to the protons from PIMA backbone. The spectrum also indicates that the two disparate resonances at 6.8 and 7.6 ppm, characteristic of two protons on the imidazole ring, are shifted to 7.20 ppm along with much lower intensity compared with the ligand alone (see Fig. 1); this shift is due to the binding to AuNPs surface and a change in its environment. In contrast, the NMR signals due to oleylamine ligands at 1.23 ppm, 1.98 ppm and 5.32 ppm in the spectrum (not shown here) collected from a solution of oleylamine capped AuNPs are essentially nonexistent in the sample of His–PIMA–PEG-capped nanocrystals. These results clearly indicate that cap exchange of oleylamine with His–PIMA–PEG is also driven by the imidazole coordination onto the surface of the AuNPs.

Quantifying the ligand coverage on the NPs

To gain additional insight into the surface-to-ligand interactions, we used the above pulsed field gradient ^{1}H-NMR data to extract an estimate of the average number of His–PIMA–PEG ligand per AuNPs. This ligand density value was extracted by comparing the molar concentrations of the polymer ligands and that of the AuNPs in the medium. The molar concentration of the polymer was extracted by comparing the integrations of the methyl-proton of the polymer backbone (labeled as a) and the α-proton of the pyridine standard (labeled as o) shown in Fig. 4B. The NP concentration was estimated from the Surface Plasmon Resonance (SPR) absorption data. Such analysis indicates that for a 5 nm radius AuNP there are about 30 polymeric ligands, which corresponds to ~500 His-anchors per AuNP.

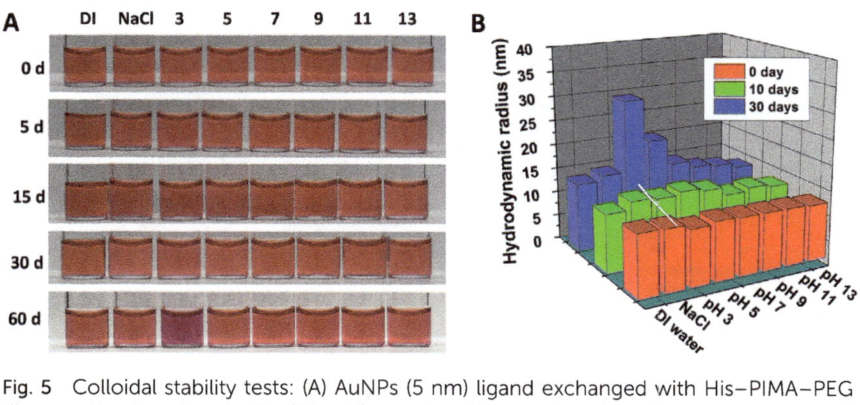

Fig. 5 Colloidal stability tests: (A) AuNPs (5 nm) ligand exchanged with His–PIMA–PEG dispersed in phosphate buffer (20 mM) over the pH range from 3 to 13 and in the presence of 1 M NaCl; (B) time-progression of the hydrodynamic radius measured from AuNP dispersions ligand exchanged with His–PIMA–PEG.

Stability test

The colloidal stability of aqueous dispersions of His–PIMA–PEG capped AuNPs (5 nm in radius) was tested in phosphate buffers at different pHs and in the presence of 1 M NaCl. The images shown in Fig. 5A indicate that dispersions of His–PIMA–PEG-capped AuNPs are colloidally stable over the pH range from 3 to 13 and in the presence of 1 M NaCl for 1 month of storage. After 2 months, a visible color change of AuNP dispersion was observed at pH 3, indicating the microscopic aggregation of NPs. Conversely, the dispersions stayed stable over the pH range 5 to 13 and in the presence of 1 M NaCl for at least 2 months of storage.

DLS measurements complemented the above data by allowing us to probe changes in the hydrodynamic size of the NPs with storage time under the various conditions shown in Fig. 5B. Data indicate that a fixed hydrodynamic size was measured for the His–PIMA–PEG-capped NPs across the full pH range and in 1 M NaCl for the first 10 day period. Also, no change was measured for dispersions at pH 7–13 after 1 month of storage. However, sizable increases in R_H were measured for the dispersions at pH 3 and pH 5, though only dispersion at pH 3 exhibited macroscopic aggregation after 2 months of storage (Fig. 5A). More precisely, the hydrodynamic size of the His–PIMA–PEG–NPs at pH 3 increased from ~11.4 nm to ~29.6 nm but to ~20 nm for dispersed at pH 5 after one month of storage. The weaker stability of His–PIMA–PEG at lower pH may be attributed to a slow protonation of the imidazole groups, reducing the NP-to-ligand affinity and thus NP stability in acidic conditions.

Surface modification of citrate-capped AuNPs and CTAB-capped AuNRs

We further applied this polymer to surface-ligating citrate-capped AuNPs and CTAB-capped AuNRs. Fig. 6A shows the absorption spectra of citrate-capped AuNPs (~6.5 nm in radius) before and after ligand exchange with His–PIMA–PEG. The spectra are essentially unchanged after ligation with His–PIMA–PEG. The hydrodynamic radius of His–PIMA–PEG–AuNP is about 15.0 nm, which is slightly larger than that measured for citrate–AuNP (~13.4 nm). This result confirms the compact coating provided by the multidentate polymer ligands. The

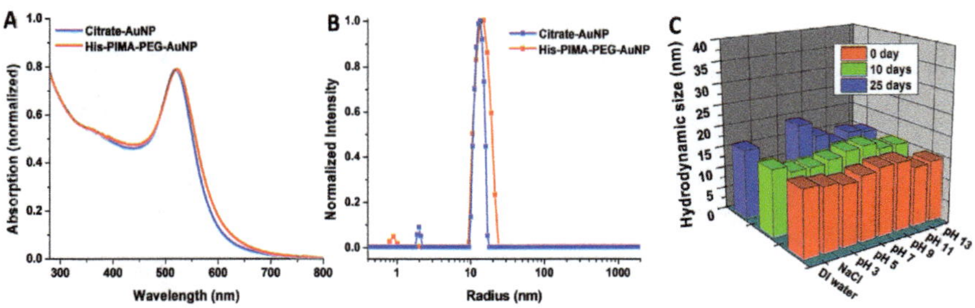

Fig. 6 (A) UV-Vis absorption spectra collected from aqueous dispersions of 13 nm (diameter) citrate-stabilized AuNPs and following ligand exchange with His–PIMA–PEG. (B) Histograms showing the distribution of the intensity *vs.* hydrodynamic radius measured for the citrate-stabilized AuNPs and His–PIMA–PEG–AuNPs. (C) Time-progression of the hydrodynamic radius of His–PIMA–PEG–AuNPs (prepared by ligand exchange of citrate–AuNPs). Values at pH 3 and in the presence of 1 NaCl are not shown after for day 25, due to sample precipitation.

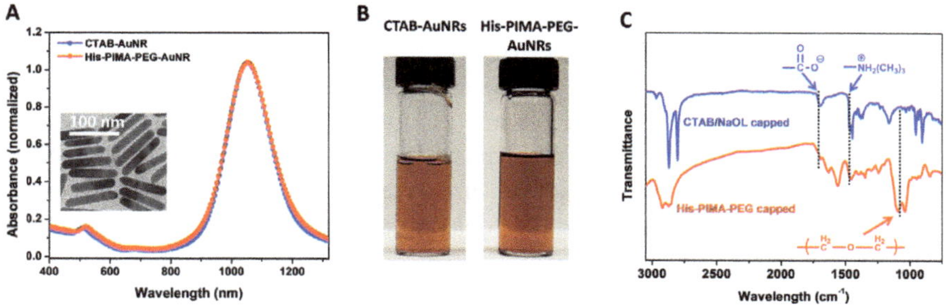

Fig. 7 (A) UV-Vis-NIR absorption spectra of aqueous dispersions of AuNRs capped with CTAB/NaOL and after ligand exchange with His–PIMA–PEG; insert shows a TEM image of the NRs. (B) White light images of CTAB/NAOL capped AuNR dispersion and His–PIMA–PEG capped AuNR dispersion in water. (C) FT-IR spectra of CTAB/NaOL- and His–PIMA–PEG-capped AuNRs.

hydrodynamic size for His–PIMA–PEG–AuNP dispersions was unchanged over the pH range 3–13 for 10 days. After ~1 month macroscopic aggregation was observed for dispersions at pH 3 and with excess salt; no reliable size measurements could be carried out for these conditions. At pH 5 R_H increased from 13 to 19 nm and a mild turbidity build up was observed. These may indicate that ligand exchange of citrate-stabilized AuNPs may not be as effective as was the case for the oleylamine–AuNPs above. This may reflect a difference in the nature of the citrate capping (*vs.* oleylamine coating) and the ligand exchange conditions used.

We further tested the ability of the His–PIMA polymer to ligate onto CTAB-coated AuNRs. For this CTAB–AuNRs ($L \times D = \sim 110$ nm × 20 nm) were mixed with His–PIMA–PEG in DI water with mild heating for several hours. Fig. 7A shows that the absorption spectra of the AuNRs before and after ligand exchange are essentially identical, with clearly defined transverse and longitudinal plasmon bands at 510 nm and 1050 nm, respectively, fully superposed for the two dispersions. The FT-IR spectra shown in Fig. 7C confirm the ligand exchange of CTAB-capped AuNRs with the His–PIMA–PEG polymer, where the bands at ~1470 cm^{-1} (ascribed to the trimethylammonium groups of CTAB) and ~1720 cm^{-1} (attributed to the carbonyl groups of NaOL) in the initial dispersions have disappeared, while a new strong peak at ~1100 cm^{-1} (due to the ether bonds of PEG chains) is measured in the spectrum of polymer-coated AuNRs.

Conclusion

We have developed a flexible amphiphilic polymer platform as a metal-coordinating ligand based on the imidazole motif. In particular, we exploited the high and specific reactivity of maleic anhydride towards amine-modified molecules *via* the nucleophilic addition reaction, and synthesized a set of imidazole- and PEG-modified polymers starting with poly(maleic anhydride) precursor. This synthetic route further allows control over the number and nature of the coordinating and hydrophilic groups introduced in the polymer platform. We showed that the imidazole-modified polymer can strongly ligate onto AuNPs and AuNRs, yielding hydrophilic metal platforms with good colloidal stability to pH changes and excess electrolytes in neutral and basic conditions. This polymer-coating of metal nanostructures based on the imidazole motif is very promising, due to the ubiquitous presence of imidazole-containing histidine in proteins expressed using bacterial vectors. These polymers can also be used to develop sensing and drug delivery platforms based on competition interactions. One idea could start with a drug modified imidazole-polymer immobilized onto AuNPs or AuNRs. These platforms could be introduced into live cells, allowing thiol-containing reducing agents, abundant in the cytoplasm, to alter the polymer coating and trigger the release of the drug. An alternative delivery mechanism could use laser induced-heating of the metal platform and partial release of the drug bearing polymer *in vivo*.

Acknowledgements

We thank FSU, the National Science Foundation (grant no. 1058957) for financial support. B.Z. was supported by a fellowship awarded by the education department of Fujian Province, China.

References

1 P. Ghosh, G. Han, M. De, C. K. Kim and V. M. Rotello, *Adv. Drug Delivery Rev.*, 2008, **60**, 1307.
2 X. Michalet, F. Pinaud, L. Bentolila, J. Tsay, S. Doose, J. Li, G. Sundaresan, A. Wu, S. Gambhir and S. Weiss, *Science*, 2005, **307**, 538.
3 H. Mattoussi, G. Palui and H. B. Na, *Adv. Drug Delivery Rev.*, 2012, **64**, 138.
4 V. Biju, T. Itoh and M. Ishikawa, *Chem. Soc. Rev.*, 2010, **39**, 3031.
5 P. Alivisatos, *Nat. Biotechnol.*, 2004, **22**, 47.
6 J. Rockenberger, E. C. Scher and A. P. Alivisatos, *J. Am. Chem. Soc.*, 1999, **121**, 11595.
7 J. Park, K. J. An, Y. S. Hwang, J. G. Park, H. J. Noh, J. Y. Kim, J. H. Park, N. M. Hwang and T. Hyeon, *Nat. Mater.*, 2004, **3**, 891.
8 S. H. Sun, H. Zeng, D. B. Robinson, S. Raoux, P. M. Rice, S. X. Wang and G. X. Li, *J. Am. Chem. Soc.*, 2004, **126**, 273.
9 W. W. Yu, J. C. Falkner, C. T. Yavuz and V. L. Colvin, *Chem. Commun.*, 2004, 2306.
10 Y. W. Jun, Y. M. Huh, J. S. Choi, J. H. Lee, H. T. Song, S. Kim, S. Yoon, K. S. Kim, J. S. Shin, J. S. Suh and J. Cheon, *J. Am. Chem. Soc.*, 2005, **127**, 5732.
11 P. Reiss, J. Bleuse and A. Pron, *Nano Lett.*, 2002, **2**, 781.
12 Y. W. Jun, J. H. Lee and J. Cheon, *Angew. Chem., Int. Ed.*, 2008, **47**, 5122.
13 F. Zhang, E. Lees, F. Amin, P. R. Gil, F. Yang, P. Mulvaney and W. J. Parak, *Small*, 2011, **7**, 3113.
14 R. Mout, D. F. Moyano, S. Rana and V. M. Rotello, *Chem. Soc. Rev.*, 2012, **41**, 2539.
15 D. Ling and T. Hyeon, *Small*, 2013, **9**, 1450.
16 B. S. Kim, J. M. Qiu, J. P. Wang and T. A. Taton, *Nano Lett.*, 2005, **5**, 1987.
17 U. I. Tromsdorf, N. C. Bigall, M. G. Kaul, O. T. Bruns, M. S. Nikolic, B. Mollwitz, R. A. Sperling, R. Reimer, H. Hohenberg, W. J. Parak, S. Forster, U. Beisiegel, G. Adam and H. Weller, *Nano Lett.*, 2007, **7**, 2422.
18 W. W. Yu, E. Chang, J. C. Falkner, J. Y. Zhang, A. M. Al-Somali, C. M. Sayes, J. Johns, R. Drezek and V. L. Colvin, *J. Am. Chem. Soc.*, 2007, **129**, 2871.

19 C. A. Lin, R. A. Sperling, J. K. Li, T. Y. Yang, P. Y. Li, M. Zanella, W. H. Chang and W. J. Parak, *Small*, 2008, **4**, 334.
20 T. Pellegrino, L. Manna, S. Kudera, T. Liedl, D. Koktysh, A. L. Rogach, S. Keller, J. Radler, G. Natile and W. J. Parak, *Nano Lett.*, 2004, **4**, 703.
21 E. E. Lees, T. L. Nguyen, A. H. A. Clayton, B. W. Muir and P. Mulvaney, *ACS Nano*, 2009, **3**, 2049.
22 X. Y. Wu, H. J. Liu, J. Q. Liu, K. N. Haley, J. A. Treadway, J. P. Larson, N. F. Ge, F. Peale and M. P. Bruchez, *Nat. Biotechnol.*, 2003, **21**, 452.
23 G. Mikhaylov, U. Mikac, A. A. Magaeva, V. I. Itin, E. P. Naiden, I. Psakhye, L. Babes, T. Reinheckel, C. Peters, R. Zeiser, M. Bogyo, V. Turk, S. G. Psakhye, B. Turk and O. Vasiljeva, *Nat. Nanotechnol.*, 2011, **6**, 594.
24 T. R. Zhang, J. P. Ge, Y. X. Hu and Y. D. Yin, *Nano Lett.*, 2007, **7**, 3203.
25 M. I. Shukoor, F. Natalio, V. Ksenofontov, M. N. Tahir, M. Eberhardt, P. Theato, H. C. Schroder, W. E. Muller and W. Tremel, *Small*, 2007, **3**, 1374.
26 M. Liong, H. Shao, J. B. Haun, H. Lee and R. Weissleder, *Adv. Mater.*, 2010, **22**, 5168.
27 M. H. Stewart, K. Susumu, B. C. Mei, I. L. Medintz, J. B. Delehanty, J. B. Blanco-Canosa, P. E. Dawson and H. Mattoussi, *J. Am. Chem. Soc.*, 2010, **132**, 9804.
28 D. Ling, W. Park, Y. I. Park, N. Lee, F. Li, C. Song, S. G. Yang, S. H. Choi, K. Na and T. Hyeon, *Angew. Chem., Int. Ed.*, 2011, **50**, 11360.
29 H. B. Na, G. Palui, J. T. Rosenberg, X. Ji, S. C. Grant and H. Mattoussi, *ACS Nano*, 2012, **6**, 389.
30 W. T. Wang, X. Ji, H. Bin Na, M. Safi, A. Smith, G. Palui, J. M. Perez and H. Mattoussi, *Langmuir*, 2014, **30**, 6197.
31 F. Aldeek, M. Safi, N. Q. Zhan, G. Palui and H. Mattoussi, *ACS Nano*, 2013, **7**, 10197.
32 B. C. Mei, K. Susumu, I. L. Medintz, J. B. Delehanty, T. J. Mountziaris and H. Mattoussi, *J. Mater. Chem.*, 2008, **18**, 4949.
33 S. Liu, G. Y. Chen, P. N. Prasad and M. T. Swihart, *Chem. Mater.*, 2011, **23**, 4098.
34 K. R. Brown, D. G. Walter and M. J. Natan, *Chem. Mater.*, 2000, **12**, 306.
35 J. Turkevich, P. C. Stevenson and J. Hillier, *Discuss. Faraday Soc.*, 1951, **11**, 55.
36 X. C. Ye, C. Zheng, J. Chen, Y. Z. Gao and C. B. Murray, *Nano Lett.*, 2013, **13**, 765.
37 T. Pons, H. T. Uyeda, I. L. Medintz and H. Mattoussi, *J. Phys. Chem. B*, 2006, **110**, 20308.
38 Y. J. Kang and T. A. Taton, *Macromolecules*, 2005, **38**, 6115.
39 W. K. Li, S. Q. Liu, R. H. Deng, J. Y. Wang, Z. H. Nie and J. T. Zhu, *Macromolecules*, 2013, **46**, 2282.

Faraday Discussions

PAPER

Three bisphosphonate ligands improve the water solubility of quantum dots†

Siti Fatimah Abdul Ghani,[a] Michael Wright,[a] Juan Gallo Paramo,[b] Melanie Bottrill,[bc] Mark Green,[c] Nicholas Long[b] and Maya Thanou*[a]

Received 11th July 2014, Accepted 1st August 2014
DOI: 10.1039/c4fd00151f

Synthesised Quantum Dots (QDs) require surface modification in order to improve their aqueous dispersion and biocompatibility. Here, we suggest bisphosphonate molecules as agents to modify the surface of QDs for improved water solubility and biocompatibility. QDs_TOPO (CdSe/ZnS–trioctylphosphine oxide) were synthesised following modification of the method of Bawendi *et al.* (*J. Phys. Chem. B*, 1997, **101**, 9463–9475). QDs surface modification is performed using a ligand exchange reaction with structurally different bisphosphonates (BIPs). The BIPs used were ethylene diphosphonate (EDP), methylenediphosphonate (MDP) and imidodiphosphonate (IDP). After ligand exchange, the QDs were extensively purified using centrifugation, PD-10 desalting columns and mini dialysis filters. Transmission electron microscopy (TEM) and fluorescent spectroscopy have been used to characterise the size and optical properties of the QDs. Cell toxicity was investigated using MTT (tetrazolium salt) and glutathione assays and intracellular uptake was imaged using confocal laser scanning microscopy and assessed by Inductively Coupled Plasma Mass Spectrometry (ICP-MS). QDs_TOPO and QDs-capped with BIPs (QDs_BIPs) were successfully synthesised. TEM showed the size and morphology of the QDs to be 5–7 nm with spherical shape. The stabilised QDs_BIPs showed significantly improved dispersion in aqueous solutions compared to QDs_TOPO. The cytotoxicity studies showed very rapid cell death for cells treated by QDs_TOPO and a minor effect on cell viability when QDs_BIPs were applied to the cells. Both EDP- and MDP-modified QDs did not significantly increase the intracellular levels of glutathione. In contrast, IDP-modified QDs substantially increased the intracellular glutathione levels, indicating potential cadmium leakage and inability of IDP to adequately cap and stabilise the QDs. EDP- and MDP-modified QDs were taken up by IGROV-1 (ovarian cancer) cells as shown by fluorescence microscopy, however, the IDP-modified QD signal was not clearly visible in the cells. Cellular uptake measured by

[a]*Institute of Pharmaceutical Sciences, Franklin-Wilkins Building, KCL, 150 Stamford Street, London, SE1 9NH, UK. E-mail: maya.thanou@kcl.ac.uk*

[b]*Department of Chemistry, Faculty of Natural Sciences, ICL, Exhibition Road, South Kensington, London, SW7 2AZ, UK*

[c]*Department of Physics, Strand Campus, KCL, London, WC2R 2LS, UK*

† Electronic supplementary information (ESI) available: Schematic of the reaction and mass spectrometry spectrum (GCMS) of ethylene diphosphonate (EDP). See DOI: 10.1039/c4fd00151f

intracellular cadmium levels using ICP-MS showed significant uptake of all three BIPs QDs. The structure of BIPs appears to play a significant role in the ability of these molecules to act as capping agents. Our findings demonstrate a novel approach to produce water-dispersible QDs through ligand exchange with certain types of BIPs molecules that can find application in bioimaging.

1. Introduction

Semiconductor QDs are nanometer-sized crystals that have unique photochemical and photophysical properties.[1] QDs are fluorescent labels that are resistant against photobleaching, with improved brightness, size-dependent fluorescence and a narrow, tunable emission spectrum. QDs provide the possibility of continuous, real-time imaging of single molecules and single cells over an extended period of time.[2] They exhibit narrow emission and wide excitation spectra; in fact, QDs can be excited at any wavelength below their emission peak.[3] The core, shell and surface compositions of semiconductor QDs nanocrystals play significant roles in their structural and optical properties, from high surface-to-volume ratios for efficient absorption and emission spectra to capping with higher band-gap shells consisting of either ZnS or ZnSe for improved luminescence, stability, aggregation and also toxicity.[4–6]

A common QD preparation uses a route involving a high-temperature solvent mixture of trioctylphosphine–trioctylphosphine oxide (TOP–TOPO), yielding hydrophobic, not aqueous dispersible, QDs. Attempts to overcome their hydrophobicity have focused on identifying suitable ligands for ligand exchange (TOPO–TOP), as well as recently developing new methods of synthesis.[4,5] Hydrophobicity also directly influences the suitability of the QDs for biological applications. The challenge in developing QDs therefore lies in the preparation method and modification of the nanoparticles towards improved biocompatibility.[7] Any phase transfer towards aqueous or water soluble QDs will require surface bio-functionalisation, or modification with hydrophilic ligands either through ligand exchange reactions or forming a micelle through hydrophobic interaction and/or silica encapsulation.[7] Ideally, surface modification or overcoating the QDs should prevent QDs from flocculating in aqueous buffers, or in storage over time, and be able to efficiently convert organic soluble/hydrophobic QDs into water soluble QDs, whilst maintaining the size of the QDs and their fluorescence properties. Surface modification also affects the hydrodynamic diameter of the QDs and this may have an impact on their application as diagnostic and therapeutic agents.[8] Several procedures for surface coating of QDs have been developed, such as cross-linking, use of peptides, avidin and amphiphilic molecules.[9–12] The latter is an example that uses a mixture of phospholipids and lipopolymers or hydrophilic polymers with hydrophobic side chains.[12]

A less investigated method of surface modification is non-covalent conjugation, which includes the interaction between biofunctionalised molecules and the metal ions (Cd, Zn) present at the surface of the QDs. This has been reported with the use of proteins or polypeptides with amino acid sequences that have an affinity for the metallic cations.[13] The mechanism of biomolecule adsorption results from ionic bonds between the carboxyl functional group (COO^-) of the ligand and Cd^{2+} or Zn^{2+}.

QD synthesis involves the formation of the QDs nanoparticles in the presence of stabilising ligands, independent of the solvents used. These stabilising ligands carry out a very important role in maintaining discrete, non-aggregated QDs, whilst also affecting the photoluminescence of the nanoparticles. This applies especially to uncapped QDs, where the surface ligand is the only barrier between the electron–hole pair in the core QD and the media or solution.[14,15] The stability of the attachment of these ligands is crucial in reducing the cytotoxicity of the QDs in biological systems.[16,17] There are studies that report that uncapped QDs give high cytotoxicity compared to capped or protected QDs and uncapped QDs have also been reported to induce apoptosis due to leaching of toxic metals from the QDs or by the formation of reactive oxygen species (ROS).[18,19] The toxicity of uncoated cadmium-containing QDs is associated with free cadmium present in the nanocrystals suspension or intracellularly released from absorbed QDs *via* surface oxidation, thus suggesting that the core could degrade in biological environment.[14]

Ligand exchange surface modification routes have been explored and employed with many different materials and protocols.[20–22] In the study presented here, we explore the ligand exchange route and surface modification in replacing the hydrophobic surface layer (TOP–TOPO) of the QDs (core/shell CdSe/ZnS) using a hydrophilic ligand containing phosphonate groups. We believe that this is the first ligand exchange method approach using this type of phosphorus-containing group as the hydrophilic ligand. BIPs or diphosphonates contain two phosphonate groups and also mimic the structure of pyrophosphate. These two phosphonate groups are linked together by one or two central carbon atoms (C) forming the P–C–P backbone. BIPs are commonly used in treating secondary breast cancer, osteoporosis and similar diseases by preventing the loss of bone mass.[23] The mechanism of action of BIPs includes Ca^{2+} chelation.[24] In this study, we aim to take advantage of this property and investigate the efficiency of BIPs as QD capping agents. We investigate the effect of structurally similar BIPs (Fig. 1) on

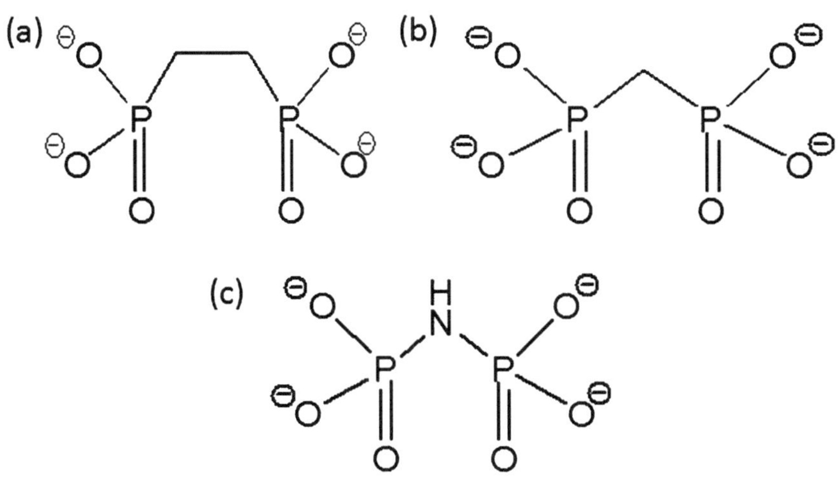

Fig. 1 Structure of BIPs used that mimic the pyrophosphate groups: (a) EDP, (b) MDP and (c) IDP.

stabilising CdSe/ZnS QDs. EDP, MDP and IDP are introduced on the surface of the QDs *via* a designed ligand exchange reaction. The nanoparticles produced are tested for their optical properties, colloidal dispersion, and photostability, as well as for their effect on cell viability and their ability to label cells.

2. Experimental materials and methods

All materials were purchased from Sigma Aldrich Corporation, St. Louis MO, (USA); Fisher Chemicals and Loughborough (UK), and Alfa Aesar, Karlsruhe (Germany); chemicals were of reagent grade and used as received. Schlenk-line techniques and inert gas nitrogen conditions have been used to carry out all synthetic preparations. $^{31}P\text{-}\{^1H\}$ NMR analytical data were obtained using a Bruker AV 400 MHz spectrometer. All isotopic solvents d_6-DMSO, D_2O, $CDCl_3$ were purchased from Cambridge Isotope Laboratories, MA USA.

For QD synthesis (CdSe/ZnS), materials used including TOPO (90%), TOP (90%), diethylzinc ($ZnEt_2$), MDP and IDP were obtained from Sigma Aldrich, MO USA. Cadmium acetate dihydrates $(CdOAc_2)\cdot 2H_2O$ and tetraethyl ethylene diphosphonate (TEDP) was purchased from Alfa Aesar. De-salting PD-10 (Sephadex G-25 Medium) columns were purchased from GE Life Sciences, UK. Absorption spectroscopy was recorded using a Lambda 25 UV/Vis spectrometer using 10 mm cuvette and emission spectra were obtained using a Varian Cary Eclipse Fluorescence Spectrometer or Tecan Infinite® 200 Pro. QDs were pre-scanned over the range of 400–700 nm.

Hydrodynamic diameters were measured using Nanoseries Nano-S Malvern (DLS) and were performed in quartz cuvettes prepared with at least three good quality data measurements analysed for each samples. Mass Spectrometry Analysis (LC MS/MS and ICP-MS) was obtained using a Perkin Elmer SCIEX ICP-Mass Spectrometer (ELAN DRC6100, USA) at the Mass Spectrometry Facility, KCL. TEM and high resolution-TEM (HRTEM) data and images were acquired using a FEI Tecnai 12 TEM operating at 120 kV with an AMT 16000M digital camera to obtain high resolution images at the Centre for Ultrastructural Imaging (CUI), KCL. Cell imaging was carried out using A1R Si Confocal Microscopy equipped with Ti-E Inverted Microscope at the Nikon Imaging Centre, KCL.

Synthesis of BIPs capped QDs

Precursor preparation. Selenium powder was added to a round bottom flask (previously purged with nitrogen) and then TOP (50 mL) was added carefully syringe-wise into the flask through a rubber septum. The mixture was stirred for 3 hours at room temperature to obtain a pale yellow, optically clear solution of the Se_TOP precursor. In a separate flask, sulfur powder was added to a Schlenk tube previously purged with nitrogen followed by TOP (5 mL) addition. The mixture was stirred overnight to make sure all the sulfur powder had dissolved completely in the TOP solvent and formed an optically clear solution to give the S_TOP precursor. A zinc and sulfur precursor solution was prepared by addition of TOP (9 mL), diethylzinc (450 μL, 1.1 M solution in toluene) and previously prepared S–TOP (480 μL) into a previously nitrogen-purged Schlenk tube. The precursor mixture was left to stir for 2–3 hours to achieve a homogenous mixture and was freshly prepared prior to synthesis.

Synthesis of core/shell CdSe/ZnS_TOPO. TOPO (5.25 g) and cadmium acetate dihydrate [$Cd(CH_3COOH)_2 \cdot 2H_2O$] (0.13 mmol; 0.045 g) were added into a round bottom flask that was purged several times with nitrogen gas. The compound was slowly heated to 220 °C and stabilised at the desired temperature with moderate stirring to obtain an optically clear solution. Selenium–TOP precursor solution was injected rapidly through the rubber septum into the hot mixture. After the injection of the precursor solution, the mixture started to change colour from pale yellow, to yellow then to orange and finally red to deep red in about 10–15 minutes. The reaction was left for 30 minutes before being cooled to 60 °C. The excess TOP–TOPO was removed using methanol and centrifuged at 4000 rpm for 30 minutes to yield a fine red precipitate. In a typical synthesis, CdSe_TOPO prepared previously was added to a three necked round-bottom flask together with TOPO (3.0 g) and the mixture was heated to 190 °C. The zinc and sulfur precursor was freshly prepared and introduced dropwise using a syringe. Following the final addition, the mixture was left for 30 minutes before cooling to 60 °C. The nanoparticles produced were then treated three times with methanol and centrifuged at 4000 rpm for 30 minutes to yield a fine red precipitate of CdSe/ZnS_TOPO. Core and core/shell quantum yields were measured compared to Rhodamine 6 G as a reference as in ref. 25.

Synthesis of EDP ligand. Tetraethyl ethylene diphosphonate (TEDP) was added into a round-bottom flask to equimolar hydrochloric acid (37%). The mixture was heated at reflux and left to stir overnight. A fine white powder formed overnight and was collected in a separating funnel. This was purified by washing the particles with hexane, dichloromethane and chloroform and then dried under vacuum and weighed to give 90% yield.

$^{31}P\text{-}\{^{1}H\}$ NMR (162 MHz, D_2O): δ 28.24 (s, P=O). ESI-MS: m/z 190 [M^+]. Result of HRMS 190.9874 (found)/190.9874 (calc.) m.p.: 210–215 °C.

Surface modification *via* ligand exchange with BIPs – general methodology. As-prepared CdSe/ZnS_TOPO was isolated from any excess TOP–TOPO by precipitation with methanol and centrifugation at 4000 rpm for 30 minutes. The nanoparticles were then suspended in hexane in a round bottom flask, with addition of 2 mg of either (a) EDP, (b) MDP; or (c) IDP, followed by insertion of penta-(ethylene)glycol (Penta_EG) *via* syringe into the mixture. The mixture was purged with nitrogen gas and left under reflux for 6 hours with vigorous stirring to ensure that the ligand exchange reactions were complete. This then yielded the nanoparticles QDs_BIPs and the precipitate was dried under vacuum. The product was weighed and a yield of 80% was noted.

Purification of water soluble QDs_BIPs – general methodology. An aqueous solution of QDs_BIPs (EDP/MDP/IDP) was sonicated for 5–10 minutes and passed through the PD-10 (Sephadex G-25 medium) de-salting column. The sample was loaded until all the solution entered the packed bed material completely. The sample was eluted with purified water and the elutes were collected for further analysis. Purification steps were repeated for all QDs_BIPs (EDP/MDP/IDP) samples. BIP-modified QDs were characterised for efficient ligand exchange using $^{31}P\text{-}\{^{1}H\}$ NMR spectroscopy.

Characterisation of QDs and QDs_BIPs

Optical measurements: core (CdSe_TOPO) and/or core/shell (CdSe/ZnS_TOPO) were dispersed in toluene while QDs_BIPs (EDP/MDP/IDP) were dispersed in deionised water at a concentration of 1 mg mL^{-1}. Absorbance and fluorescence intensities of all solutions were obtained using a 10 mm quartz cuvette scanning from 400 nm to 700 nm in a UV/Vis spectrometer (Perkin Elmer) and fluorescence spectrometer (Varian Cary Eclipse).

Colloidal stability and photostability studies of QDs in aqueous media. Purified QDs_BIPs were prepared in water, media and 10%, 20%, 50% FBS (Foetal Bovine Serum) in cell culture media. All these solutions were assessed using absorbance/fluorescence measurements for photostability and colloidal stability over 4 hours. Measurements were taken using Tecan Platereader Infinite® 200 Pro Series spectrometer. Absorbance values at 550 nm wavelength were recorded for all solutions to measure the turbidity values of the solutions as an indication for QD aggregation.[26]

Nanoparticle characterisation. Both diluted core and core/shell (CdSe/ZnS_TOPO) in hexane and aqueous suspensions of QDs_BIPs were dropcast using a pipette onto holey carbon TEM grids (300 mesh, copper grids), horizontally placed with an inverted tweezer and left to dry for 5 minutes. Filter paper was used to passivate the solution by placing the sharp edge of the filter paper by the side of the grids. These nanoparticles were then analysed under the microscope for size and distribution. High resolution-TEM was also used for determination of the nanocrystals lattice spacing using FEI Tecnai G2 T20 FE TEM at a 200 kV accelerating voltage. BIP-modified QDs were characterised for their ζ potential using a Malvern Nanoseries.

Energy dispersive X-ray spectroscopy (EDAX) analysis. QDs_TOPO and QDs_BIPs nanoparticles were prepared in 1 mg mL^{-1} solution in hexane and in water, respectively. 5 µL of these solutions were pipetted onto holey titanium grids horizontally with an inverted tweezer and left to dry under an argon lamp for 5 minutes. These nanoparticles were then analysed under the microscope for image and elemental analysis.

Cell viability assay (MTT viability assay and glutathione assay). Human ovarian adenocarcinoma cells (IGROV-1) were cultured in Roswell Park Memorial Institute (RPMI 1641) media supplemented with 2 mM glutamine, 10% v/v fetal bovine serum (FBS), 1% penicillin–streptomycin (pen–strep). Cells were cultured in a humidified atmosphere containing 5% CO_2, at 37 °C. For MTT cell viability studies, the 3-(4,5-dimethylthiazol-2-yl)-2,5-diphenyltetrazoliumbromide (MTT) reduction assay and glutathione assay (determination of reactive oxygen species-induced toxicity) were selected to monitor the cell viability of the proliferation of the QDs_TOPO and QDs_BIPs in media. For MTT assays, cells were seeded into 96 well plates overnight, and were then incubated with media containing QDs_TOPO and QDs_BIPs in triplicate. In each well, 10 µL of MTT reagent was added 24 hours after addition of the nanoparticles, and the plate was then incubated for 4 hours and DMSO added to each well to solubilise the formazan crystals after removing the media. Results for the MTT viability assay were expressed as:

$$\text{Percent viability} = \frac{[\text{cells} \times (\text{treated cells}) - \text{background}]}{[\text{cells} \times (\text{untreated cells}) - \text{background}]} \times 100$$

The glutathione assay was carried out to determine the reactive oxygen species-induced toxicity in cells. Total concentration of intracellular glutathione (tGSH) was determined biochemically using a Sigma glutathione assay kit. Typically, cells were counted, incubated and then lysed with the nanoparticles. Total glutathione was determined biochemically following the assay protocol by constructing the standard curve against total glutathione in each sample. It was calculated using the equation below;

$$[\text{Total GSH}] = \left(\frac{(\text{absorbance at 412 nm}) - (y\ \text{intercept})}{\text{slope}}\right)$$

QDs_BIPs cell uptake. Adherent cells (IGROV-1) were grown and counted manually using haemocytometer and seeded into an 8-well chamber plate. The chamber plate was previously primed with 10% w/v porcine skin gelatine in water overnight at 4 °C. The cells were incubated for 24 hours to let them adhere to the chamber and were then dosed with RPMI media containing QDs_BIPs. Subsequently, the cells were washed with phosphate buffered saline (PBS) and fixed using 1% w/v paraformaldehyde (PFA in PBS). The cells were washed with PBS before staining them with 4′,6-diamidino-2-phenylindole (DAPI) to label the nucleus. The cells were washed again and stored in PBS before imaging with confocal laser scanning fluorescent microscopy.

ICP-MS. The level of QDs_BIPs uptake was characterised by assessing intracellular levels of Cd^{2+} following the 24 hours incubation with the cells. QDs_TOPO (coarse dispersion in cell medium) and QDs_BIP (fine dispersion) were applied on IGROV-1 cells and were incubated for 24 hours. At the end of the experiment the supernatant was removed and cells were washed with PBS and lysed with lysis buffer. Quantification of Cd^{2+} concentration in cell uptake and in standard solution was carried out by ICP-MS. Each biological sample was digested with 10 times mass or volume matter with nitric acid, HNO_3 at 60 °C overnight until all matter was digested completely. Each sample was then diluted with Mili-Q water (18.2 Ω) to give an acid concentration of ~15%. Yttrium was used as the internal standard and was spiked in all samples at the same concentration. Solution samples of the same kind used in the cellular uptake were analysed following the same procedures described above to correlate with the Cd^{2+} concentration. All samples were analysed using the same ICP-MS initialisation and set up.

3. Results and discussions

Core and core/shell QDs were synthesised using a modification of published procedures.[27–31] Quantum yields for both core and core/shell species were measured to be 0.17 and 0.32 respectively using Rhodamine 6 G (reference QY: 0.95) as the reference fluorescence dye. The aim was to prepare colloidally stable QDs in aqueous media using the following BIPs: EDP , MDP and IDP (Fig. 1). Most of the frequently used ligands for phase transfer utilise thiol or carboxylic containing capping agents.[32,33] However, in 2004, Green *et al.* reported that nucleotides such as ATP could be used as passivating agents for QDs.[34] Based on this report, we considered investigating bisphosphonate molecules as capping agents for water soluble QDs. MDP and IDP are commercially available BIPs, whereas EDP is prepared from tetraethylene-1,2-diphosphonate under acidic conditions.

Analysis with mass spectrometry indicated that EDP was prepared successfully (see ESI†). These BIPs were used to modify CdSe/ZnS_TOPO QDs, which were selected due to their frequently reported superior optical properties. The pioneering synthesis methods of Bawendi *et al.* and Peng *et al.* [ref. 21 and 44] were followed to prepare the initial QDs coated with TOPO to further undergo phase transfer.[27,28] The results in Fig. 2 indicate the optical characteristics of CdSe/ZnS_TOPO and CdSe_TOPO, as well as size characteristics *via* TEM, lattice spacing *via* HRTEM and energy dispersive X-ray analysis. These data are in agreement with those in previously reported studies using these methods of synthesis.[30,35]

Several structurally different molecules and polymers have been suggested to modify the surface of QDs to provide improved biocompatibility and hydrophilicity.[7,26,36] The ligand exchange reactions used for EDP, MDP and IDP were designed based on well-documented phase transfer reported studies.[33,37,38] After the ligand exchange reactions, the QDs were characterised for their photoluminescence and size (TEM), and the efficiency of the ligand exchange reaction was monitored. Fluorescence spectroscopy of the modified QDs showed a decrease in photoluminescence; however this also indicated successful phase transfer resulting in biocompatible/water soluble QDs (Fig. 3).

TEM indicated that the QDs kept their size characteristics while $^{31}P\text{-}\{^{1}H\}$ NMR showed that ligand exchange was successful and no traces of TOPO or TOP could

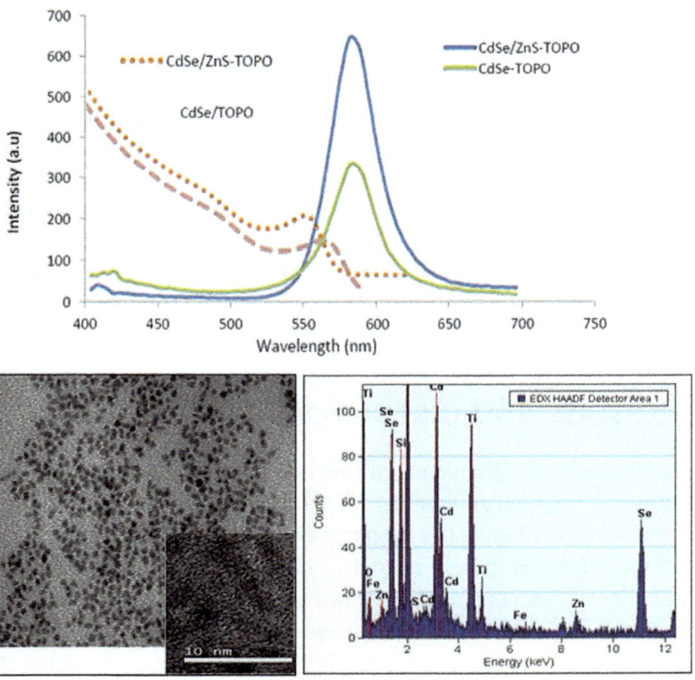

Fig. 2 Characteristics of QDs core/shell CdSe/ZnS (top) absorbance and fluorescence spectrum of core CdSe in hexane; absorbance and fluorescence spectrum of core/shell CdSe/ZnS in hexane; (bottom, left) TEM image of QDs and HRTEM image indicating confirmation the lattice spacing for nanocrystals, and (bottom, right) EDAX.

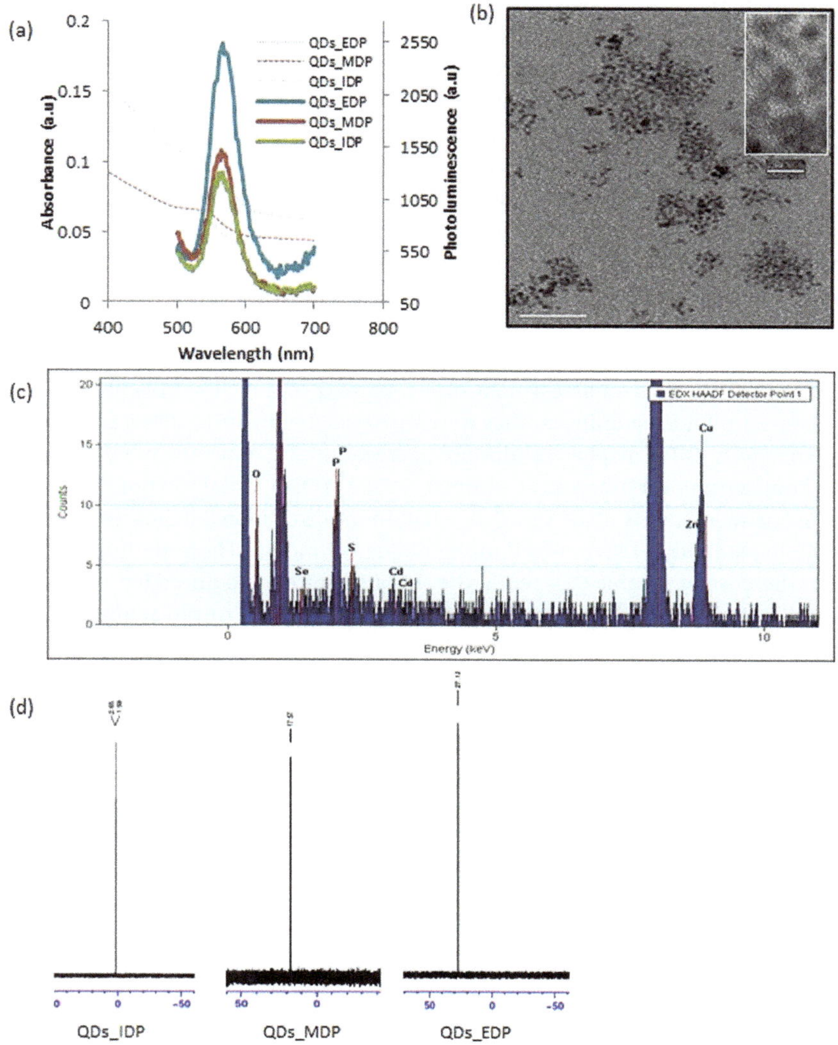

Fig. 3 QDs_BIPs characteristics: (a) emission/absorbance spectra of QDs_BIPs in water at RT after the ligand exchange; (b) TEM and HRTEM images of QDs_BIPs showing size and distribution in water (scale bar: 100 nm, inset: 5 nm); (c) EDAX spectra of QDs_BIPs showing presence of elements related to QDs_BIPs; (d) ^{31}P NMR spectrum of QDs_BIPs indicating no impurities detected in the precipitate after purification (no signal recorded for TOP and TOPO). Pure TOPO in chloroform-d ^{31}P NMR shows a typical signal at ~50 ppm whilst free TOPO in chloroform-d signal at ~52 ppm.

be found on the surface of the QDs.[39] QDs_BIP were also characterised for their zeta potential and this was found to be within the range −0.50 mV to 0.05 mV for all QDs_BIP (~pH 7.0 at room temperature). The neutral readings (−10 mV to +10 mV) indicated that the nanoparticles were well-dispersed in water and were suitable for further cell studies without cationic charge-related cytotoxicity.

The photostability of the modified QDs in water was investigated, as well as in aqueous solution with different pH values. The results (Fig. 4) indicate the photostability of BIPs-coated QDs after repeated irradiation and show that 80–85% of the fluorescence intensity remains after 120 scans. Fluorescence intensity also remains unchanged at different pH values (Fig. 4). This is important for bioimaging, as QDs may diffuse through cell compartments with different pH values and, therefore, their initial photoluminescence might be affected. It has been reported that low pH values can induce dissociation of thiolate ligands from cadmium chalcogonide nanocrystals with an effect on their photoluminescence.[40] Studies have also shown that the photostability of surface-modified QDs is negatively affected after ligand exchange.[41] In our studies, BIPs-modified QDs show minimum change in photoluminescence at different pH values.

The QDs_BIPs also show good colloidal stability over time in media containing FBS. We measured the effect of QDs_BIPs on the transparency of serum containing solutions as an assessment of the stability of the QDs_BIPs in biological systems. Colloidal stability studies were carried out with 10%, 20% and 50% FBS media. QDs_BIPs display minimum aggregation in aqueous media with or without serum over 4 hours, in contrast to QDs_TOPO, which do not disperse in aqueous media. QDs_EDP and QDs_MDP (non-nitrogenous) exhibited excellent colloidal stability in serum-containing biological media. These findings suggest that these water soluble QDs retain the colloidal stability required for application in cell imaging. QDs_IDP and QDs_EDP were slightly more turbid at higher serum concentration (Fig. 5) than QDs_MDP, thus indicating the structure of the bisphosphonate ligand has an effect on the colloidal stability of the QDs. Stabilised BIP-capped QDs have shown good dispersion profile in aqueous solutions and retained their fluorescence characteristics.[26]

There is concern over the potential toxicity of QDs containing cadmium within the core of the nanoparticles. The potential cytotoxicity of the semiconductor nanoparticles still remains unclear. Studies have been carried out investigating cell viability to determine the level of toxicity or exposure limits of these QDs.[17,42]

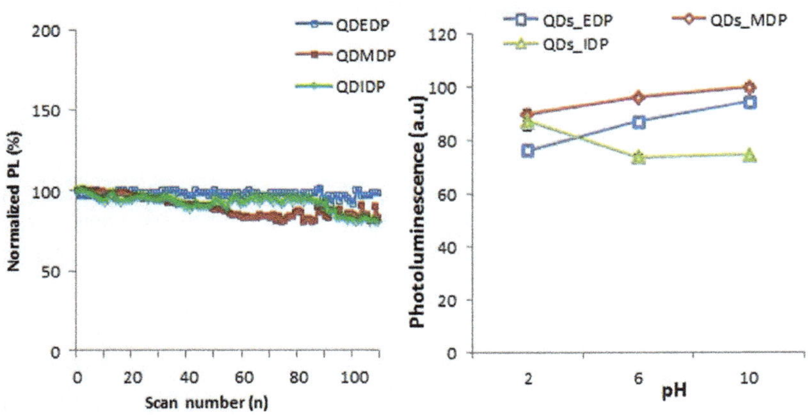

Fig. 4 Photostability of aqueous dispersion of QDs_BIPs. (Left) Photostability was measured at emission wavelength of 560 nm by consecutive excitation at 365 nm at room temperature. The fluorescence intensity was normalised to the initial value of photoluminescence. Finding indicate that 80–85% initial fluorescence intensity remain after 120 measurements/illuminations. (Right) QDs_BIPs remain soluble in water under acidic to basic conditions with no decrease in photoluminescence from pH 2.0–10.0.

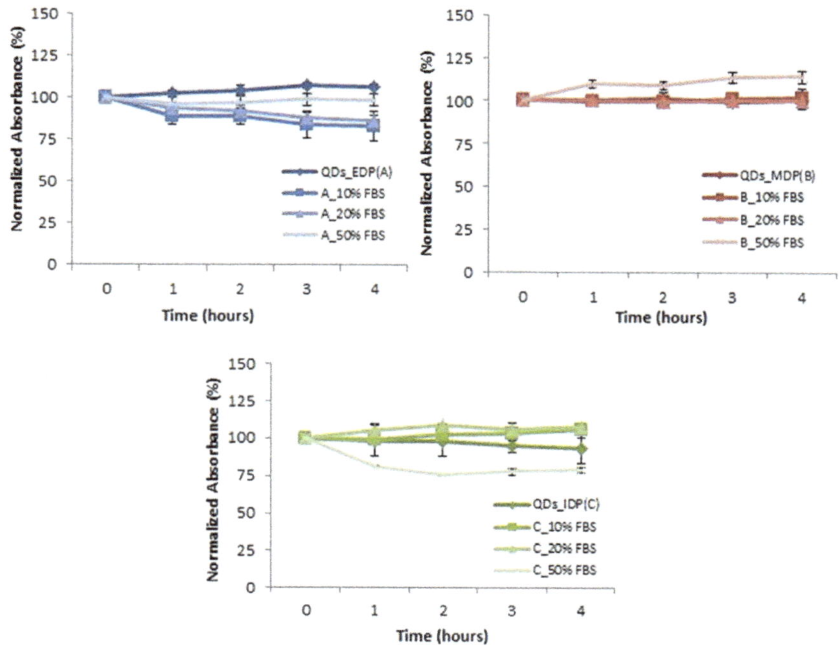

Fig. 5 Stability studies of QDs_BIPs for 10%, 20% and 50% FBS in media. QDs_BIPs have shown stability over 4 hour time studies for in media containing these percentages of FBS. Overall, QDs_MDP showed better dispersion and stability for all percentages of FBS, indicating minimal changes throughout the 4 hour time studies. In contrast, QDs_EDP and QDs_IDP showed nearly a 25% change in absorbance values from initial fluorescence as the percentage of FBS in the media increased. All stabilised water soluble QDs_BIPs have shown good dispersion profiles in aqueous solutions containing up to 50% FBS and also retained their fluorescence characteristics.

QDs_TOPO and QDs_BIPs underwent MTT assays to measure their effect on cell viability after 4 hours incubation, whereas the effect on reactive oxygen species was assessed using the glutathione assay. Fig. 6c shows that QDs_BIPs induced a minimal effect on cell viability whereas QDs_TOPO caused significant cytotoxicity at both low and high concentrations. These data are expected due to rapid cell uptake and leakage of the heavy metals into the cells.[43] Excellent viability with QDs_BIPs is observed even at the higher doses tested. This suggests that capping with bisphosphonate ligands has produced water soluble QDs with a minimal effect on cell viability.

Determination of Cd^{2+} leakage-induced ROS was carried out through glutathione assays using IGROV-1 cell lines. The QDs_EDP/MDP induce a small and dose dependent increase of intracellular total level of glutathione, and in particular, QDs_EDP induces a minimum increase of total glutathione (non-significant difference to non-treated cells at 5 μg mL^{-1}). QDs_IDP show a significant increase of total glutathione for both concentrations compared to the other QDs_BIPs (EDP and MDP). This may indicate that IDP-coated QDs have inferior stability, thus leading to intracellular Cd^{2+} leakage. Even at the lowest concentrations (5 μg mL^{-1}) QDs_IDP show elevated levels of total glutathione relative to untreated cells (control) and exhibit the highest total induced

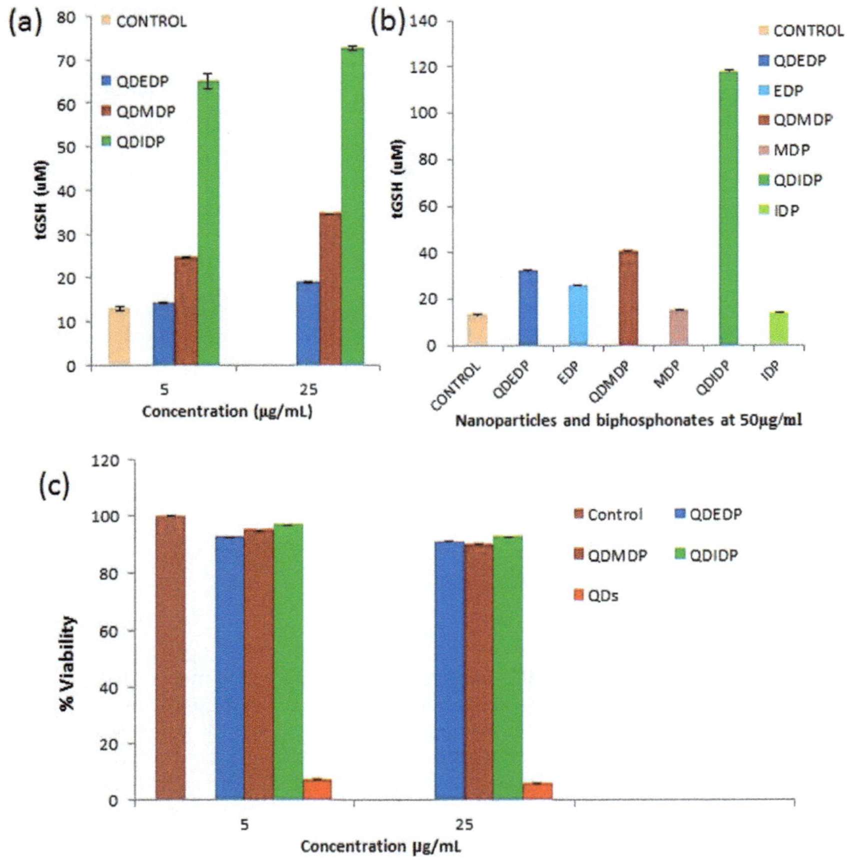

Fig. 6 Cytotoxicity studies of QDs_BIPs using MTT and glutathione assays to measure cell proliferation. (a and b) Glutathione assay to measure ROS in IGROV-1 cells line for BIPs alone (EDP/MDP/IDP) and QDs_BIPs. (c) MTT assays of QDs_BIPs and QDs_TOPO where QDs_TOPO show acute cytotoxicity on IGROV-1 cells. IGROV-1 cell toxicity studies exhibited a dose–response relationship on cell viability induced by the water soluble QDs_BIPs.

glutathione. We also measured the total glutathione levels induced by the BIP molecules (*i.e.* EDP, MDP and IDP) and their contributions towards the overall levels of total glutathione produced in the cells. Fig. 6b shows that; BIPs do not affect the intracellular glutathione levels even at high concentrations (higher concentrations compared to the amount capping the QDs). It is possible that the release of Cd^{2+} ions intracellularly by the QDs_IDP is the main factor that contributes to cell impairment, cell death or the formation of ROS, as the effect of BIPs is minimal (Fig. 6b).

Reported studies carried out with mercaptopropionic acid-coated cadmium telluride (CdTe) by Maysinger *et al.* showed that unmodified or 'naked' core CdTe can induce formation of ROS leading to damage to multiple organelle and cell death. They also showed in the same study that effects associated with ROS formation can be prevented either with antioxidant treatment *i.e.* *N*-acetylcysteine (NAC-cysteine) or by further surface modification.[44] Niemeyer *et al.* have carried out systematic investigation on the generation of free radicals in aqueous solution by photoirradiation of stabilised CdS, CdSe and CdSe/ZnS with mercaptoacetic acid (MAA) through two methods (EPR spectroscopy and highly sensitive, radical-specific fluorimetric assays to measure generated radical species).

CdS generated both hydroxyl and superoxide free radicals, whilst CdSe generated only hydroxyl free radicals, which contrasts with CdSe/ZnS, which did not induce formation of free radicals under the condition used in the study.[45] Over-coating of a ZnS shell prevents the production of both free radicals with higher band-gap energy barriers, as compared to core CdSe only.[46] Finally, Green and Howman *et al.* have reported generation of free radicals from CdSe and CdSe/ZnS QDs that are likely responsible for DNA nicking, both when incubated in the dark and when exposed to UV irradiation.[47]

Many studies have been carried out to determine the level of toxicity *i.e.* investigating levels of QD concentrations, exposure times of the nanoparticles, different cell types and also different surface coatings. From some of these studies, it has been clearly indicated that bare core QDs (either CdSe or CdTe) have the most effect on cell viability.[48] QDs capped with another layer or multi-layer of wider band gap ZnS or ZnSe feature good capping material that contributes to the nanoparticles having longer exposure times in living cells and showing less toxicity compared to naked core CdSe or CdS.[49,50] Moreover, it has also been demonstrated that surface modifications, either through ligand exchange or encapsulation in liposomes or micelles, greatly improves the biocompatibility of the nanoparticles for *in vitro* and *in vivo* experiments.

In order to assess the intracellular uptake of the nanoparticles, fluorescence microscopy experiments were carried out. IGROV-1 cells were co-incubated with QDs_BIPs at 50 μg mL^{-1} for 4 hours. Fluorescence microscopy images indicated significant intercellular QDs uptake. Images in Fig. 7 show evidence that the nanoparticles are taken up by the cells. The amount of endocytosed QDs is sufficient to allow diffusion into the cellular cytoplasm. Together with the nuclear counter staining using DAPI, it can be seen that the nanoparticles QDs_MDP are found around the nucleus, while the QDs_EDP label the cytosol. In contrast,

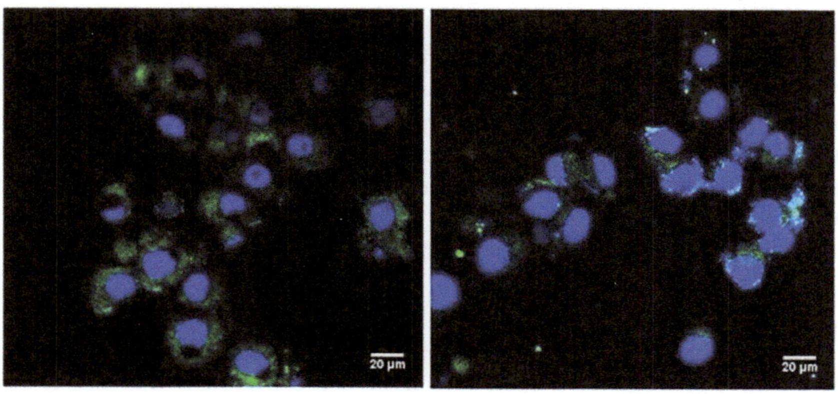

Fig. 7 Confocal microscopy cell images of IGROV-1 cells labelled with DAPI (blue) and QDs_BIPs (green): QDs_EDP (right) and QDs_MDP (left). Imaging cells treated with QDs_IDP showed no fluorescent signal related to the QDs.

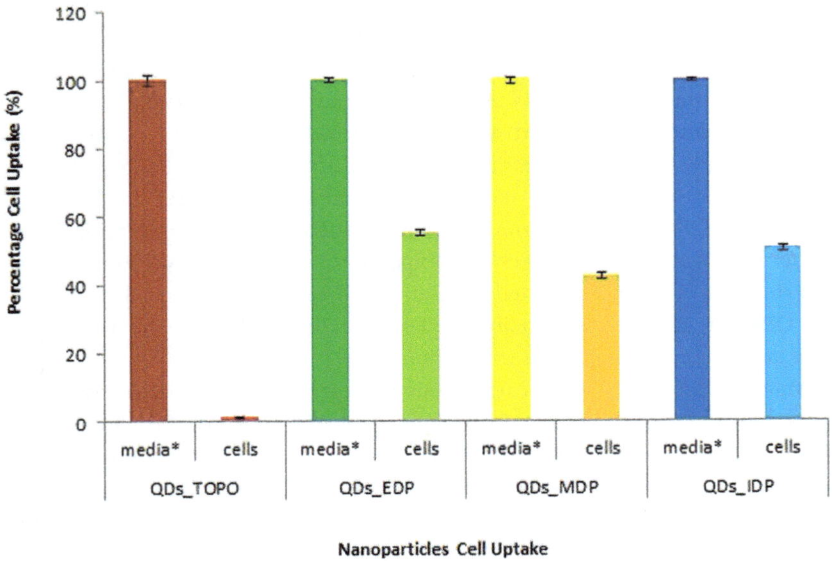

Fig. 8 Intracellular uptake of QDs_BIPs measured through ICP-MS. Intracellular Cd^{2+} cell measurements of QDs_TOPO and QDs_BIPs treated cells (24 hours incubation time; IGROV-1 cell lines). *media labelled in graph refers to media solution containing nanoparticles at the same concentration used for cell uptake.

QDs_IDP showed minimum labelling of the cells, supporting the hypothesis that IDP is a poor capping agent that allows QD degradation in biological environments.

Data of cell uptake by ICP-MS in Fig. 8 indicated that up to 60% of the applied QDs were taken up by the cells during 24 hours incubation studies, whilst there were minimal levels of Cd^{2+} in QDs_TOPO treated cells (less than 5% uptake) due to rapid cell death (cells were lysed and therefore most Cd^{2+} was in the cell lysate). The stabilised or as-prepared QDs_IDP have shown high cellular levels of Cd^{2+}, even with relative colloidal aggregation (that would prevent their cell uptake). It is possible that, out of the three BIPs studied, IDP has weak or limited capping interaction between the BIPs/IDP with the surface of QDs and that this phenomena leads to a shedding off of the QDs layer leading to cell damage. These findings correlated with the lack of imaging in cell labelling experiments, where no labelling or fluorescence signals were observed under confocal microscopy on QDs_IDP treated cells. There is also evidence that suggests that aggregations formed in the QDs_BIPs/IDP system can trigger Cd^{2+} release, due to shedding of the capping layer and generation of Cd^{2+} ions in the biological solution that can contribute to the loss of fluorescence intensity.

4. Conclusions

Hydrophilic BIPs ligands offer an attractive route to produce biocompatible and water soluble QDs. Here we show that BIPs have successfully produced water-soluble QDs that are colloidally stable in biological media even up to 50% serum. These QDs_BIPs are photostable and their fluorescence is pH independent and therefore can be used for cell imaging. The structure of the bisphosphonate appears critical for their colloidal and chemical stability. MDP- and EDP-modified QDs had minimal effect on cell viability and they show adequate cell internalisation and labelling of cells. Capping QDs with bisphosphonate ligands lead to colloidally stable fluorescent particles that can be used in bioimaging without compromising cell viability. In this study we have used the most widely explored QDs (CdSe/ZnS), and we hypothesise that selected non-nitrogenous BIPs can act as biofunctionalisation agents on a wide range of imaging nanoparticles that require colloidal stability for imaging cells and tissues *in vivo*.

Acknowledgements

We acknowledge Dr Gema Vizcay-Barrena and Prof. Alice Warley, Centre for Ultrastructural Imaging, KCL for TEM assistance; Mr Andrew Cakebread and Mr Tobias Krams, Mass Spectrometry Facility, KCL for mass spectrometry analysis (ESI/ICP/MS) also Dr John Harris and Dr Jan Soetaert, Nikon Imaging Centre, KCL for the training and guidance in confocal microscopy imaging. We also gratefully acknowledge EPSRC (Theranostic Nanoparticles) and the Ministry of Science and Technology (MOSTI), Malaysia and the University Science Malaysia (USM) for the support and funding of the research.

References

1 A. P. Alivisatos, W. Gu and C. Larabell, *Annu. Rev. Biomed. Eng.*, 2005, **7**, 55–76.
2 W. C. Chan and S. Nie, *Science*, 1998, **281**, 2016–2018.
3 M. A. Hines and P. Guyot-Sionnest, *J. Phys. Chem.*, 1996, **100**, 468–471.
4 N. Erathodiyil and J. Y. Ying, *Acc. Chem. Res.*, 2011, **44**, 925–935.
5 K. T. Yong, W. C. Law, I. Roy, Z. Ling, H. J. Huang, M. T. Swihart and P. N. Prasad, *J. Biophotonics*, 2011, **4**, 9–20.
6 R. E. Bailey, A. M. Smith and S. M. Nie, *Phys. E.*, 2004, **25**, 1–12.
7 M. Geszke-Moritz and M. Moritz, *Mater. Sci. Eng., C*, 2013, **33**, 1008–1021.
8 E. E. Lees, M. J. Gunzburg, T. L. Nguyen, G. J. Howlett, J. Rothacker, E. C. Nice, A. H. A. Clayton and P. Mulvaney, *Nano Lett.*, 2008, **8**, 2883–2890.
9 D. Gerion, F. Pinaud, S. C. Williams, W. J. Parak, D. Zanchet, S. Weiss and A. P. Alivisatos, *J. Phys. Chem. B*, 2001, **105**, 8861–8871.
10 B. C. Lagerholm, M. M. Wang, L. A. Ernst, D. H. Ly, H. J. Liu, M. P. Bruchez and A. S. Waggoner, *Nano Lett.*, 2004, **4**, 2019–2022.
11 E. R. Goldman, E. D. Balighian, H. Mattoussi, M. K. Kuno, J. M. Mauro, P. T. Tran and G. P. Anderson, *J. Am. Chem. Soc.*, 2002, **124**, 6378–6382.
12 B. Dubertret, P. Skourides, D. J. Norris, V. Noireaux, A. H. Brivanlou and A. Libchaber, *Science*, 2002, **298**, 1759–1762.
13 F. Pinaud, D. King, H. P. Moore and S. Weiss, *J. Am. Chem. Soc.*, 2004, **126**, 6115–6123.
14 A. M. Derfus, W. C. W. Chan and S. N. Bhatia, *Nano Lett.*, 2004, **4**, 11–18.
15 W. W. Yu, E. Chang, R. Drezek and V. L. Colvin, *Biochem. Biophys. Res. Commun.*, 2006, **348**, 781–786.
16 A. Hoshino, K. Fujioka, T. Oku, M. Suga, Y. F. Sasaki, T. Ohta, M. Yasuhara, K. Suzuki and K. Yamamoto, *Nano Lett.*, 2004, **4**, 2163–2169.

17 A. Shiohara, A. Hoshino, K. Hanaki, K. Suzuki and K. Yamamoto, *Microbiol. Immunol.*, 2004, **48**, 669–675.

18 J. Lovric, S. J. Cho, F. M. Winnik and D. Maysinger, *Chem. Biol.*, 2005, **12**, 1227–1234.

19 J. M. Tsay and X. Michalet, *Chem. Biol.*, 2005, **12**, 1159–1161.

20 Y. Zhang and A. Clapp, *Sensors*, 2011, **11**, 11036–11055.

21 M. Xie, H. H. Liu, P. Chen, Z. L. Zhang, X. H. Wang, Z. X. Xie, Y. M. Du, B. Q. Pan and D. W. Pang, *Chem. Commun.*, 2005, 5518–5520.

22 I. L. Medintz, H. T. Uyeda, E. R. Goldman and H. Mattoussi, *Nat. Mater.*, 2005, **4**, 435–446.

23 M. Gnant, *Cancer Treat. Rev.*, 2014, **40**, 476–484.

24 F. H. Ebetino, A. M. L. Hogan, S. T. Sun, M. K. Tsoumpra, X. C. Duan, J. T. Triffitt, A. A. Kwaasi, J. E. Dunford, B. L. Barnett, U. Oppermann, M. W. Lundy, A. Boyde, B. A. Kashemirov, C. E. McKenna and R. G. G. Russell, *Bone*, 2011, **49**, 20–33.

25 M. Grabolle, M. Spieles, V. Lesnyak, N. Gaponik, A. Eychmüller and U. Resch-Genger, *Anal. Chem.*, 2009, **81**, 6285–6294.

26 F. Dubois, B. Mahler, B. Dubertret, E. Doris and C. Mioskowski, *J. Am. Chem. Soc.*, 2007, **129**, 482–483.

27 C. B. Murray, D. J. Norris and M. G. Bawendi, *J. Am. Chem. Soc.*, 1993, **115**, 8706–8715.

28 Z. A. Peng and X. Peng, *J. Am. Chem. Soc.*, 2002, **124**, 3343–3353.

29 B. O. Dabbousi, J. Rodriguez-Viejo, F. V. Mikulec, J. R. Heine, H. Mattoussi, R. Ober, K. F. Jensen and M. G. Bawendi, *J. Phys. Chem. B*, 1997, **101**, 9463–9475.

30 M. J. Murcia, D. L. Shaw, H. Woodruff, C. A. Naumann, B. A. Young and E. C. Long, *Chem. Mater.*, 2006, **18**, 2219–2225.

31 M. Green, P. Rahman and D. Smyth-Boyle, *Chem. Commun.*, 2007, 574–576.

32 V. Biju, *Chem. Soc. Rev.*, 2014, **43**, 744–764.

33 M. Green, *J. Mater. Chem.*, 2010, **20**, 5797–5809.

34 M. Green, D. Smith-Boyle, J. Harries and R. Taylor, *Chem. Commun.*, 2005, 4830–4832.

35 D. V. Talapin, I. Mekis, S. Gotzinger, A. Kornowski, O. Benson and H. Weller, *J. Phys. Chem. B*, 2004, **108**, 18826–18831.

36 A. F. Hezinger, J. Tessmar and A. Gopferich, *Eur. J. Pharm. Biopharm.*, 2008, **68**, 138–152.

37 B. K. Pong, B. L. Trout and J. Y. Lee, *Langmuir*, 2008, **24**, 5270–5276.

38 S. Kim and M. G. Bawendi, *J. Am. Chem. Soc.*, 2003, **125**, 14652–14653.

39 J. Wang, J. Xu, M. D. Goodman, Y. Chen, M. Cai, J. Shinar and Z. Lin, *J. Mater. Chem.*, 2008, **18**, 3270–3274.

40 J. Aldana, N. Lavelle, Y. J. Wang and X. G. Peng, *J. Am. Chem. Soc.*, 2005, **127**, 2496–2504.

41 Q. F. Ma, J. Y. Chen, X. Wu, P. N. Wang, Y. Yue and N. Dai, *J. Lumin.*, 2011, **131**, 2267–2272.

42 M. P. Jain, A. O. Choi, K. D. Neibert and D. Maysinger, *Nanomedicine*, 2009, **4**, 277–290.

43 D. Maysinger, J. Lovric, A. Eisenberg and R. Savic, *Eur. J. Pharm. Biopharm.*, 2007, **65**, 270–281.

44 J. Lovrić, S. J. Cho, F. M. Winnik and D. Maysinger, *Chem. Biol.*, 2005, **12**, 1227–1234.
45 B. I. Ipe, M. Lehnig and C. M. Niemeyer, *Small*, 2005, **1**, 706–709.
46 X. Michalet, F. F. Pinaud, L. A. Bentolila, J. M. Tsay, S. Doose, J. J. Li, G. Sundaresan, A. M. Wu, S. S. Gambhir and S. Weiss, *Science*, 2005, **307**, 538–544.
47 M. Green and E. Howman, *Chem. Commun.*, 2005, 121–123.
48 J. Lovric, H. S. Bazzi, Y. Cuie, G. R. Fortin, F. M. Winnik and D. Maysinger, *J. Mol. Med.*, 2005, **83**, 377–385.
49 F. M. Winnik and D. Maysinger, *Acc. Chem. Res.*, 2013, **46**, 672–680.
50 C. Kirchner, T. Liedl, S. Kudera, T. Pellegrino, A. Munoz Javier, H. E. Gaub, S. Stolzle, N. Fertig and W. J. Parak, *Nano Lett.*, 2005, **5**, 331–338.

Faraday Discussions

PAPER

The properties and performance of a pH-responsive functionalised nanoparticle

Sandra Ast, Peter J. Rutledge and Matthew H. Todd*

Received 14th May 2014, Accepted 22nd July 2014
DOI: 10.1039/c4fd00110a

We report fluorescence measurements of three quantum dots (QDs) of different sizes functionalised with the same pH responsive naphthalimide dye. QD size strongly influences energy transfer between dye and dot. Using QDs with an emission maximum of 570 nm gives rise to an interesting transfer of energy from dye to dot, while QDs with an emission maximum at 670 nm give unexpected enhancement of the dye emission. Titrations of QDs with the dye provide a means to establish the loading and hence an approximation of the surface dye density, which varies in proportion to QD size. Quenching effects are observed beyond the loading limit, and may indicate non-specific interactions between the excess dye and the nanoparticle. Attachment of the dye to the QD core is achieved by a thiol/disulfide exchange process that has been interrogated with Raman spectroscopy. The stability of these QD–dye conjugates over time and across a physiological pH range has been investigated to provide an assessment of their performance and robustness.

Introduction

Fluorescent nanoparticles, and specifically quantum dots (QDs), have found widespread application in biomedical imaging in recent years.[1–4] Much of this work involves tracking the localisation of nanoparticle reporters functionalised with a ligand that binds some cognate biological molecule, and much of the interest is focused on *intra*cellular measurements. We have become interested in two less well-explored areas of sensing that we feel could be addressed with appropriately built fluorescent nanoparticles: (1) monitoring analyte concentrations and dynamics in the *extra*cellular (*inter*cellular) space; and (2) developing nanoparticles that are responsive to analytes, in that their fluorescent output changes with the (varying) local concentration of an analyte – responsive QDs (RQDs).

Our initial approach to sensing using nanoparticles targets pH. *Inter*cellular pH regulates a wide range of poorly understood processes including cell adhesion,[5] motility,[6] myelin repair,[7] and cancer progression.[8] We felt that proof of

School of Chemistry, The University of Sydney, NSW 2006, Australia. E-mail: matthew.todd@sydney.edu.au; Fax: +61 2 9351 3329; Tel: +61 2 9351 2180

concept for a reversible sensing scaffold would be more easily attained with a pH-sensitive system than a prototype based on the metal-responsive probes we have developed recently.[9–11] While there is some precedence for pH-responsive nanoparticles[12–16] there is considerable scope for the development of new responsive probes that can be simply prepared.

To this end we recently reported an easily-constructed functionalised nanoparticle that responds to pH but which also exhibits an enhanced QD photoluminescence (PL) (Fig. 1).[17] This system was designed so that the dye emission overlaps with the absorption band of the nanoparticle, leading to energy transfer and an enhanced PL signal of the QD (Fig. 1B). The conjugate thus exhibits energy transfer from *dye-to-dot*, in contrast to the majority of the quantum dot literature in which energy transfer is designed and observed to occur to the dye from the excited dot. Dye-to-dot transfer was sought, but not conclusively found, by Mattoussi in 2005,[18] who proposed that in their system energy transfer in this direction was precluded by competition between the dye's fast radiative decay pathways and the slower non-radiative Förster Resonance Energy Transfer (FRET) pathways, as well as strong, unavoidable direct excitation of the nanoparticle itself. However, three more recent reports have described dye-to-dot transfer in QDs directly functionalised with simple organic dyes[19,20] or a photosynthetic light harvesting complex.[21] None of these systems respond to analytes. Our conjugate consists of a naphthalimide dye attached to a QD (*via* a passive disulfide exchange reaction[22]) to give a system in which the resulting emission is sensitive to the pH. The conjugate exhibits dye-to-dot energy transfer that is pH-dependent, constituting the first *responsive* dye-to-dot system. The fluorophore is (2-(dimethyl)ethyl) amino-naphthalimide, a pH-sensitive probe that is switched off at basic pH due to a photoinduced electron transfer (PET) from the non-protonated dimethylamino-group.[23] When this kind of dye is attached to the nanoparticle, pH-induced changes to the properties of the dye are communicated to the QD, modulating the intensity of the QD emission.

The signal changes observed as a function of pH are significant but not as large as had been anticipated. Further investigation is required to understand the mechanism behind this dye-to-dot fluorescence enhancement. Several aspects of this system in particular require further elucidation. First, we wished to confirm

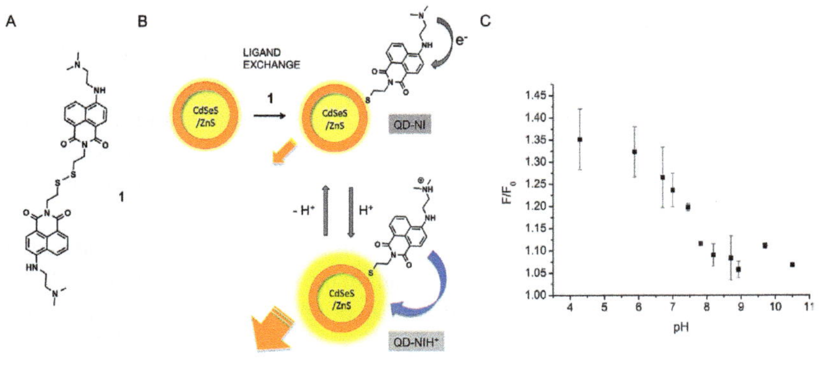

Fig. 1 Schematic representation of the previously reported pH-responsive QD–dye conjugate using disulfide 1 and water-soluble, core/shell type CdSeS/ZnS QDs.[17]

that the spectroscopic performance of the system can indeed be attributed to the spectroscopic overlap between dye and dot as envisaged in the original conjugate design. Second, it was important to ascertain whether the dye interacts directly with the surface of the core of the particle (and not with the outer polymer coating added to the QD to aid aqueous solubility) and if so to what extent, *i.e.*, how many molecules of dye could be installed on the particles. Third it was important to assess the performance and stability of the conjugate with respect to time and changes in pH.

Results and discussion

1 The dot–dye energy transfer changes dramatically with particle size

In our previous studies we found that there was an enhancement of the dye emission as well as the QD PL signal, that might arise from energy transfer of different origin going in both directions, *i.e.*, dye-to-dot as well as dot-to-dye energy transfer. To understand these effects further, QDs of different sizes (QD-460 and QD-670) were employed to complement the originally-used QDs (QD-570) (*vide infra* for further discussion of the particle sizes).

The three types of dot were exposed to increasing concentrations of disulfide **1** under the same conditions (HEPES buffer, pH 7.4, room temperature). The resulting PL spectra for QD-570 (Fig. 2a) show the enhancement of the QD fluorescence at 570 nm, and the growth of the 'shoulder' next to the QD arising from the fluorophore emission, which, by virtue of the conjugation to the QD, is blue-shifted to 516 nm with respect to the emission of **1** alone (the lower emission

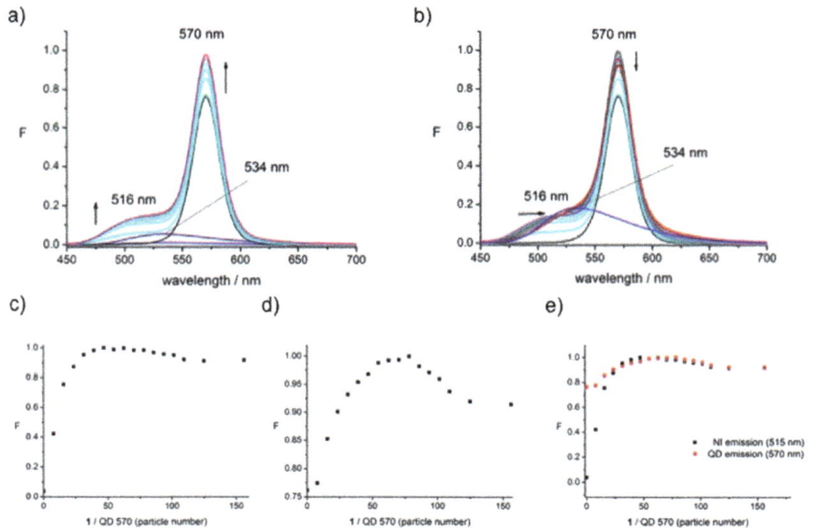

Fig. 2 Normalised PL emission spectra and plots of emission signal changes as a function of particle number (*i.e.*, ratio of **1** : QD-570) in HEPES buffer, excitation at 435 nm. (a) Addition of **1** up to saturation of fluorescence emission increase (up to 1 μM), (b) addition of **1** beyond emission signal saturation (up to 4 μM) and **1** alone at the same concentration (4 μM), (c) plot of fluorophore emission intensity at 516 nm, (d) plot of QD emission intensity at 570 nm, and (e) overlay of plots of emission intensities at 516 and 570 nm.

maximum centred at 534 nm). Clearly this blue-shift is accompanied by a substantial increase in PL intensity of the fluorophore emission. A maximum is reached for the QD fluorescence output after the addition of a given amount of **1**, beyond which the intensity decreases (Fig. 2b). The spectra are notable in that they show enhancement of both signals.

Plotting the intensity at the emission maximum as a function of the ratio of numbers of dye molecules over number of QDs ("Particle Number", Fig. 2c – calculated as described below) shows that the PL signal approaches an asymptote around 50, implying further conjugation (or even interaction) of dye and dot is not taking place above a certain concentration of the added dye. This is a strong indicator that the surface of the dot is covered with the maximum amount of dye; each particle thus accommodates *ca.* 100 fluorophore molecules (since the assumption is made that each molecule of the disulfide **1** delivers two molecules of the fluorophore). The absence of further QD–dye interactions is indicated by the increasing red-shift as the titration of **1** is continued (Fig. 2b) (up to 4 μM). However, the signal intensity arising from the fluorophore remained essentially unchanged in this subsequent titration, *i.e.*, a small increase in dye emission was expected (because more dye is being added) but this was not observed, suggesting some non-specific interaction between the dye and some other component (likely the polymer) of the nanoparticle. This unexpected phenomenon becomes more apparent when comparing the final spectrum (in Fig. 2b) with the emission of **1** alone taken at the concentration that would be expected in the absence of the QD (4 μM, blue line) where the intensities are seen as being approximately equivalent indicating there is no energy transfer from the dot to the dye beyond the surface saturation point.

The plot of the QD's PL signal intensity as a function of particle number shows the same behaviour as that of the dye emission (Fig. 2e) with the only difference being the weaker changes in signal intensity, providing supporting evidence that we are looking at a saturation of the dye on the surface of the dot. We conclude that the communication, presumably FRET, operates from dye to dot. There is no donor quenching in this system, precluding a simple calculation of FRET efficiency. The origin of the fluorescence enhancement of the fluorophore emission in the presence of the dot was unclear. To examine this further, QD-460 and QD-670 were employed in equivalent experiments. The spectra obtained when **1** was titrated into solutions of QD-460 (Fig. 3) and QD-670 (Fig. 4) show very different behaviour.

In the case of QD-460, with an emission maximum at 459 nm, the combination with **1** (absorption maximum at 435 nm) implies that the nanoparticle is the energy donor, and the dye the acceptor, when the sample of the conjugate is excited at 400 nm. Thus adding **1** to QD-460 results in an immediate decrease in QD PL intensity (unlike the case for QD-570) accompanied by a large enhancement of fluorophore emission (Fig. 3a). The blue-shifted dye emission is centred at 515 nm (pink line), as seen for QD-570.

Interestingly, further addition of dye **1** to the QD–**1** conjugate led to a decrease in fluorophore emission intensity and a red-shifted emission signal (Fig. 3b, grey lines). This emission signal experienced a second saturation at a number of particles per dot roughly twice that seen for the first saturation, with the emission centred at 534 nm (the emission maximum of the dye **1** itself), followed by a subsequent increase in emission intensity. The signal increase was accompanied

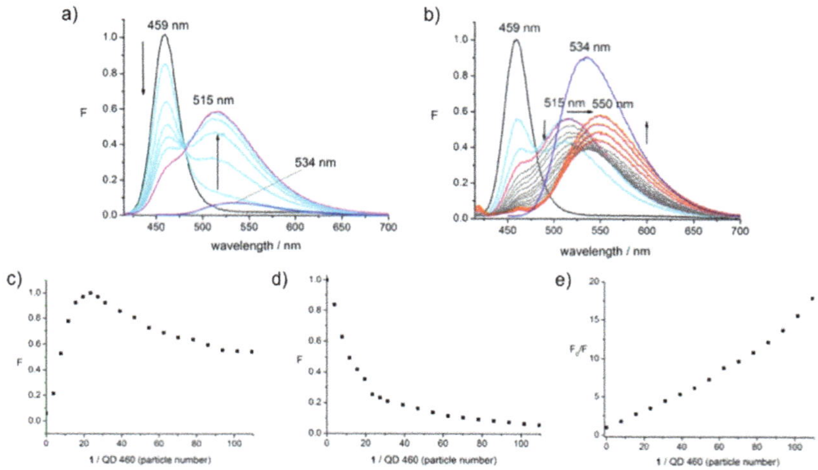

Fig. 3 Normalised PL emission spectra and plots of emission signal changes as a function of particle number (*i.e.*, ratio of **1** : QD-460), in HEPES buffer (pH 7.4), excitation at 400 nm. (a) Addition of **1** up to saturation of fluorophore emission increase of (up to 0.5 μM) and **1** alone at this same concentration (0.5 μM, blue line), (b) addition of **1** beyond emission signal saturation (up to 4 μM) and **1** alone at same final concentration (4 μM), (c) plot of fluorophore emission intensity at 515 nm, (d) plot of QD emission intensity decrease at 459 nm and (e) Stern–Volmer plot (F/F_0) of fluorescence emission intensity at 459 nm.

by a further red-shift to 550 nm, suggesting that the free dye is experiencing some interaction with the QD or its surroundings, again potentially with the polymer outer layer. The total increase in dye emission intensity (up to the maximum seen at 515 nm during the initial addition of **1**, Fig. 3a) is of about 6 orders of magnitude, as judged by the final intensity compared with the intensity of **1** alone at the same concentration. The quenching of QD emission during addition of **1** is high, with the signal disappearing almost completely after further dye addition (Fig. 3b); the further decrease in intensity at this wavelength for this conjugate was observed to be a partly time-dependent phenomenon (*vide infra*, Fig. 9a).

From the decrease in QD donor PL intensity, the FRET efficiency (E) can be determined using eqn (1).

$$E = 1 - \frac{F_{\mathrm{D}}^{\mathrm{q}}}{F_{\mathrm{D}}} \tag{1}$$

The FRET efficiency is thus defined as the ratio of the fluorescence intensity of the donor in the presence of the quencher ($F_{\mathrm{D}}^{\mathrm{q}}$) over the fluorescence intensity of the donor in the absence of the quencher (F_{D}). Here the FRET efficiency is 80% at maximal QD coverage and approaches 96% in the course of further dye addition. Plotting the fluorescence intensity at 515 nm as a function of the number of dye molecules covering the QD, shows that PL signal increase of the dye emission signal is maximal around a dye-to-dot ratio of 25 (Fig. 3c). The same number is obtained by plotting the decrease in QD PL intensity at 459 nm (Fig. 3d) at 80% FRET efficiency.

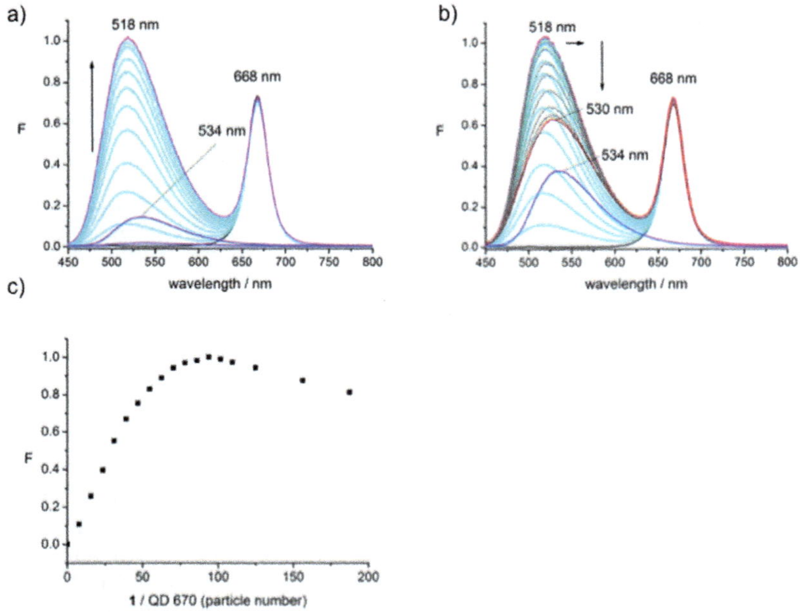

Fig. 4 Normalised PL emission spectra and plots of emission signal changes as a function of particle number of **1** per single QD-670 in HEPES-buffer (pH 7.4), excitation at 435 nm. (a) Addition of **1** up to saturation of fluorophore emission increase and **1** alone at the same concentration (1.8–2.2 μM), (b) addition of **1** beyond emission signal saturation (up to 6.6 μM) and **1** alone at the same final concentration (6.6 μM) and (c) plot of fluorophore emission intensity at 518 nm.

From the Stern–Volmer plot[24] (Fig. 3e), where the rate of the fluorescence intensity quenching (F/F_0) is plotted as a function of the quencher concentration (as particle number) it is obvious that a linear relationship only holds at low dye density, *i.e.*, in the concentration range up to the fully covered QD where static quenching is possible. Addition of further dye continues the linear trend up to a particle number of around 50 (corresponding to the second saturation limit) with subsequent addition leading to an upward curving plot indicative of combined static and dynamic quenching processes; one would expect dynamic quenching with dye in solution. The linear range between particle number 25 and 50 is interesting in that this may be supporting evidence for the association of the dye with the polymer coating, as opposed to direct interaction with the dot or free dye in solution.

Also shown (Fig. 3b) is the emission intensity of the dye alone (*i.e.*, the disulfide **1** at a concentration of 4 μM) at the concentration equal to the final concentration reached following addition of **1** into the QD solution, which is observed to be significantly higher (blue line) than in the presence of the QD. A quenching effect appears to operate on the dye even after the surface of the QD is fully covered with dye molecules. However, the increase in dye emission after the point of saturation indicates that these processes are starting to be suppressed and radiative processes of the fluorophore start to dominate. We were not able to observe this for the QD-570 system, but would expect similar behaviour to be operating.

The PL spectra of QD-670 in the presence of disulfide dye **1** (Fig. 4) were acquired. The absorption of the first exciton band is centred around 660 nm and the emission maximum of the QD is centred at 668 nm. Addition of **1** to solutions of QD-670 resulted in a large increase of the fluorophore emission band and a blue shifted emission signal to 518 nm as before, but the QD emission remained unchanged. These results are clearly surprising, since the emission of the QD is far outside the fluorophores' absorption, which stretches out to only 570 nm and thus should not be able to function as an energy donor. It was expected that a QD of this size would have limited capacity for FRET due to FRET's strong distance dependence[25] and these QDs were not chosen for our previous studies for precisely this reason. However, the blue-shifted emission suggests very similar interactions to the ones found for the smaller QDs, *i.e.*, energy transfer from the QD donor. However, such a conclusion would need to be verified with, for example, fluorescence lifetime measurements.

Plotting the emission intensity as a function of dye-to-dot density shows that maximal fluorescence enhancement of the dye is achieved at a ratio in the range 80–100. Further addition of **1** led to a decrease in intensity of the dye signal (Fig. 4b) which reached an asymptote (red line) around twice the number of molecules of **1** (200) (Fig. 4c) in a similar way to the QD-460 system. Comparison with the signal intensity of **1** alone (blue line) at 518 nm at this final concentration shows that the final intensity of the dye emission in the conjugate is higher than the dye signal alone; thus the signal of the fluorophore is enhanced in the presence of the QD and remains enhanced in the presence of an excess of the dye. As before it is possible that the first 100 molecules of added **1** interact with the inorganic core of the QD, with the remaining dye interacting with the outer polymer matrix in a non-specific manner.

The Förster radius (R_0) represents the distance for 50% of energy transfer and normally falls within the core/shell radius for larger dots due to the 6th power relationship in eqn (2).[25,26] However the large number of dye molecules attached to these larger dots appears to change this situation significantly, since the FRET efficiency (E) is dependent on the number of dye molecules (n) as given by eqn (2).[27]

$$E = \frac{nR_0^{\,6}}{nR_0^{\,6} + r^6} \tag{2}$$

Thus it would appear that the more populous "antenna" of fluorophores surrounding the dot is able to increase the FRET efficiency to allow for energy transfer despite the sub-optimal donor–acceptor distance and spectroscopic overlap.

A question remained: why did the PL of the QD remain unchanged through the addition of the dye? That the dot-to-dye transfer is not accompanied by a quenching in QD PL implies that fluorescence enhancement is competing (perfectly) with fluorescence quenching, leaving the QD signal unchanged – a net zero change of energy transfer at the particle though this would need to be confirmed by independent lifetime measurements of the QD.

From these studies of all three dot–dye conjugates we conclude that attachment of dye **1** to QD-570 (the medium sized particle previously reported) leads to a system involving energy transfer from dye-to-dot. Conjugates based on smaller

QDs give instead the expected dot-to-dye FRET because the reverse process is energetically disfavoured. Large QD conjugates only give clear FRET from dot-to-dye but the dye-to-dot transfer is assumed from the lack of any observed QD quenching when the dot transfers energy to the dye, and this may be a feature of FRET efficiency being enhanced by the larger number of dye molecules attached to the core; there remains the possibility that electronic or even molecular mechanisms, rather than FRET, could explain these observations.

2 Raman spectroscopy can probe the nature of dye–dot linkage

The QDs employed in these studies are core shell QDs covered with a capping layer, surrounded by a layer of amphiphilic polymer, functionalised with carboxylic acids. The assumption was made in our previous work that the conjugate is assembled *via* sulfide exchange between **1** and the "inner" inorganic surface of the dot, requiring the disulfide to diffuse into the particle (through the polymer, without reacting with it) for such exchange to occur. We had previously performed an experiment in which a model compound **2**, lacking disulfide functionality, was added to solutions containing QD-570 (Fig. 5a).[17] No apparent modification of the QD PL, and certainly no enhancement of the QD fluorescence output, was observed, strongly suggesting that covalent attachment of the dye was necessary for the optical effects we had observed in the conjugate. Additionally, attempts to isolate a potential QD–**2** conjugate failed, while **1** formed such an isolable conjugate (Fig. 5b).

It remained possible that the disulfide was engaging in an exchange reaction with the outer polymer layer of the particle, even though such a reaction would not *prima facie* lead to any PL effects based on FRET due to the inevitably larger distances between donor and acceptor. Nevertheless we had no direct evidence of the formation of new bonds on the particle. We therefore measured Raman spectra of the QD, dye and conjugate to attempt to observe changes in the relevant

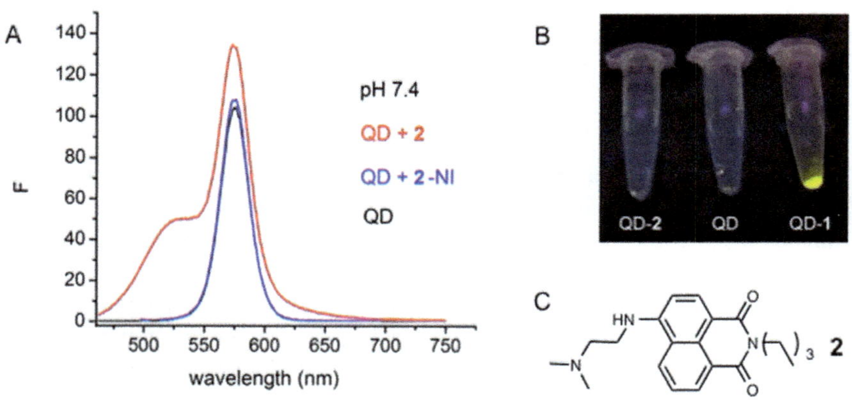

Fig. 5 Emission spectra of QD-570 showing attempts to isolate a QD-conjugate with the model compound 2. (a) Emission spectra of QD-570 (black line), QD + 2 (red line) and deconvoluted PL spectrum of QD after substraction of fluorophore component (blue line) in HEPES-buffer, (b) pictures of attempted isolation of the conjugates with 2 and 1 and (c) structure of 2. This figure is adapted from our previous work and is reproduced here for clarity.[17]

regions of the spectra. The disulfide S–S stretching vibration occurs in the range of 500–530 cm^{-1}, while the C–S stretching is typically found between 715 and 579 cm^{-1}.[28]

The individual Raman spectra (overlay, Fig. 6) were acquired from solid samples, using a *MultiRAM* FT-Raman-spectrometer irradiating at 1064 nm. The QD–dye conjugate was prepared as described previously,[17] by precipitation and washing of the pellet and subsequent drying under a stream of nitrogen. The spectrum of the dye **1** (black line) shows distinct peaks indicative of the disulfide bond as well as the aromatic core of the amino-naphthalimide. The spectrum of the QD itself (red) shows less resolved peaks and vibrations originating from the nanoparticle and the encapsulating polymer. The QD–**1** conjugate represents the linear combination of the two spectra, except for the disulfide region, which lacks the disulfide vibration peaks at 500 and 660 cm^{-1}. Though not conclusive proof of the attachment of the dye *via* sulfide exchange on the inner surface of the particle, these results do suggest the expected covalent attachment to the QD in the isolated conjugate.

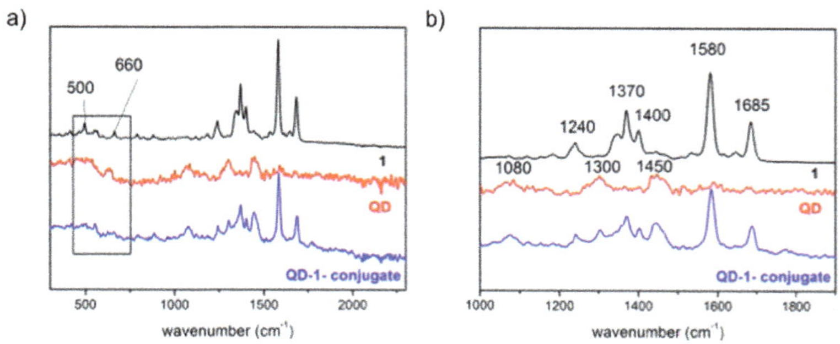

Fig. 6 Overlay of the Raman spectra of the dye **1** alone (black), QD-570 (red) and the isolated conjugate QD-570–**1** (blue). (a) The whole range from 2300 to 300 cm^{-1} (representative of the disulfide and aromatic vibrations) and (b) the region from 1900 to 1000 cm^{-1} (representative of the aromatic region).

3 Titration experiments permit calculation of the "loading" of the dye on the particle

The concentration and dimensions of the QDs used in the present experiments are of interest, for the purpose of understanding the loading calculations described above, *i.e.*, to answer the question: how many dye molecules are installed on each particle? The concentration of each QD solution, as provided by the supplier, Ocean NanoTech, was equal to 8 μM. The default volume of QD solution used in each experiment (8 μL, mixed with 3 mL HEPES buffer) implies approximately 3.9×10^{13} particles per experiment, though without weighing the sample this does not permit an estimation of the molecular weight of each particle (and we are unaware of any reports of mass spectrometry applied to quantum dot solutions that might give empirical measures of molecular weights, though this is not an unreasonable experiment). These aqueous-compatible QDs are covered with a layer of polymer that significantly enhances their hydrodynamic radii and which presents a presumably porous barrier between the bulk medium and the covalently functionalised surface of the core particle. The supplier-provided hydrodynamic radii (Table 1) are clearly far larger than the radii normally associated with QDs without the polymer layer encapsulating the particle (1–5 nm), though the hydrodynamic radius may be the value for a polymeric particle encapsulating more than one QD core.[29]

The experiments described above (Fig. 2–4) combined with these data provide a means to establish an approximate surface density or loading of the dye on the surface of the conjugate. For each of the three dots a maximum signal is reached at a certain concentration, where this maximum is taken as evidence that the surface of the dot has accepted as many copies of the dye as it is able – though it is unclear whether this limit arises from (i) addition to "vacant" sites on the particle surface, (ii) exchange with a limited number of labile ligands on the surface or (iii) a genuine limit that arises from, *e.g.*, coulombic repulsion between the polar/charged dye molecules installed. From the concentration at the maximal signal change in the sample volume the molarity was calculated, which then allowed for the determination of the number of dye particles.

Assuming perfect correspondence between the empirical limit of fluorescence increase and covalent attachment to a surface (*i.e.*, rather than an equilibrium process of disulfide exchange), the number of dye molecules per dot was calculated: approximately 40–50 dye molecules for QD-460, 80–100 for QD-570 and 170–200 for QD-670. These numbers match well with the comparative surface area of the dots derived from the hydrodynamic radii (Table 1), *i.e.*, with QD-570 having double the surface area, it is able to accommodate twice as many dye molecules. The hydrodynamic radius is, however, likely to represent the overall particle size that includes the polymeric outer layer, rather than the radius of the core QD, which should be much lower. Literature values for estimates of the number of added ligands assembling on the surface of QDs are around the same order of magnitude as those shown above;[27,30] these values, derived from QDs without an outer polymeric layer, suggest that the exchange processes occurring in the present work are indeed those taking place on this inner QD surface, rather than the far larger outer polymeric structure, not least because covalent addition to the outer polymeric structure would be expected to give a value for number of dye molecules per particle that is much higher than that calculated here.

Table 1 Physical parameters and loading of the QDs with disulfide dye **1**

	QD-460	QD-570	QD-670
Hydro. rad./nm	4.8	6.85	8.75
Surface area/nm^2	290	590	962
Moles in exp.^{a,b}/mol	$1.25–1.5 \times 10^{-9}$	$2.5–3.0 \times 10^{-9}$	$5.5–6.5 \times 10^{-9}$
Number of part. in exp.^{c}	$7.5–9.0 \times 10^{14}$	$1.5–1.8 \times 10^{15}$	$3.3–3.9 \times 10^{15}$
Ratio of dye **1** : QD^{d}	19.5–23.4	40–46.9	85.9–101.5
No. dye per part.^{e}	40–47	80–94	172–203

^{a} 1 mM stock solution in DMSO. ^{b} **1** (0.5 μL) was added to 8 μL QD stepwise into 3 mL aqueous buffer. ^{c} Obtained with Avogadro number. ^{d} Particle number of dye over particle number of QD. ^{e} Size of the dye over surface area.

If we take an estimated radius for the inner core (CdSe/ZnS) of the QD-570 particle as 2.5 nm,[31] we would expect the 100 fluorophore molecules to be spread over an area of 80 nm^2. If we (crudely) estimate the cross-sectional area of the dye molecule to be 0.9 nm long by 0.5 nm wide (the approximate dimensions of the naphthalimide dye), each fluorophore would occupy a static cross-sectional area of about 0.5 nm^2 meaning this value of loading seems reasonable (coverage of 50 nm^2 of the available 80 nm^2). If one assumes that each molecule is moving and is solvated, this may imply that the surface of the core has become completely covered with the added dye, which may explain the fairly sharp saturation limit observed above upon addition of the disulfide.

4 The conjugate is moderately robust with respect to time and pH change

In our previous work we were able to show that the QD emission signal varies with pH when the dye is attached. These signal changes were measured as the relative fluorescence signal increase with respect to the QD alone and gave a sigmoidal relationship when plotted as a function of pH. The present studies aimed to address several features of the conjugate performance.

a) pH sweeping. Fluorescence output was measured in two separate experiments, screening the response of QD-570 and its conjugate with **1** to pH changes starting at neutral pH (7.4) but moving to either acidic or alkaline values through the addition of HCl (0.5 M) or NaOH (0.5 M). For the conjugate, the QD : dye ratio was employed that had been found (above) to provide a fully covered nanoparticle.

The measurements for the acidic pH region (Fig. 7) show that the intensity of the QD signal itself is significantly affected by the presence of protons reducing in intensity towards lower pH values (Fig. 7a, black to grey). The emission signals of the conjugate also decrease as the pH decreases. At the lowest pH value assessed, 4.5 (Fig. 7a, blue line), the conjugate's signal intensity has dropped below the intensity of the initial QD emission (Fig. 7a, black line). These effects could be attributed to protonation of the polymer surrounding the dot. Plotting the fluorescence intensity at the signal maximum *vs.* pH (Fig. 7b) strongly indicates that the conjugate experiences the same effects as the QD. This becomes even more

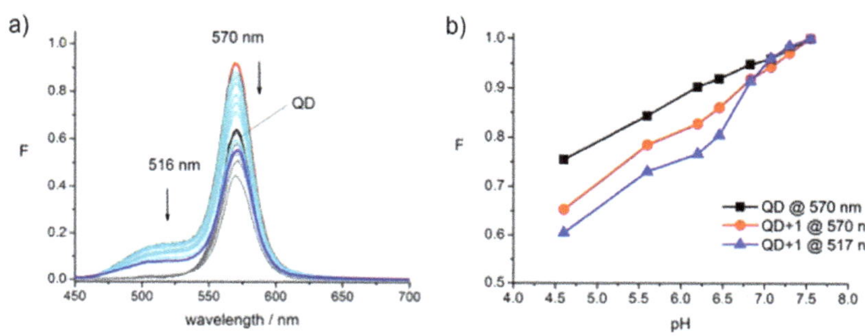

Fig. 7 pH-sensitivity of QD-570 and its conjugate with dye **1** in the pH range 4.6 to 7.4. (a) Normalised PL spectra of QD 570 (black and grey lines) and of QD-570–1 conjugate (red, then cyan lines) and (b) plot of normalised intensity at 570 nm *vs.* pH, sweeping from 7.5 to 4.6.

apparent when looking at the fluorophore emission at 517 nm in the conjugate (Fig. 7b, blue), which follows the same trend as the QD (Fig. 7b, black).

This situation changes completely when sweeping from neutral to alkaline pH (Fig. 8). In this range the signal from the QD remains perfectly stable while the emission intensity signals (at 516 and 570 nm) of the conjugate decrease towards higher pH values (Fig. 8b). The decrease is expected: it reflects the response of model fluorophore **2** during pH-measurements[23] and has been shown by us before for the QD-570–**1** conjugate (Fig. 1C).[17] Thus, the QD signal changes measured in the alkaline region can be attributed to changes at the dye. Attempts to reverse the pH-sweep were less successful and also appeared to be dominated by a change in QD signal intensity (Fig. 8c and d). Thus, in the reversed sweep, the QD signal of the conjugate (Fig. 8c, pink) shows a trend similar to the signal of the QD alone (Fig. 8c, blue). The signal of the fluorophore at 516 nm is affected in the reverse sweep as well (Fig. 8d, red) and does not retain its original signal intensity (Fig. 8d, black).

The fact that the fluorescence signal of the conjugate is reduced at acidic pH would appear to preclude its application as a switch-on probe in that region. However, the pK_a of dye **2**, reported to be 7.8,[23] narrows the window of practical application down to a range between pH 7 and 10, the region where the QD-570 was found to be responsive and stable in the present experiments.

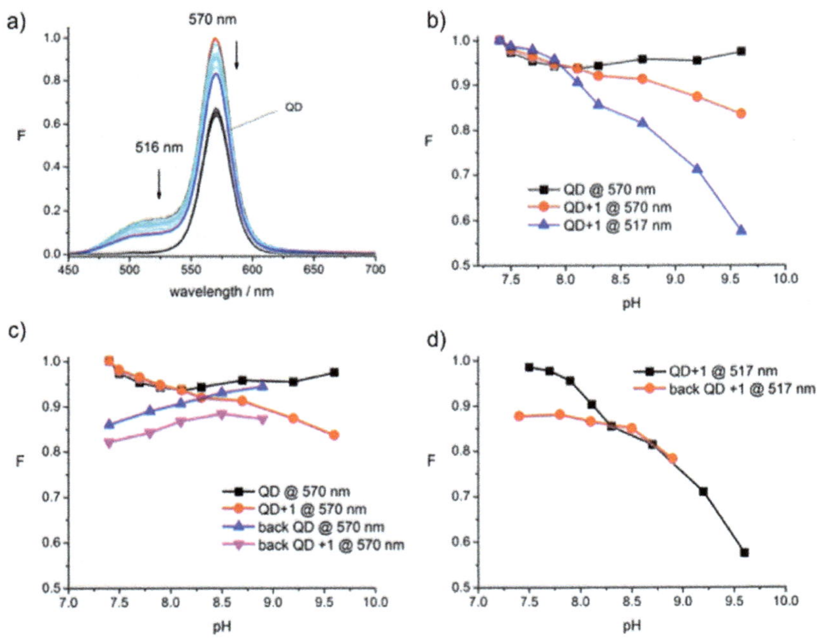

Fig. 8 pH-sensitivity of QD-570 and its conjugate with dye **1** in the pH range 7.5–9.6. (a) Normalised PL spectra of QD-570 (black and grey lines) and of QD-570–**1** conjugate (red, then cyan lines) and, (b) plot of normalised intensity at 570 nm *vs.* pH, sweeping from 7.5–9.6, (c) plot of normalised intensity at 570 nm *vs.* pH, back-sweep from 9.6 to 7.4 (blue and pink) and (d) plot of normalised intensity at 515 nm *vs.* pH, back-sweep (red) from 9.6 to 7.4.

b) Time. The fluorescence output of the QD-570–**1** conjugate has been found to be reasonably stable over the hour-long periods involved in previous experiments (data not shown). However, small changes in output intensity have been observed immediately following the combination of QD and dye. In the case of the smaller QD-460 particles very rapid formation of the conjugate upon addition of the disulfide is observed, with small subsequent changes (decrease in dot emission, increase in dye emission) as time passes (Fig. 9a) that could be attributed in part to the disulfide exchange reaction reaching a position of equilibrium in the minutes following combination of dot and dye, but which may also be the result of an inherent decrease in fluorescence of these dots over time that may be observed in a sample only of the QD (Fig. 9b). The QD-570–**1** conjugate exhibits smaller changes immediately following combination, with a slight decrease in the QD emission in the 16 minutes following conjugate synthesis that may be attributed to chemical exchange processes completing on the QD surface, since the QD on its own emits a very stable intensity over the same timescale (Fig. 9d).

There is an obvious explanation as to why chemical exchange processes may take a few minutes to complete in these systems: diffusion of the dye through the polymeric outer layer. Indeed it should be remembered that the aqueous solubility of the particle arises from this charged outer layer, while the inner particle itself remains fairly hydrophobic. Though this was not addressed as part of the current study, measurements of pH are made on the assumption that the solvent environment between the QD core and the outer polymeric layer reflects that of the bulk medium, but this has yet to be established in the present case.

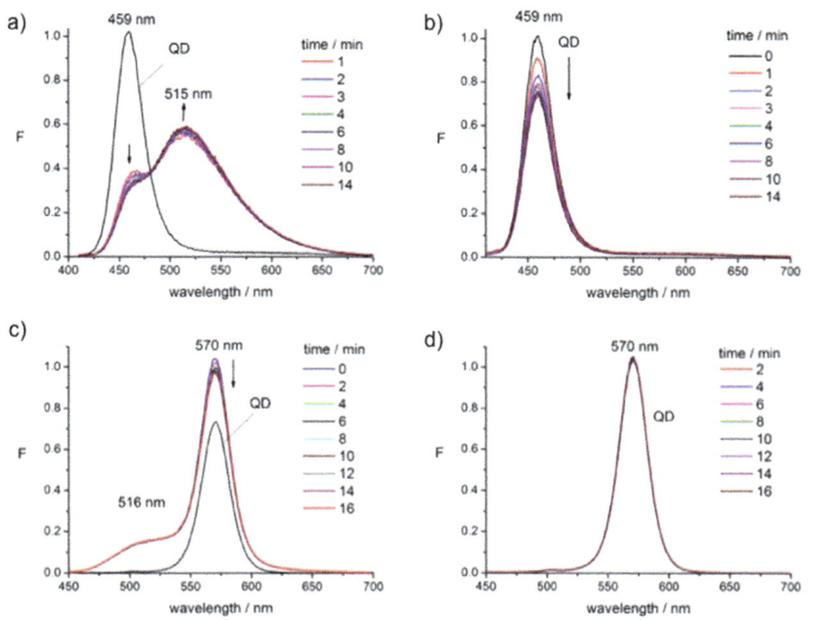

Fig. 9 Normalised PL spectra of QD-460 and QD-570 and their conjugates with **1** over time (14–16 min). (a) QD-460 and QD-460–**1** conjugate, (b) QD-460 alone, (c) QD-570 and QD-570–**1** conjugate and (d) QD-570 alone.

c) Absorption changes with pH. In our previous studies we were able to show that the QD signal intensity changes as a function of the excitation wavelength (around the maximum for the dye), an effect that led us to conclude that an energy transfer-type mechanism was operating since such a mechanism would be strongly dependent on the number of excited dye-molecules. We showed that the highest fluorescence intensity was realised at the wavelengths around the centre of the maximal absorption (440 nm), providing an emission intensity from the dot 25–30% higher than those obtained when exciting at 400 nm.

To confirm this behaviour, and examine the mechanism of the energy transfer, absorption measurements were undertaken of the QD-570–**1** conjugate at varying pH, but these results became unclear at the higher concentrations required for such measurements, possibly arising from issues of solubility. Thus for the absorption measurements the model dye **2** was employed. A solution of **2** (3.33×10^{-5} M) was prepared in buffer and the pH was changed by addition of small amounts of HCl (0.5 M) or NaOH (0.1 M). The resulting UV/vis absorption spectra (Fig. 10a) show that the lowest energy transition around 440 nm changes significantly with pH, leading to both intensity changes and a wavelength shift. Notably, no isosbestic point is found. When absorption at a particular wavelength is plotted as a function of pH (Fig. 9b and c) it is clear that the absorption maxima change significantly depending on the excitation wavelength and that especially large differences are found when moving from neutral to alkaline pH. In the fluorescence studies described above excitation at 435 nm was chosen since the absorption at this wavelength varies least with pH.

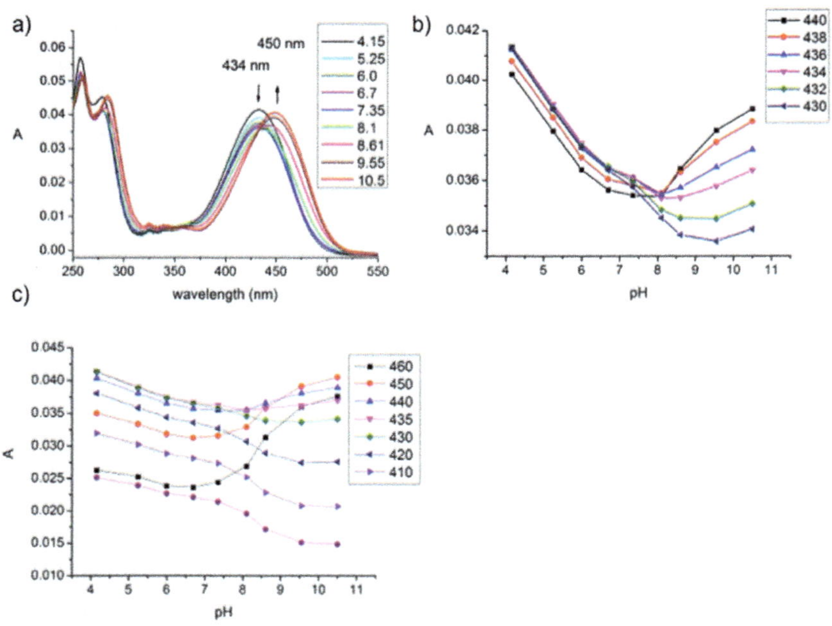

Fig. 10 UV/vis absorption measurements of model dye 2 in solutions of different pH and plots of the absorbance *vs.* pH at different absorption maxima. (a) Absorption spectra in the pH-range 4.15–10.5, (b) plot of the absorbance at wavelengths between 430 and 440 nm and (c) plot of the absorbance at wavelengths between 410 and 460 nm.

The absorption of the model dye therefore clearly changes with pH, and if extrapolated to the conjugate, this could influence the eventual communication between dye and dot and therefore be a possible explanation for the pH responsiveness of the conjugate (as opposed to a feature of the dye's fluorescence changes). However, the proportional absorption change of *ca.* 10–15% (in Fig. 9b) would seem unable to account for the factor of *ca.* 30% change observed in the original QD *vs.* QD–**1** conjugate emission. Thus there is clearly communication between dye and dot that is responsive to pH, and a component of this could arise from absorption changes in the dye employed as the pH is altered.

Conclusions

The photophysical measurements described here for three different QDs functionalised with the same naphthalimide dye demonstrate how the nature of energy transfer within such conjugates depends on nanoparticle size. The QD-570 conjugate exhibits clear energy transfer between the dye and dot, an elusive and hitherto rarely reported phenomenon. Titration experiments between QDs and the dye demonstrate a saturation limit with a maximum number of dye molecules accommodated per particle, and dye densities that suggest complete surface coverage. The greater the number of dye molecules attached to the larger dots may be promoting FRET that would not otherwise be expected.

The water-soluble QDs used in this study have hydrodynamic radii that are far larger than the core particle size owing to a covering of a solubilising polymer. This coating is assumed to be porous to the dye, and Raman spectra offer some evidence for the loading of the naphthalimide dye directly onto the core surface through disulfide exchange. Temporal changes following addition of the dye suggest that this reaction, though fast, can take several minutes to complete in some cases, consistent with the dye journeying through the polymer coating to reach the dot surface. When greater amounts of the dye are added, at levels above the saturation limit of the core, there appear to be non-specific interactions that lead to quenching of the dye signal.

The QD–dye conjugates (like the QDs themselves) are unstable at low pH, rendering these probes unsuitable for measurements at acidic pH. The conjugates perform well at physiological pH, and in the alkaline region around the pK_a of the naphthalimide dye, although the fluorescence output is not perfectly reversible as pH is swept. Towards the goal of biomedical application, it will be of interest to assess the performance of these conjugates in more realistic (*i.e.*, serum-based) media as a precursor to cell-based work, though clearly for these functionalised nanoparticles to be applied in such environments quantitatively, ratiometric probes would be needed. However, of greater interest perhaps is to further characterise the nature of the fluorescence enhancement processes, using intermediate-sized dots to optimise the energy transfer, or acquiring fluorescence decay time measurements to explore the possibility of "antenna-like" FRET behaviour in more detail.

Acknowledgements

This work was supported by a Discovery Grant from the Australian Research Council (DP120104035). We would like to thank Dr Elizabeth Carter of the

University of Sydney Vibrational Spectroscopy Facility for assistance with the Raman spectroscopy.

References

1 P. Zrazhevskiy, M. Sena and X. Gao, *Chem. Soc. Rev.*, 2010, **39**, 4326–4354.
2 I. L. Medintz, H. T. Uyeda, E. R. Goldman and H. Mattoussi, *Nat. Mater.*, 2005, **4**, 435–446.
3 X. Michalet, F. F. Pinaud, L. A. Bentolila, J. M. Tsay, S. Doose, J. J. Li, G. Sundaresan, A. M. Wu, S. S. Gambhir and S. Weiss, *Science*, 2005, **307**, 538–544.
4 C. M. Tyrakowski and P. T. Snee, *Phys. Chem. Chem. Phys.*, 2014, **16**, 837–855.
5 R. K. Paradise, D. A. Lauffenburger and K. J. Van Vliet, *PLoS One*, 2011, **6**, e15746.
6 C. Stock, F. T. Ludwig, P. J. Hanley and A. Schwab, *Compr. Physiol.*, 2013, **3**, 59–119.
7 A. Jagielska, K. D. Wilhite and K. J. Van Vliet, *PLoS One*, 2013, **8**, e76048.
8 B. A. Webb, M. Chimenti, M. P. Jacobson and D. L. Barber, *Nat. Rev. Cancer*, 2011, **11**, 671–677.
9 S. Ast, P. J. Rutledge and M. H. Todd, *Eur. J. Inorg. Chem.*, 2012, 5611–5615.
10 Y. H. Lau, J. R. Price, M. H. Todd and P. J. Rutledge, *Chem.–Eur. J.*, 2011, **17**, 2850–2858.
11 E. Tamanini, A. Katewa, L. M. Sedger, M. H. Todd and M. Watkinson, *Inorg. Chem.*, 2008, **48**, 319–324.
12 C. M. Lemon, P. N. Curtin, R. C. Somers, A. B. Greytak, R. M. Lanning, R. K. Jain, M. G. Bawendi and D. G. Nocera, *Inorg. Chem.*, 2013, **53**, 1900–1915.
13 P. T. Snee, R. C. Somers, G. Nair, J. P. Zimmer, M. G. Bawendi and D. G. Nocera, *J. Am. Chem. Soc.*, 2006, **128**, 13320–13321.
14 A. M. Dennis, W. J. Rhee, D. Sotto, S. N. Dublin and G. Bao, *ACS Nano*, 2012, **6**, 2917–2924.
15 T. Jin, A. Sasaki, M. Kinjo and J. Miyazaki, *Chem. Commun.*, 2010, **46**, 2408–2410.
16 R. C. Somers, R. M. Lanning, P. T. Snee, A. B. Greytak, R. K. Jain, M. G. Bawendi and D. G. Nocera, *Chem. Sci.*, 2012, **3**, 2980–2985.
17 S. Ast, P. J. Rutledge and M. H. Todd, *Phys. Chem. Chem. Phys.*, 2014, **16**, 25255–25257.
18 A. R. Clapp, I. L. Medintz, B. R. Fisher, G. P. Anderson and H. Mattoussi, *J. Am. Chem. Soc.*, 2005, **127**, 1242–1250.
19 H. Xu, X. Huang, W. Zhang, G. Chen, W. Zhu and X. Zhong, *ChemPhysChem*, 2010, **11**, 3167–3171.
20 Y. Chang, S. Yueming, S. Bo, T. Wenwen, Q. Qi, Z. Yingping, D. Yunqian and J. Wei, *Nanotechnology*, 2013, **24**, 435704.
21 M. Werwie, X. Xu, M. Haase, T. Basché and H. Paulsen, *Langmuir*, 2012, **28**, 5810–5818.
22 W. Ma, L. X. Qin, F. T. Liu, Z. Gu, J. Wang, Z. G. Pan, T. D. James and Y. T. Long, *Sci. Rep.*, 2013, **3**, 1537.
23 A. P. de Silva, H. Q. N. Gunaratne, J.-L. Habib-Jiwan, C. P. McCoy, T. E. Rice and J.-P. Soumillion, *Angew. Chem., Int. Ed.*, 1995, **34**, 1728–1731.

24 J. R. Lakowicz, *Principles of Fluorescence Spectroscopy*, Kluwer Academic/ Plenum, New York, 1999.

25 I. L. Medintz, J. H. Konnert, A. R. Clapp, I. Stanish, M. E. Twigg, H. Mattoussi, J. M. Mauro and J. R. Deschamps, *Proc. Natl. Acad. Sci. U. S. A.*, 2004, **101**, 9612–9617.

26 A. R. Clapp, I. L. Medintz, J. M. Mauro, B. R. Fisher, M. G. Bawendi and H. Mattoussi, *J. Am. Chem. Soc.*, 2003, **126**, 301–310.

27 M. Hardzei, M. Artemyev, M. Molinari, M. Troyon, A. Sukhanova and I. Nabiev, *ChemPhysChem*, 2012, **13**, 330–335.

28 G. Socrates, *Infrared and Raman Characteristic Group Frequencies*, Wiley, 2001.

29 A. M. Smith, H. Duan, M. N. Rhyner, G. Ruan and S. Nie, *Phys. Chem. Chem. Phys.*, 2006, **8**, 3895–3903.

30 M. Gonzalez-Bejar, M. Frenette, L. Jorge and J. C. Scaiano, *Chem. Commun.*, 2009, 3202–3204.

31 B. O. Dabbousi, J. Rodriguez-Viejo, F. V. Mikulec, J. R. Heine, H. Mattoussi, R. Ober, K. F. Jensen and M. G. Bawendi, *J. Phys. Chem. B*, 1997, **101**, 9463–9475.

Faraday Discussions

PAPER

In vitro study of the interaction of heregulin-functionalized magnetic–optical nanorods with MCF7 and MDA-MB-231 cells

A. Lesniak,†[a] D. Kilinc,†[a] Suad A. Rashdan,[ab] A. von Kriegsheim,[c] B. Ashall,[d] D. Zerulla,[d] W. Kolch[c] and G. U. Lee*[a]

Received 21st May 2014, Accepted 21st July 2014
DOI: 10.1039/c4fd00115j

Multifunctional nanoparticles that actively target specific cells are promising tools for cancer diagnosis and therapy. In this article we review the synthesis and surface chemistry of Fe–Au nanorods and their characterization using microscopy. The diameter of the rods used in this study was selected to be 150–200 nm so that they did not enter the cells. The 80 nm-long Au tips of the nanorods were functionalized with heregulin (HRG), and the micron-long Fe portion was coated with a poly(ethylene glycol) monolayer to minimize non-specific interactions. Nanorods functionalized with HRG were found to preferentially bind to MCF7 cells that express high levels of the receptor tyrosine-protein kinase ErbB2/3. Magnetic tweezers measurements were used to characterize the kinetic properties of the bond between the HRG on the rods and ErbB2/3 on the surface of the cells. The strong magnetization of Fe–Au nanorods makes them excellent candidates for *in-vitro* and *in-vivo* imaging, and magnetic therapeutic applications targeting cancer cells in circulation.

A. Introduction

Our capacity to synthesize nanoparticles of specific shape and composition has advanced rapidly over the last decade due to the significant interest that nanotechnology has received. It is now possible to design and synthesize multi-segmented and multi-layered nanoparticles of defined composition from two or more inorganic, organic, or biological materials.[1,2] Novel nanoparticles have been developed for biomedical and biotechnological applications where they play an

[a]*Bionanotechnology Group, Conway Institute and School of Chemistry, UCD, Belfield, Dublin, Ireland. E-mail: gil.lee@ucd.ie*
[b]*University of Bahrain, Kingdom of Bahrain*
[c]*Systems Biology Ireland, UCD Conway Institute, UCD, Dublin, Ireland*
[d]*School of Physics, UCD, Dublin, Ireland*
† Equal contribution.

important role in *in-vitro* diagnostics, *in-vivo* imaging, vaccines, and therapeutics, due to their strong interaction with electromagnetic fields and their ability to interact with cells and biomolecules.[3] Of particular interest are multifunctional nanoparticles that act as highly sensitive contrast agents for cancer diagnosis as well as carriers of small molecule or biomolecular therapeutic agents. For example, gold nanoparticles can be readily functionalized with antibodies and polynucleotides, and act as contrast agents for X-ray based computer tomography and near infrared imaging when excited at their plasmon resonance frequencies. Even more interestingly, coating Au nanoparticles on dielectric spheres has been found to enhance their optical properties.[4] Superparamagnetic iron oxide nanoparticles (SPIONs) have also been used for targeting and imaging of specific cancers *in vivo*,[3] and the fact that these nanomaterials also absorb magnetic radiation has led to their use as agents for hypothermia therapy.[5,6]

Targeting of specific cell types *in vivo* for diagnostic and therapeutic purposes has remained a challenge due to the numerous macromolecular components of blood that adsorb onto nanoparticles and the body's multiple mechanisms for clearing nanoparticles from the circulation.[7,8] Nanoparticle surfaces are typically coated with a polymer monolayer, such as dextran or poly(ethylene glycol) (PEG), to improve their stability and biocompatibility, and prolong their blood half-life. These polymer coatings decrease the amount of protein that adsorbs on the nanoparticle surface and increase the hydrodynamic size of the particles sparing the smaller particles from rapid elimination in the reticuloendothelial system. To target specific cell types, nanoparticles need to be further functionalized with agents that recognize specific molecules presented on the target tissues. Nanoparticles have been synthesised that are capable of targeting a variety of proteins that are over-expressed on certain types of solid tumours, *i.e.*, folate receptors *via* folic acid,[9] integrin $\alpha_v\beta_3$ *via* RGD peptide,[10] and mutant epidermal growth factor (EGF) receptor EGFRvIII *via* a specific antibody.[11] Unfortunately, these receptors are rarely unique to cancerous tissues and this leads to the targeting of the nanoparticle to additional regions in the body. Studies of the interaction of nanoparticles with cultured cell lines indicate that they also non-specifically interact with cells that have poorly regulated endocytic pathways. For example, the enhanced permeability and retention (EPR) effect has been used to target cancers with nanoparticles based on particle size.[12–14] Ideally, interaction of the nanoparticles with culture cells *in vitro* could be rapidly screened to determine the mechanism of particle–cell interaction before testing in animal models.

In this article we present the results of a study of the interaction of Fe–Au nanorods ~150–200 nm in diameter composed of segments of defined length with cultured cells using an optical microscope capable of tracking individual rods. The Fe–Au nanorods were functionalized with a PEG monolayer and heregulin (HRG), which is known to bind to a set of receptors that are over-expressed in some breast cancers, *i.e.*, the ErbB2/3 receptors. A novel microfluidics device is described that allowed the dynamics of the nanoparticle interactions with the cells to be studied in very small volumes under defined flow conditions in conjunction with measurements using magnetic tweezers. This technique has been used to characterize the interaction of HRG-functionalized nanorods (NR-HRG) with specific regions of two cell lines with known levels of expression of ErbB2/3. Magnetic tweezers have been used to characterize the interaction of the nanorods with the cells, which took place at the cell membrane since the particles appear not to be internalized. We suggest that these tools provide a powerful means to characterize nanoparticle–cell interactions *in vitro*, and the strong MRI contrast of the nanorods suggests that these studies can be translated to *in-vivo* animal studies.

B. Methods

Synthesis of Fe–Au two segmental nanorods

The porous anodic alumina (PAA) templates were synthesized by a three-step anodization method described previously.[14] A high-purity (99.999%) piece of aluminium was electropolished in a mixture of 5 vol% sulfuric acid, 95 vol% phosphoric acid, and 20 g L^{-1} chromium oxide at 20 V and 60 °C for 45 s. This polished piece of aluminium was anodized in 0.42 M oxalic acid at 40 V and 0.9 °C for 11 h. The alumina layer was removed from the sample using an aqueous mixture of chromium oxide, phosphoric acid, and water (45 g L^{-1}/3.5 vol%/96.5 vol%) at room temperature for at least 5 h. A second 6-h anodization and 5-h stripping step was conducted under the same conditions as described above. A third 5-h anodization step produced the highly ordered PAA template. The PAA was removed by immersing in saturated mercury chloride aqueous solution for 4 h. The as-prepared PAA membrane was then immersed in phosphoric acid solution for pore widening for 10 min resulting in a pore diameter of approximately 60 nm.

The synthesis of segmented metal nanorods was based on the method developed described previously.[14] Briefly, an Au layer (*ca.* 200 nm) was sputtered onto one side of the PAA template to serve as a working electrode in an electrochemical setup. An Ag/AgCl reference electrode and a Pt wire counter electrode were utilized to form a three-electrode configuration. The electrodeposition was controlled with a potentiostat/galvanostat. An Ag layer (*ca.* 2 μm) was first electrodeposited into the nanopores from a silver plating solution at a constant potential of −0.8 V (*vs.* Ag/AgCl). The Au and Fe layers were deposited from a commercial electroplating solution and freshly made iron sulfate solution ($FeSO_4 \cdot 7H_2O$, 120 g L^{-1}; boric acid, 45 g L^{-1}; ascorbic acid, 1 g L^{-1}), respectively, at −0.9 V (*vs.* Ag/AgCl). The segmental length of each element was controlled by monitoring the charge during deposition, and 100 nm segments of Au and Fe required 0.003 and 0.009 Coulombs, respectively. The PAA membrane and Ag segment were removed with sodium hydroxide aqueous solution and fuming nitric acid, respectively. Dispersed Fe–Au nanorods were obtained after surface modification as follows: the nanorods were washed thoroughly with sodium hydroxide solution, deionised water, and ethanol. In a subset of experiments, Fe–Au rods 203.4 ± 6.1 nm in diameter were synthesized using a commercial Al_2O_3 membrane filter with pore back-side diameters of ~200 nm as described above using the following parameters: 125 mC for Au and 3 C for Fe segments. Fig. 1A presents a scanning electron micrograph of the rods in which the gold sections can be easily identified due to their higher propensity to scatter electrons.

Surface chemistries

Fe–Au nanorods 203.4 ± 6.1 nm in diameter, composed of 77.9 ± 5.4 nm-long Au and 1.35 ± 0.07 μm-long Fe segments, were coated with self-assembling PEG

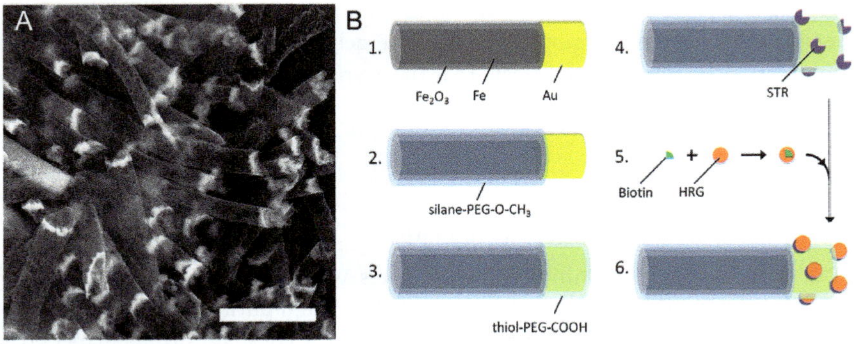

Fig. 1 (A) Scanning electron micrograph showing synthesized Fe–Au nanorods with highly uniform size and element ratio. Scale bar = 1 µm. (B) Cartoon depicting surface modifications: (1) Fe segment is coated with PEG-methoxy (2); Au segment is coated with PEG-carboxyl (3) and conjugated with streptavidin (4). Free HRG was modified with biotin (5). Biotinylated HRG was bound to STR at the gold tips (6).

monolayers (Fig. 1B). First, the Fe segment was coated with methoxy-PEG *via* silane chemistry; then, the Au segment was coated with carboxyl-PEG *via* thiol chemistry. The Fe segment was coated with 1 mg mL^{-1} methoxy-PEG-silane (2 kDa; Creative PEGWorks) dissolved in a 2 : 1 mixture of toluene and triethylamine (Et_3N). Following reaction overnight at RT, pentane was added to precipitate the nanorods and the mixture was centrifuged. The precipitation step was repeated 3× . The Au segment was then coated with COOH-PEG-SH (3.4 kDa; Creative PEGWorks) dissolved in EtOH (1 mg mL^{-1}). Following reaction overnight at RT, the nanorods were rinsed 3× with EtOH. The nanorods were dispersed in 2-(*N*-morpholino)ethanesulfonic acid (MES) buffer and incubated with a mixture of 1-ethyl-3-[3-dimethylaminopropyl]carbodiimide hydrochloride (EDC; Thermo Scientific) and *N*-hydroxysulfosuccinimide (Sulfo-NHS; Thermo) for 30 min, both 10 g L^{-1}, dissolved in MES buffer. After 3× PBS rinsing, the particles were incubated with 0.6 g L^{-1} streptavidin (STR) in PBS for 2 h at RT and overnight at 4 °C. STR-conjugated nanorods (NR-STR) were rinsed 3× with and dispersed in PBS containing 0.02% Tween20. Recombinant HRG (R&D Systems) was biotinylated using sulfo-NHS-biotin at a 20-fold excess of the latter. Excess biotin was removed using a spin column (Thermo). Biotinylated HRG (5 µM) was mixed with NR-STR in a rotating wheel for 1 h, after which the nanorods were rinsed thoroughly with PBS.

Lab-on-a-chip and magnetic tweezers experiments

In-vitro cell-targeting experiments were conducted on cells cultured in a novel microfluidic device.[15] Polydimethylsiloxane (PDMS) microfluidic devices that have 4 mm-long, 750 µm-wide, and 75 µm-high grooves were fabricated using photolithography and replica molding processes. Inlet and outlet holes were punched out of the PDMS and the devices were permanently bonded to glass coverslips using oxygen plasma. MCF7 and MDA-MB-231 cells were seeded in UV-sterilized microfluidic devices at low confluence. Cells were starved prior to experiment by replacing full DMEM with serum-free medium at least for 1 h. The inlet and outlet wells of the channel were filled with 20 μL fresh medium, representative groups of cells were located and imaging was started. Set volumes of HRG or STR-conjugated nanorod suspensions (10 μL) were added to the channels and flowed over the cells until the volume difference in the inlet and outlet reservoirs diminished. Imaging was conducted at 1 fps (frame per second). Nanorods with a velocity lower than 15 $\mu m\ s^{-1}$ and with a trajectory that traverses the cell area were considered for cell-targeting analysis. For HRG competition experiments, MCF7 cells in microfluidic devices were incubated with medium containing soluble HRG at varying concentrations for 15 min. HRG-conjugated rods suspended in the same medium were introduced and cell targeting was quantified as described.

A custom-built electromagnet was used to apply forces on the HRG-conjugated particles bound to MCF7 cells. Forces acting on the nanorods were calibrated by measuring their drag velocity in fluids of defined viscosity.[16] The force is a function of the distance between the tip of the electromagnet core and the nanorod, as well as the current passing through the coil. The magnitude of the applied force for a given horizontal and vertical distance from the tip and for a given current was calculated through interpolation. The acting force per nanorod in each experiment was estimated by interpolating the force values measured at known distances and currents. The multilayer design of the microfluidics chips allows us to maintain an electromagnet tip–nanorod distance of no more than 500 μm, which is critical for maximizing the force in the cell-stretching experiments. MCF7 cells cultured in these modified devices were targeted with HRG-conjugated nanorods as described. Live cell imaging was conducted at 1.5 fps while applying a periodic stretch (1 s on, 1 s off) at different force magnitudes for 15 min.

C. Results and discussion

Optical and magnetic properties of the nanorods

We have recently characterized the optical signature of the Fe–Au nanorods using transmission polarization microscopy.[14] Nanorods of length ~2 μm (~1 μm Fe and ~1 μm Au) and ~300 nm in diameter could be visualized in both bright-field

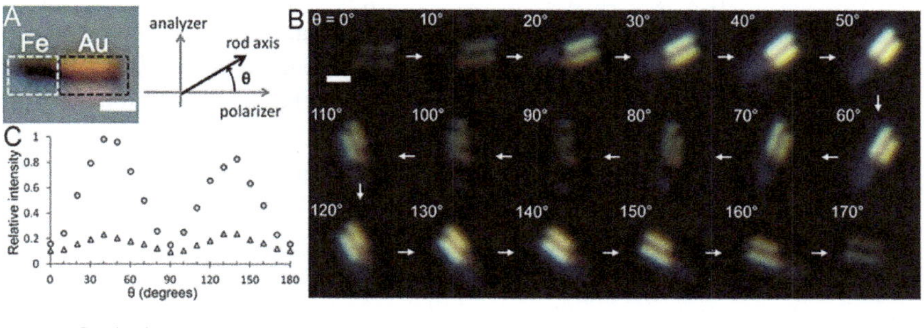

Fig. 2 Optical signature of iron–gold composite nanorods. (A) Bright-field image of a 295 nm diameter nanorod. (B) A series of crossed polarisation microscopy images of the same rod (right panel) analyser where *θ* represents the angle between the incident polarisation plane and the nanorods orientation. Scale bars = 500 nm. (C) Scattered light intensity from small (triangles) and large (circles) rods changes as a function of the angle (*θ*) between the polarizer and the rod axis.

microscopy (Fig. 2A) and crossed polarized optical microscopy (Fig. 2B), where the latter was achieved by using an orthogonally-aligned polarizer/analyzer pair. The sequential images in Fig. 2B show how the scattered light intensity depends on the orientation of the rod, where θ is the angle between the polarization axis and the rod's longitudinal axis. The optical signature of the gold nanorod was characterized by the polarization-dependent change in the scattering intensity of the two bright stripes, which is a result of the optical effects associated with the cylindrical geometry of the nanorod, the plasmonic modes excited on the Au section of the nanorod, the ultra-high collection angle of the 1.49 NA objective lens, and the illumination geometry. The dark stripe along the middle of the rod's longitudinal axis resulted from the shadow produced by the cylindrical rod when imaged in the transmission direction. In contrast, light scattered from plasmonic modes, which have an altered polarization direction and scatter diffusely from the edges of the rod, can be detected by the wide-angle collection of the objective lens and pass through the analyzer.

Fig. 2C presents the intensity of the optical signature of the rods that was measured from the images in Fig. 2B as a function of θ. The minimum intensity was observed when the illumination polarization was oriented parallel to the longitudinal and transverse axes, *i.e.*, θ of 0°, 90°, 180°, or 270°, and the maximum intensity was observed when the illumination polarization was orientated halfway between the longitudinal and transverse axes, *i.e.*, θ of 45°, 135°, 225°, or 315°. The plasmonic contribution to this behaviour can be understood by considering three exemplary orientations. First, when the polarization was aligned parallel to the longitudinal axis of the nanorod, *i.e.*, θ of 0° or 180°, the longitudinal axis plasmon was excited at a maximum level. However, the scattering was polarized along the longitudinal axis and thus was almost completely rejected at the analyzer resulting in the observed intensity minimum. Second, when the polarization was aligned perpendicular to the longitudinal axis of the nanorod, *i.e.*, θ of 90° or 270°, the transverse axis plasmon was excited. Analogous to the previous case, upon scattering, this light had a polarization orientation aligned in the transverse direction that was rejected by the analyzer resulting in the observed minimum. Third, consider now the case where the polarization was aligned halfway between the longitudinal and transverse orientations of the nanorod. In these cases, both the longitudinal and transverse axis plasmonic modes were partially excited. However, as the resulting scattered light maintained the polarization defined by the plasmonic modes, the polarization lies at an angle of 45° or 135° to the polarizer and analyzer. When this light was analyzed a relatively large portion of the light passed through the analyzer maximizing the observed intensity. The relative intensity of light transferred through the polarization microscope provides a powerful technique that enabled the orientation of sub-wavelength nanorods to be defined in the far-field without visualization of the Fe section of the rod.

The magnetic properties of the Fe–Au nanorods were characterized using a SQUID magnetometer (Quantum Design, MPMS XL 7T), and the magnetization measurements were performed on the nanorod arrays in the PAA membrane with the magnetic field oriented parallel and perpendicular to the longitudinal axis of the rods. The hysteresis loops indicated that the nanorods were magnetically soft and had an easy axis that is oriented along their longitudinal direction. The longitudinal coercive magnetization (H_c), remanent magnetization (M_r), and

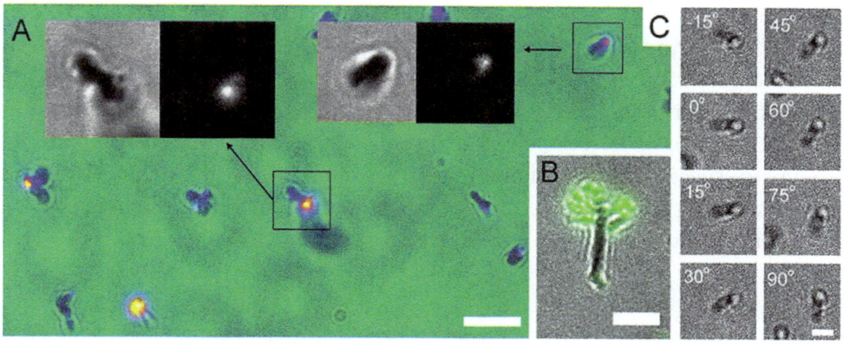

Fig. 3 (A) Streptavidin-conjugated nanorods (NR-STR) showing their fluorescent tips after incubation with fluorescein-biotin. The fluorescent image is pseudo-colored and superposed with the phase contrast image. Bright-field and fluorescent images of boxed areas are 2× magnified. (B) NR-STR incubated with biotinylated fluorescein-DNA. (C) NR-STR rotates freely around its gold tip attached to the PEG-biotin surface. Bars = 5 μm (A) and 2 μm (B and C).

saturation magnetization (M_s) of the iron section of the nanorods were 40, 14.7 ± 0.7, and 1457.7 ± 73 emu cm^{-3}, respectively. The transverse H_c, M_r, and M_s of the iron section were 40, 17.7 ± 0.9, and 1600 ± 80 emu cm^{-3}, respectively. The magnetic properties of the Fe nanorods indicate that their saturation magnetization was quite close to that of bulk Fe, which is 1750 emu cm^{-3} or 220 emu g^{-1}.[17] An important feature of the magnetic properties of the Fe segment of the nanorods was that they were magnetically soft, which allowed the nanorods to be repeatedly used for magnetic separation without acquiring a residual magnetic moment. The hardness of the Fe nanorods is expected to be defined by the size of the iron domains formed during the electrodeposition and the size of the nanorods.[18] The size dependence of the magnetization has been characterized for different sized rods, *i.e.*, diameters of 60 and 295 nm, by their squareness, *i.e.*, the ratio between the remanent and saturation magnetization, M_r/M_s.[19] Smaller rods have a higher level of squareness than the larger rods, but the relatively low magnitude of squareness values for all our nanorods suggests that the Fe rods are composed of multi-domain particles. This has been confirmed by high-resolution electron microscopy (results not shown). The strong magnetic susceptibility, soft magnetic behaviour, and relatively large size of the class I rods make them an ideal material for magnetic tweezers and diagnostic applications.

Magnetic nanoparticles developed for drug delivery or theranostic applications, such as FePt capsules[20] and Fe_2O_3 tubes,[21] may also be optimized for application as magnetic tweezers like those presented in this article. Other potential candidates are the soft composite materials with programmable shape changes: stripes that twist or bend,[22] and tubes that collapse upon exposure to weak magnetic fields[23] can be further exploited for their use as magnetic tweezer probes.

Targeting of ErbB2/3 receptors in cancer cell lines and magnetic tweezers measurements

Fig. 3A illustrates that streptavidin (STR) was successfully localized to the Au tips of the nanorods using fluorescently-labeled biotin. STR also bound to bare Au, but not to unmodified carboxyl groups on a PEG monolayer. Localization of STR to the Au tips was further verified by biotinylated fluorescent DNA (Fig. 3B). When the NR-STR were reacted with PEG-biotin surfaces, the Fe segments were free to follow the external magnetic field around the Au tips that acted as pivot points (Fig. 3C). These observations cumulatively suggested that the two-step surface chemistry successfully functionalized Fe–Au nanorods with specific ligands for active targeting, yet they minimized non-specific interactions which can lead to the premature clearance and mis-targeting of nanoparticles *in vivo*.

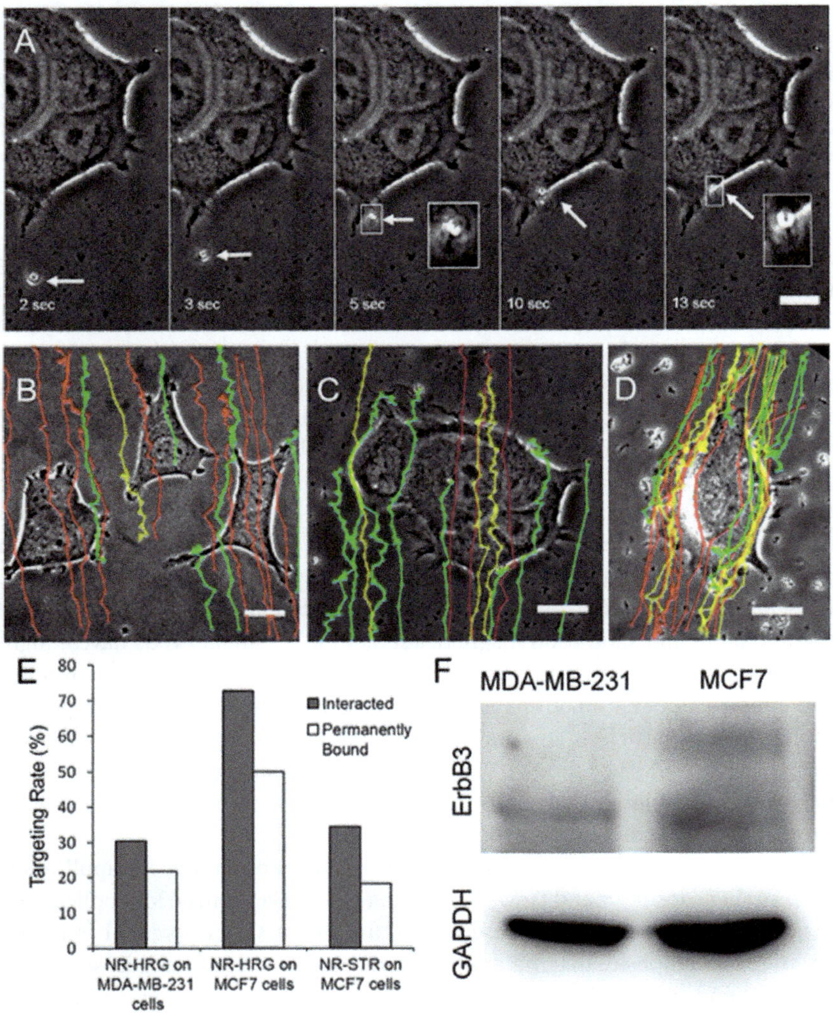

Fig. 4 (A) Time-lapse images of MCF7 cells targeted with individual heregulin-conjugated Fe–Au nanorods (NR-HRG). Scale bar = 5 μm. Exemplary trajectories of permanently bound (green), interacting (yellow), and non-interacting (red) nanorods: NR-HRG on MDA-MB-231 (B) and MCF7 (C) cells, and STR-conjugated nanorods on MCF7 cells (D). Scale bars = 5 μm. NR-HRG targeting rates (E) of MDA-MB-231 and MCF7 cells correlate with their ErbB3 expression levels shown by Western blots made using equal amounts of MDA-MB-231 and MCF7 cells with antibodies against ErbB3 and GAPDH (F). Bars = 5 μm. Boxed areas are 2.5× magnified.

ErbB-2 and -3 receptors on two different human breast adenocarcinoma cell lines were targeted with NR-HRG. ErbB2 expression levels on both cell types were quantified by using immunofluorescence followed by fluorescence-activated cell sorting (FACS, data not shown) and ErbB3 expression was quantified through Western blotting (Fig. 4F). In agreement with previous studies,[24] ErbB3 expression was higher in the MCF7 cells (Fig. 4F). When flowed over cultured cells, the NR-HRG successfully targeted MCF7 cells resulting in higher interaction rates compared to NR-HRG flowed over MDA-MB-231 cells and NR-STR flowed over MCF-7 cells (Fig. 4). Nanorod targeting rates for MCF7 relative to MDA-MB-231 cells correlated with their relative ErbB-2 and -3 expression levels. Fig. 4B and 4C present the paths of HRG-conjugated nanorods flowing over MCF7 and MDA cells, respectively, in microfluidic channels. HRG-conjugated rods that were reacted with the MCF7 cells exhibited significantly higher binding rates than those reacted with the MDA cells (Fig. 4E). There was also a significant level of interaction of STR-conjugated nanorods with MCF7 and MDA cells, which we attribute to non-specific protein–protein interactions. A detailed microscopic analysis of particle targeting experiments revealed the non-uniform distribution of bound nanorods on cell surfaces. 'Active zones', or dynamic extrusions in the cell periphery, attracted both HRG-conjugated rods and, to a lesser extent, STR-conjugated rods. HRG-conjugated rods targeted active zones six times more often than the rest of the cell surface (30.7 ± 7.9 *vs.* 4.9 ± 0.8 rods per 1000 μm^2, $p < 0.005$), whereas STR-conjugated rods targeted active zones $3\times$ more often than the rest of the cell surface (18.8 ± 4.5 *vs.* 6.3 ± 0.8 rods per 1000 μm^2; $p < 0.01$). These results suggest that cells expressing specific surface receptors can be targeted with Fe–Au nanorods whose tips are functionalized with appropriate ligands.

In order to determine whether targeting of HRG-conjugated nanorods was influenced by soluble HRG in binding to MCF7 cells, the in-flow cell-targeting experiment was repeated by exposing the cells to increasing concentrations of soluble HRG. Particle binding rates were not affected by the presence of free HRG up to 100 nM, a concentration that is two orders of magnitude higher than that needed to induce prolonged activation of the mitogen-activated protein kinase/extracellular signal-regulated kinase 1/2 (MAPK/ERK) pathway in these cells.[25] The local apparent concentration of HRG on the gold tips of the rods was calculated by dividing the estimated number of HRG molecules by the contact volume, which is the contact area multiplied by the estimated diameter of the molecule. The number of molecules was estimated by adjusting the previously-measured surface coverage of a similarly sized heterobifunctional PEG coated on spherical particles.[16] The local apparent density was determined to be 300 μM, which may explain the lack of competition between HRG-nanorods and soluble HRG in these experiments. Considering the (hetero) dimerization of ErbB-2 and -3 receptors[26] and their relative HRG dissociation rates,[27] it appears that the very high local concentration of HRG at the nanorod tips results in the formation of strong bonds with ErbB2/3 dimers.

MCF7 cells specifically targeted with NR-HRG were mechanically stimulated by using a custom-built magnetic tweezers coupled to a modified microfluidic device that minimized the gap between the cells in the flow channel and the electromagnet core. Cyclic stretching forces at a constant magnitude of 4.1, 8.3 and 12.7 pN were applied to cells targeted with individual rods over a fixed period of time.

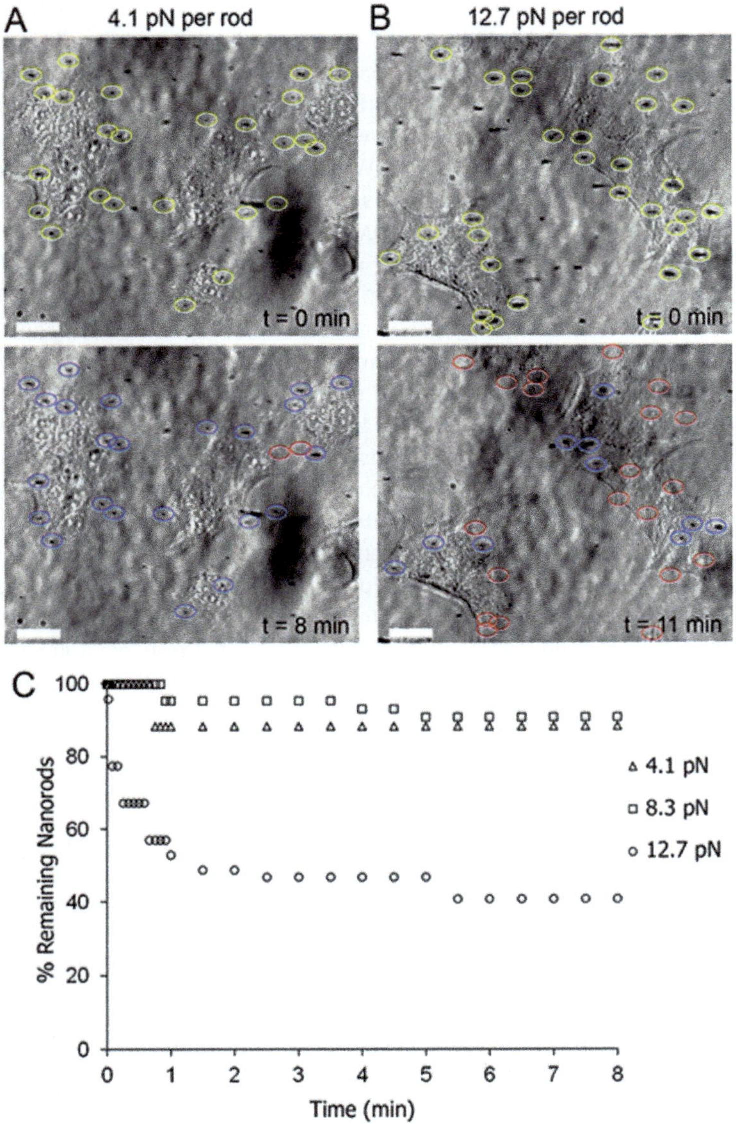

Fig. 5 Exemplary snapshots of MCF7 cells targeted with NR-HRG which were mechanically stimulated by using a custom-built electromagnet. Yellow circles indicate bound nanorods immediately prior to force application (time = 0), and red and blue circles indicate escaped and remaining nanorods, respectively, at the indicated time points. Scale bars = 20 μm. The plot shows the percentage of nanorods remaining on cell surface during periodic force application as a function of time. Only single nanorods were included in this analysis.

Receptor–ligand bonds withstand cyclic stretching at low force magnitudes but gradually fail at higher force magnitudes (Fig. 5). These results demonstrate that magnetic nanorods targeting specific membrane receptors form bonds that can withstand stretching forces. Due to the small size of the Au tips, the force will be distributed to a limited number of ErbB2/3 receptors.

The kinetic and energetic properties of protein–protein interactions can be analyzed through analysis of the magnetic tweezers results.[16,28] The energy landscape of a chemical reaction may contain one or more sharp activation barrier that needs to be overcome for the reaction to occur. External force application increases the unbinding rate by skewing this energy landscape.[29] According to Bell's model, the off-rate of a single, specific intermolecular interaction depends on the externally-applied constant force F,

$$k_{\text{off}} = k_{\text{off},0}\exp(F/f) \tag{1}$$

where $k_{\text{off},0}$ is the dissociation rate of the interaction in the absence of an external force and f is the scaling force, characteristic of that interaction.[30] This model implies that weak, non-covalent bonds have a limited lifetime and will break under any level of force if pulled on for a sufficient time period. An extension of this model is proposed[31] for multiple parallel bonds of the same type where the off-rate also depends on the number of bonds, n:

$$k_{\text{off}} = k_{\text{off},0}\left(\sum_{i=1}^{n}\exp\left(\frac{(-x/(if))}{i}\right)\right)^{-1}. \tag{2}$$

In this study, constant magnetic forces were applied of 4, 8 and 13 pN at a frequency of 0.5 Hz to characterize the HRG-ErbB2/3 intermolecular bonds. Clearly, very few of the particles were displaced from the cells at the lower forces, but a significant number of particles were displaced at a force of 13 pN. This behaviour is consistent with the formation of a small number of slip bonds between the nanorods and the cells.

D. Conclusions

Active cell targeting with multifunctional nanoparticles has recently gained considerable attention. Core–shell[32,33] and dumbbell-shaped[34] structures have been proposed to obtain nanoparticles that contain both Au and Fe_3O_4. We have recently described the synthesis of Fe–Au nanorods that have a high saturation magnetization, low remanence, uniform size and strong plasmonic activity.[14,35] In this study, we targeted specific cancer cell receptors using large ($\varnothing = 200$ nm) magnetic nanorods with heregulin-functionalized Au tips. Nanorods functionalized with heregulin using a PEG monolayer selectively bound to the MCF7 target cells that have high levels of ErbB2/3 but did not enter the cells. These nanorods bind to cells displaying a low density of ErbB2/2 (*i.e.*, MDA-MBA-231 cells) with a specificity similar to background levels as measured from the interaction of streptavidin-functionalized nanorods with MCF7 cells. Magnetic tweezers measurements indicate that the rods bind to the cells through a small number of receptors. This approach promises to allow us to develop a detailed understanding of the kinetics of interaction of nanoparticles with cells *in vitro*. More generally, the template synthetic approach employed here may also be scaled down to smaller ($\varnothing < 50$ nm) nanorods, which are suitable for cell internalization *in vivo*. The anisotropic nature of Fe–Au nanorods also renders them potential agents for cell destruction therapy, where heating acting upon internalized particles has been shown to induce apoptosis. The polycrystalline nature of the Fe segment makes these rods magnetically soft; yet their M_s is >10× higher than of SPIONs, which are widely used as MRI contrast agents.[36] Pure Fe NPs require a >20× smaller concentration to generate the same amount of heat as SPIONs of similar size[37] and hence are more suitable for magnetic hyperthermia.[38] Due to these superior qualities, Fe–Au NRs also have the potential to replace SPIONs in targeted molecular imaging and/or hyperthermia, where high particle concentrations are needed.[38,39] Thus, this study confirms that the physical and chemical properties of the nanorods can be tuned to selectively bind to cells expressing a high level of receptors, and suggests that targeting metastasized cells in circulation for diagnostic applications could be highly selective.

Acknowledgements

We would like to acknowledge Manuel DaSilva, Timothy D. Sands, and Yong Zhang for their collaboration on the development of the Fe–Au nanorod synthesis techniques. This article is based on works supported by the Science Foundation Ireland under grants 08/RP1/B1376 (G.U.L.), 08/IN1/B2072 (G.U.L.), 06/CE/B1129 (A.v.K. and W.K.), Erasmus Mundus Gulf Countries Postdoctoral Fellowship (S.A.R.), and Marie Curie Intra-European Fellowship (D.K.). We thank the Conway Institute core facilities for providing the instrumentation.

References

1 J. Della Rocca, D. Liu and W. Lin, *Acc. Chem. Res.*, 2011, **44**, 957–968.

2 M. A. Mahmoud, R. Narayanan and M. A. El-Sayed, *Acc. Chem. Res.*, 2013, **46**, 1795–1805.

3 D. Yoo, J. H. Lee, T. H. Shin and J. Cheon, *Acc. Chem. Res.*, 2011, **44**, 863–874.

4 S. Link and M. A. El-Sayed, *J. Phys. Chem. B*, 1999, **103**, 4212–4217.

5 Q. A. Pankhurst, N. T. K. Thanh, S. K. Jones and J. Dobson, *J. Phys. D: Appl. Phys.*, 2009, **42**, 224001.

6 K. Maier-Hauff, R. Rothe, R. Scholz, U. Gneveckow, P. Wust, B. Thiesen, A. Feussner, A. von Deimling, N. Waldoefner, R. Felix and A. Jordan, *J. Neuro-Oncol.*, 2007, **81**, 53–60.

7 C. H. Choi, J. E. Zuckerman, P. Webster and M. E. Davis, *Proc. Natl. Acad. Sci. U. S. A.*, 2011, **108**, 6656–6661.

8 Q. Dai, C. Walkey and W. C. Chan, *Angew. Chem., Int. Ed.*, 2014, **53**, 5093–5096.

9 F. Sonvico, S. Mornet, S. Vasseur, C. Dubernet, D. Jaillard, J. Degrouard, J. Hoebeke, E. Duguet, P. Colombo and P. Couvreur, *Bioconjugate Chem.*, 2005, **16**, 1181–1188.

10 J. Xie, K. Chen, H. Y. Lee, C. Xu, A. R. Hsu, S. Peng, X. Chen and S. Sun, *J. Am. Chem. Soc.*, 2008, **130**, 7542–7543.

11 C. G. Hadjipanayis, R. Machaidze, M. Kaluzova, L. Wang, A. J. Schuette, H. Chen, X. Wu and H. Mao, *Cancer Res.*, 2010, **70**, 6303–6312.

12 F. Li, L. Zhang and R. M. Metzger, *Chem. Mater.*, 1998, **10**, 2470–2480.

13 H. Maeda, J. Wu, T. Sawa, Y. Matsumura and K. Hori, *J. Controlled Release*, 2000, **65**, 271–284.

14 Y. Zhang, M. DaSilva, B. Ashall, G. Doyle, D. Zerulla, T. D. Sands and G. U. Lee, *Langmuir*, 2011, **27**, 15292–15298.

15 D. Kilinc, J. M. Peyrin, V. Soubeyre, S. Magnifico, L. Saias, J. L. Viovy and B. Brugg, *Neurotoxic. Res.*, 2011, **19**, 149–161.

16 D. Kilinc, A. Blasiak, J. J. O'Mahony, D. M. Suter and G. U. Lee, *Biophys. J.*, 2012, **103**, 1120–1129.

17 D. L. Huber, *Small*, 2005, **1**, 482–501.

18 J. M. Baik, M. Schierhorn and M. Moskovits, *J. Phys. Chem. C*, 2008, **112**, 2252–2255.

19 R. M. Cornell and U. Schwertmann, *The Iron Oxides: Structure, Properties, Reactions, Occurrences and Uses*, Wiley-VCH Verlag GmbH & Co. KGaA, Weinheim, Germany, 2003.

20 T. Fuchigami, Y. Kitamoto and Y. Namiki, *Biomatter*, 2012, **2**, 313–320.

21 N. Kang, J. H. Park, J. Choi, J. Jin, J. Chun, I. G. Jung, J. Jeong, J. G. Park, S. M. Lee, H. J. Kim and S. U. Son, *Angew. Chem., Int. Ed.*, 2012, **51**, 6626–6630.

22 J. W. Tavacoli, P. Bauer, M. Fermigier, D. Bartolo, J. Heuvingh and O. du Roure, *Soft Matter*, 2013, **9**, 9103–9110.

23 R. Fuhrer, C. M. Schumacher, M. Zeltner and W. J. Stark, *Adv. Funct. Mater.*, 2013, **23**, 3845–3849.

24 G. Bowers, D. Reardon, T. Hewitt, P. Dent, R. B. Mikkelsen, K. Valerie, G. Lammering, C. Amir and R. K. Schmidt-Ullrich, *Oncogene*, 2001, **20**, 1388–1397.

25 J. V. Thottassery, Y. Sun, L. Westbrook, S. S. Rentz, M. Manuvakhova, Z. Qu, S. Samuel, R. Upshaw, A. Cunningham and F. G. Kern, *Cancer Res.*, 2004, **64**, 4637–4647.

26 Y. Yarden and M. X. Sliwkowski, *Nat. Rev. Mol. Cell Biol.*, 2001, **2**, 127–137.

27 M. Hiroshima, Y. Saeki, M. Okada-Hatakeyama and Y. Sako, *Proc. Natl. Acad. Sci. U. S. A.*, 2012, **109**, 13984–13989.

28 D. Kilinc and G. U. Lee, *Integr. Biol.*, 2014, **6**, 27–34.

29 C. Bustamante, Y. R. Chemla, N. R. Forde and D. Izhaky, *Annu. Rev. Biochem.*, 2004, **73**, 705–748.

30 G. I. Bell, *Science*, 1978, **200**, 618–627.

31 P. M. Williams, *Anal. Chim. Acta*, 2003, **479**, 107–115.

32 J. Kim, S. Park, J. E. Lee, S. M. Jin, J. H. Lee, I. S. Lee, I. Yang, J. S. Kim, S. K. Kim, M. H. Cho and T. Hyeon, *Angew. Chem., Int. Ed.*, 2006, **45**, 7754–7758.

33 L. Wang, J. Bai, Y. Li and Y. Huang, *Angew. Chem., Int. Ed.*, 2008, **47**, 2439–2442.

34 C. Xu, J. Xie, D. Ho, C. Wang, N. Kohler, E. G. Walsh, J. R. Morgan, Y. E. Chin and S. Sun, *Angew. Chem., Int. Ed.*, 2008, **47**, 173–176.

35 Y. Zhang, B. Ashall, G. Doyle, D. Zerulla and G. U. Lee, *Langmuir*, 2012, **28**, 17101–17107.

36 J. E. Rosen, L. Chan, D. B. Shieh and F. X. Gu, *Nanomedicine*, 2012, **8**, 275–290.

37 C. G. Hadjipanayis, M. J. Bonder, S. Balakrishnan, X. Wang, H. Mao and G. C. Hadjipanayis, *Small*, 2008, **4**, 1925–1929.

38 B. Mehdaoui, A. Meffre, J. Carrey, S. Lachaize, L.-M. Lacroix, M. Gougeon, B. Chaudret and M. Respaud, *Adv. Funct. Mater.*, 2011, **21**, 4573–4581.

39 L. M. Lacroix, N. F. Huls, D. Ho, X. Sun, K. Cheng and S. Sun, *Nano Lett.*, 2011, **11**, 1641–1645.

Faraday Discussions

PAPER

Studies of the dynamics of biological macromolecules using Au nanoparticle–DNA artificial molecules

Qian Chen,[abc] Jessica M. Smith,[ab] Haider I. Rasool,[bd] Alex Zettl[bd] and A. Paul Alivisatos*[ab]

Received 9th July 2014, Accepted 29th July 2014
DOI: 10.1039/c4fd00149d

The recent development of graphene liquid cells, a nanoscale version of liquid bubble wrap, is a breakthrough for *in situ* liquid phase electron microscopy (EM). Using ultrathin graphene sheets as the liquid sample container, graphene liquid cells have allowed the unprecedented atomic resolution observation of solution phase growth and dynamics of nanocrystals. Here we explore the potential of this technique to probe nanoscale structure and dynamics of biomolecules *in situ*, using artificial Au nanoparticle–DNA artificial molecules as model systems. The interactions of electrons with both the artificial molecules and the liquid environment have been demonstrated and discussed, revealing both the opportunities and challenges of using graphene liquid cell EM as a new method of bio-imaging.

Introduction

It has been a challenging goal in microscopy to visualize, and potentially modulate, the transformative kinetics of biomolecules during biological reactions at the nanometer resolution in their native liquid environment.[1–13] Such dynamics involving morphological details are critical to the functions of individual biological molecules, the feedback network of biological processes, and biomedical applications. For example, the folding and unfolding of proteins have inspired extensive theoretical and simulation efforts[2] to understand the molecular mechanisms of the exposing and hiding of active sites or binding pockets. In contrast, experimental efforts that can probe the conformational dynamics of biomolecules are largely limited, due to the lack of an imaging tool that can resolve *in situ* dynamics at the nanoscale.

[a]*Department of Chemistry, University of California, Berkeley, CA 94720, USA. E-mail: alivis@berkeley.edu*
[b]*Materials Sciences Division, Lawrence Berkeley National Laboratory, Berkeley, CA 94720, USA*
[c]*Miller Institute for Basic Research in Science, University of California, Berkeley, CA 94720, USA*
[d]*Department of Physics, University of California, Berkeley, CA 94720, USA*

The previous focus in the development of experimental approaches has been on two strategies, both of which have had great successes, as well as limitations: pushing the spatial resolution for dynamic imaging techniques, such as super-resolution optical microscopy,[5,14,15] and pushing the dynamic capability for high spatial resolution techniques, such as cryo EM[1] and X-ray crystallography.[3,4] Super-resolution optical microscopy, in its various forms, can have a record spatial resolution of ~15 nm, achieved usually by fitting the point spread function of the emission of individual fluorescent labels. This mechanism of imaging already has its own limitation: the morphology of a biomolecule can only be generated when the whole biomolecule is covered with labeling agents, which is not experimentally easy and also poses complications in possible interference effects from labeling agents. On the other hand, cryo EM has developed a suite of imaging and analysis techniques that can generate the three-dimensional (3D) shape of biomolecules at sub nm resolution. However, TEM (transmission electro microscopy) was considered to be incompatible with the existence of a liquid sample, and thus dynamics in liquid, due to its high vacuum working condition. In cryo EM techniques, therefore, the biological samples are embedded in vitrified ice, leaving only snapshots of the conformational states of biomolecules along their dynamic trajectory during reactions. X-ray crystallography shares similar limitations, with the additional constraint that it can work only on the crystalline forms of biomolecules. This constraint is more pronounced when the relevant functional states of the biomolecules are not necessarily their bulk crystalline forms. For example, membrane proteins are still presented as a great challenge for X-ray crystallography, since they function and transform in close proximity with the cellular membrane.

Here we investigate the opportunities and challenges of a third strategy towards the elusive nanoscale dynamics in biological systems: graphene liquid cell TEM, with the unique advantage of retaining a native liquid environment and the capability of capturing continuous dynamics at nanometer resolution.[13,16] In principle, graphene liquid cell TEM integrates the powers of the previous two approaches and can thus probe a large sample space that has not been approachable before in bio-imaging. It allows samples to remain in a liquid environment, like super-resolution optical microscopy, but can also reveal biomolecular morphological details. It has ultrahigh spatial resolution, just like cryo EM and X-ray crystallography, but now the samples are in their native liquid environment. In comparison to the conventional thick-walled liquid phase EM strategies using Si_3N_4 to hold liquid against high vacuum,[17–21] graphene liquid cells wrap a liquid samples using two graphene layers and seal the sample *via* the van der Waals force between two graphene layers. In this liquid cell configuration, we can minimize the loss of imaging electrons from the one-atom thick graphene window. The atomic resolution imaging capability of EM can thus be maintained, despite the liquid molecules surrounding the sample, which is potentially crucial for the observation of low atomic number biological samples in liquid. The van der Waals sealing between the graphene layers can also minimize possible contamination resulting from adhesive spacers.

Meanwhile, the traditionally identified complication with high spatial resolution imaging techniques can persist here: the probing beams, electron beam or X-ray, have high energy that interacts with, and sometimes damages, the sample being imaged. This complication is an important concern in the broad use of liquid phase EM in imaging nanoscale dynamics of biomolecules. However, graphene has its own unique properties that make it more advantageous than other window materials for liquid phase EM. Our earlier work,[13] as well as other recent independent research,[9,22,23] has observed that graphene can help increase the electron beam tolerance of samples, such as an inorganic MoS_2 lattice and ferritin molecules, by wrapping the samples inside. While detailed mechanisms are still under investigation, the speculation is that the ability of graphene to rapidly conduct heat and charges can play a role.

The model system we chose to image here is 3D shaped Au–DNA artificial molecules, to study and quantify the effect of electron beam on the DNA molecules during the imaging process. Such artificial molecules can demonstrate: (1) the effect of the electron beam on DNA molecules, detected from the easily visible motions of Au nanoparticles (NPs); (2) determination of the 3D shape and dynamics of biomolecules that can be generalized to other small and low-contrast biomolecules. In addition, Au NPs inside the artificial molecules have plasmonic coupling that is sensitive to the construction of the artificial molecules. They have been extensively used as "plasmon rulers"[24–26] to detect the dynamics and activity of biomolecules, but their real 3D configuration in solution could not be easily probed. Imaging these artificial molecules in their working liquid environment is also crucial for optimizing the optical properties of their design for bio-sensing.

Experimental procedures

Gold NP preparation

Citrate-stabilized 5 nm gold NPs (83 nM) were purchased from Ted-Pella, Inc. (Redding, CA). We first stirred such NP solutions with an excess (1 mg BSPP per 1 mL deionized water) of bis(*p*-sulfanatophenyl)phenyl-phosphine (BSPP from Strem Chemicals, Inc., Newburyport, MA) overnight. Later, we collected the NPs, now coated with BSPP, by centrifuging the solution at 3000 rpm for 15 min. We then removed the supernatant solution and resuspended the NPs in BSPP solution (1 mg BSPP per 1 mL deionized water). Our stock solution was concentrated to 2 μM, and was kept at ambient temperature when not in use.

Preparation of single-stranded (ss) DNA

We obtained our ssDNA of desired lengths from Integrated DNA Technologies Inc. (Coralville, IA). The 5′ ends were modified with hexylthiol disulfide linkers, and then ssDNA were resuspended in 10 mM Tris buffer containing 0.5 mM EDTA in a ~100 μM concentration.

Preparation of Au–ssDNA conjugates

In one reaction, we mixed 70 μL stock solution of Au NPs, 1.5 μL ssDNA stock solution, 10 μL BSPP (5 mg mL^{-1} in deionized water), with 10 μL 500 mM NaCl solution, and had them react overnight without stirring at room temperature. The reaction mixture was then mixed with 10 μL of 20 mM poly(ethylene glycol) methyl ether thiol (MPEG) and reacted for 2 hours to replace the BSPP ligands on the Au NPs. The solution then went through high performance liquid chromatography (HPLC) to collect Au NPs with one and two ssDNA on the surface separately, depending on the desired nanoconjugate structure.

Preparation of trimers and pyramids

For trimers, Au NPs conjugated with two ssDNA were mixed and hybridized with Au–ssDNA monoconjugates twice in amount at room temperature overnight. Then reacted solution went through an Agarose gel (3 wt% in deionized water) in electrophoresis to obtain Au–ds DNA conjugates, trimers, from unreacted Au–ssDNA monoconjugate. The electrophoresis was done in 0.5× tris-borate-EDTA (TBE) buffer. The gels were run for 1 hour at 125 V, after which several distinct bands were visible. We extracted dimers, or trimers, by cutting the gel at the end of the proper band and immersing the gel piece in 0.5× TBE buffer and 200 mM NaCl solution. The obtained Au–double-stranded (ds) DNA nanoconjugate solution was stocked in the fridge when not in use. For pyramids, we followed the procedure in ref. 25, but used a DNA strand design such that one edge of the pyramids was shorter than the others.

Preparation of graphene covered TEM grids

The base TEM grids are Quantifoil R1.2/1.3300 Mesh Au holey carbon TEM grids purchased from SPI supplies. These TEM grids were pressed against graphene grown on Cu foil *via* chemical vapor deposition. A drop of isopropanol was cast on the Cu foil to immerse the pressed TEM grids to bind them with graphene as isopropanol evaporates. The Cu foil, with attached TEM grids facing up towards the air, was then placed on the surface of a sodium persulfate (Sigma Aldrich) solution (10 mg mL^{-1}), which etched away the Cu foil, leaving TEM grids coated with graphene. These TEM grids were subsequently rinsed with deionized water three times to wash off the etching solution, and were left to dry for further use.

Loading liquid sample into graphene liquid cell

We laid one graphene coated TEM grid onto a hard substrate such as a glass slide with the graphene side facing upwards. One tiny droplet (diameter ¼–$^1/_3$ of the TEM grid diameter) of the aqueous solution was deposited onto the center of the TEM grid. The other graphene coated TEM grid was then positioned on top to cover the bottom TEM grid and the droplet. During the slight evaporation of water after 10 min in the air, the top and bottom TEM grids were brought into close contact and kept the rest of the solution intact and sealed.

Microscopy

The TEM imaging was performed on LaB6 Tecnai G2 S-TWIN TEM at 200 kV. An accelerating voltage of 200 kV was used to obtain enough contrast and also to balance between avoiding damaging the graphene layer and lowering radiation damage. We set 0.1 seconds of exposure time and 0.9 seconds of read out time, therefore giving 1 second for the entire frame time, to acquire time-serial images.

Results

Graphene liquid cell imaging of 3D pyramidal artificial molecules

EM imaging of 3D structure and dynamics of biomolecules has two prerequisites: first, to be able to visualize low contrast, or sometimes invisible, biomolecules in morphological detail under EM, and second, to be able to obtain the 3D structure of biomolecules from their 2D projections. Cryo EM developed staining methods using heavy metal ions to enhance the contrast, and thus the visibility, of biomolecules under EM. In addition, cryo EM relies on the homogeneity of biomolecules at a particular functional state to reconstruct the 3D structure of biomolecules from snapshots of millions of randomly oriented biomolecules. Using 3D shaped artificial molecules, we demonstrated our proposed strategies towards solving these two issues. The biomolecules, dsDNA, in the artificial molecules are too thin to be seen directly under TEM, and we show how the decoration of a specific staining agent, tiny Au NPs, can help capture the status of dsDNA and enable us to detect its conformational change. As to the 3D structure, instead of generating one structure from millions of molecules, we are able to obtain a series of 2D projections of one particular artificial molecule and to construct the 3D model accordingly.

Our 3D model artificial molecule is a pyramidal shape, with four Au NPs positioned at the vertices and dsDNA as the bridges. These pyramids are of the simplest geometric shape that extends into 3 dimensions, enabling multiplexed probing of distance or reaction dynamics along different directions. Moreover, the pyramidal arrangement mimics the atomic backbone of glyceraldehyde, the smallest commonly used chiral molecule. Theoretical calculations have predicted that such pyramidal artificial molecules, once their symmetry is broken, can generate circular dichroism signals in addition to the geometry-dependent back scatterings from coupled plasmons. Many experimental strategies were suggested to introduce such chirality in pyramids,[25] including using Au NPs of different sizes, or different dsDNA determined edge lengths. To allow the practical use of these artificial molecules beyond their use as model systems, we designed their DNA linker sequence such that the tetrahedral DNA constructs, if assembled and base-paired successfully, have one bond shorter than the others (see the red line in Fig. 1). This strategy, if successful, can serve as the first step towards more variations in DNA linker lengths that eventually generate chirality in the structure.

Conventional dry TEM imaging of the final assembled artificial molecules can show whether the Au NPs are in proximity, but cannot resolve their 3D arrangement in their native liquid environment, which is crucial to their optical properties and applications as biosensors. The backbone structure of the artificial molecules can be distorted during dry TEM sample preparation, due to the strong capillary force from the meniscus of the evaporating solvent. As shown in Fig. 2,

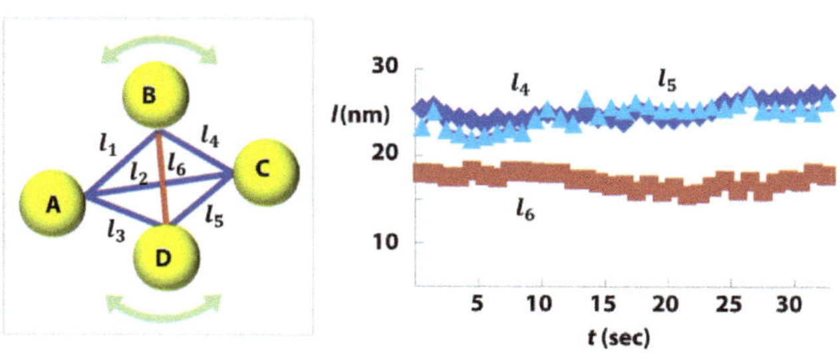

Fig. 1 3D remodeling of artificial pyramidal molecules. Left: the designed pyramidal molecule, with the red bond shorter than the other five blue bonds. Right: the calculated bond lengths for all the different frames (x axis).

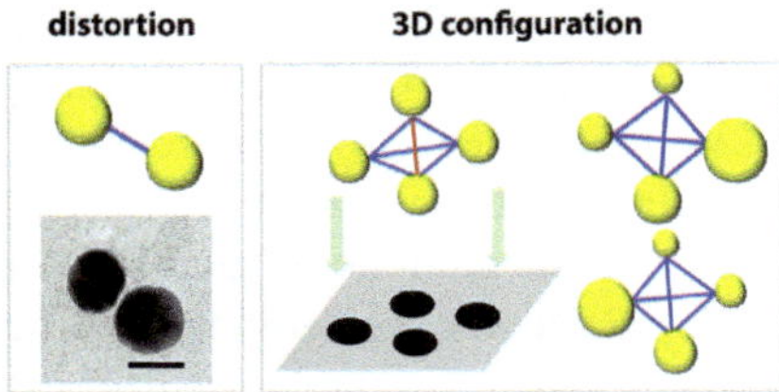

Fig. 2 Limitations of dry TEM imaging for artificial molecules. Left: the collapse of a dimeric artificial molecule. Right: the flattening of a pyramidal artificial molecule. Scale bar: 20 nm.

the component Au nanocrystals of a dimeric molecule were dragged closer together to a spacing of ~5 nm, despite the expected 15 nm dsDNA linker in between. Note that previous studies using small angle X-ray scattering[26] clearly demonstrated that, in the solution phase, the bond length of the same model dimeric molecules closely follows their linker length. It is therefore speculated that the capillary force can overwhelm the rigid linker and collapse their otherwise well-defined geometric configuration. Likewise, 3D artificial molecules get flattened onto the TEM grid, leaving hardly any information of their 3D configuration (see Fig. 2). This loss of 3D information poses a great challenge, since small angle X-ray, the alternative technique to probe the bond length between component artificial atoms, has difficulty interpreting 3D conformation data, not to mention that small angle X-ray can probe only the ensemble interparticle distances.

Using graphene liquid cell EM, we reconstructed the 3D configuration of pyramidal shaped artificial molecules in their native liquid environment. We sandwiched the solution of pyramids between two graphene layers and captured their dynamics under TEM. The Au NPs we used are small, around 5 nm in diameter, in order to minimize their effect on biomolecules as the "staining" agent. These small Au NPs can easily be seen under TEM, in comparison with the lower spatial resolution obtained from a thick-walled Si_3N_4 liquid chamber. We see this advantage of graphene liquid cell EM as crucial for the imaging of low-contrast biomolecules, as well as biomolecules stained with small high contrast agents. Here we watched one pyramid tumbling, due to the surrounding liquid environment, and recorded its continuous rotations as a TEM movie. Every snapshot in the movie has one 2D projection of the pyramid, and one set of six measurable projected distances between the four component Au NPs. This is our raw data, which can be obtained independently for every snapshot. We then employ the iterative optimization method described in detail in an earlier report from this laboratory.[13] In short, we assume three bond lengths as the inputs (see l_1, l_2, l_3 in Fig. 1), and calculate the 3D coordinates of particles B, C, D relative to particle A based upon the inputs and measured projected distances. Now that we have the 3D configuration of the pyramids, we further calculated the other unknown bond lengths (see l_4, l_5, l_6 in Fig. 1). We did the same calculation for each frame in a continuous movie, and optimized the three inputs until the variances of the three calculated bond lengths reach their minimums, which means the 3D configurations obtained from all the snapshots can converge into one structure.

Several observations can be concluded from this remodeling. The calculated bond lengths are consistent with our DNA design. As shown in the graph in Fig. 1 and representative TEM and 3D configurations shown in Fig. 3, one bond is indeed distinctly shorter than the other two bonds. The two longer bonds are also not only comparable with each other in length, but also comparable with the three input lengths. This consistency demonstrates that our solution phase-based DNA directed assembly into artificial molecules was able to introduce a bond length difference in the pyramidal shapes. More work is currently in progress to design a DNA sequence that can base pair into chiral pyramids for their unique optical properties. More importantly, this convergence into one 3D configuration of the pyramid indicates the geometric configuration of the pyramid stays intact during this time window of imaging. Otherwise, the pyramids will be distorted to different extents and to different shapes during the imaging process. This unique feature indicates the compatibility of graphene liquid cell TEM with imaging of other samples involving biomolecules and, eventually, even biomolecules themselves. The principle of imaging 3D shapes of biomolecules, although demonstrated here in the manner of "coarse-grained" staining, is shown in this example, and can be generalized to other biomolecules.

Electron beam effect

Interaction with the artificial molecules. Now that we have shown how a projected image relates to the 3D configuration of the pyramids, we can visualize the structural damage to the pyramids induced by electron beam in real time. During TEM imaging of biomolecules, the electron beam has been found to induce dramatic structure changes through ionization and consequently bond

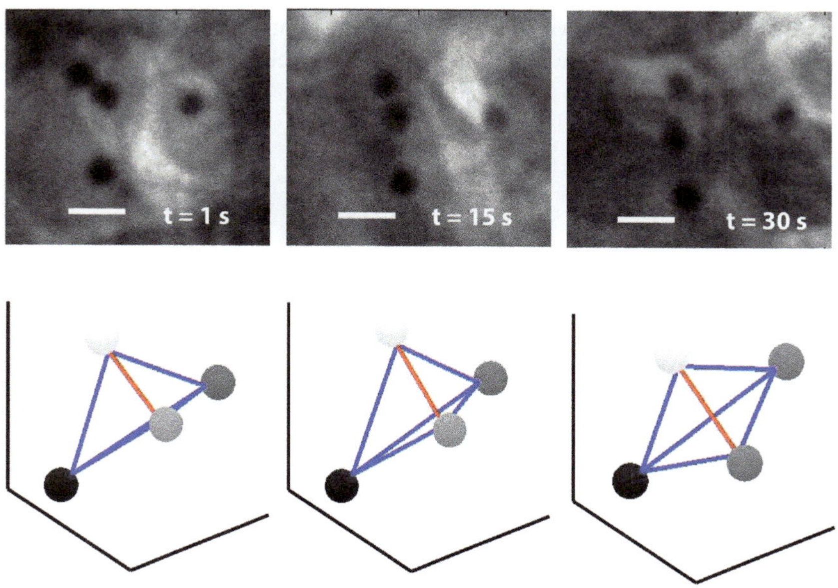

Fig. 3 Snapshots of a continuous TEM movie (top) and the corresponding 3D models (bottom). Scale bars are 10 nm. The spheres in the 3D models are color coded according to their *z* height.

breakage and bond rearrangements. Extensive studies have been done for biomolecules at cryogenic temperatures, showing that radiation damage is reduced significantly when biomolecules are embedded in a layer of vitrified ice, with the limitation that the samples are also locked in place and become stationary.

Recent EM imaging of a liquid phase sample has started to suggest new mechanisms for lowering such radiation damage and allowing imaging of biomolecules at higher spatial resolution and for longer observation time. Reports using Si_3N_4 liquid cell EM techniques[6,8,10,12] argued that the native liquid environment significantly attenuates the structural distortion of bio-samples, including both proteins and living cells, under continuous exposure to the electron beam. The use of graphene in this context for static yet hydrated ferritin[9] and bacteria[7,11] has also shown lowering of radiation damage, possibly by rapid conductance of heat and charges. Due to the complicated origins of electron beam damage, the mechanisms for such a reduction of the electron beam effect have not been thoroughly investigated. In some cases, seemingly controversial observations were reported about the nature of radiation damage, dose dependent or dose rate dependent, and about the state of the liquid environment under electron beam, dried out or truly hydrated. Our observation of a dynamic pyramid in the graphene liquid cell can serve as a good control system for further investigation into the electron beam effect.

We followed the pyramidal artificial molecule used in our 3D remodeling for a prolonged time and found it suffered dramatic structural deformation after 40 seconds of exposure to the electron beam. As highlighted in Fig. 4, the distance between particles A and B became abruptly large; particle B later diffuses away from the other three component Au NPs. This observation has several indications. The connection between particles A and B was broken once the accumulated electron dose reached a threshold value. Such dose dependent radiation damage is consistent with the previous studies. This dose threshold is larger than that determined from cryogenic TEM imaging of biomolecules,[1] but comparable with

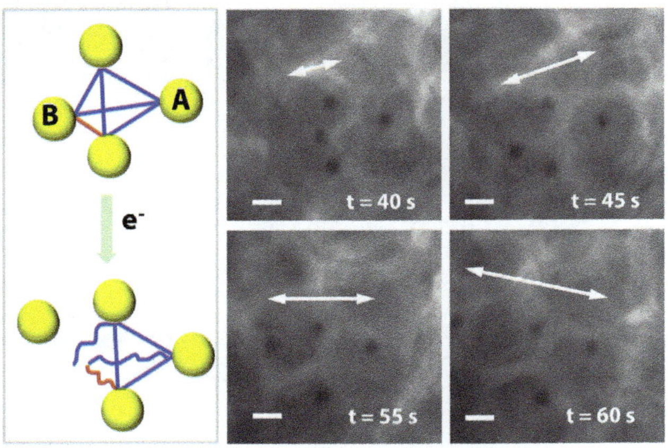

Fig. 4 The breakage of DNA linkage after imaging the pyramid shown in Fig. 2 and 3 for a longer time. Left: schematics. Right: TEM series, where the arrows highlighted the distance between particles A and B. Scale bar is 10 nm.

the reported increased thresholds in both Si_3N_4 and graphene liquid cell experiments.[6–8,10–12] Meanwhile, this transition from a converged 3D pyramidal shape to a dissembled structure proves in return that the pyramidal structure was indeed in liquid and that the positions of the Au NPs fully described the 3D conformation of the pyramids.

It is still an open question if the deformed pyramids truly indicate in real time the structural damage to dsDNA in the graphene liquid cell. It is very probable that the breakage of individual bonds in DNA molecules occurs prior to the loss of the connecting capability of the DNA chains, therefore prior to the observed phenomenon of Au nanocrystals. Several reasons can lead to this possible gap between the apparent critical dose and the real critical dose for biomolecules. Nanoscale objects have been repeatedly found to have slowed motions[18,21] within the liquid cell configuration. This sluggish motion may effectively act like a "cage" that keeps the structure from diffusing apart altogether, even after the linkage is broken. For the imaging of artificial molecules, this mechanism is efficient enough, as we shown earlier in the text, for us to remodel their 3D conformation.

Of course our use of visual markers, the Au NPs, makes our probing of the DNA molecules indirect. A more vigorous and quantitative experiment will be the direct visualization of high-contrast biomolecules, to quantify and capture the critical dose at which the structural details of biomolecules start to blur or to degrade. Both high-contrast biomolecules, ribosomes for example, and heavy metal ion-stained biomolecules can be good candidate systems to show if there is a quantitative improvement in the dosage tolerance of biomolecules when using graphene liquid cells.

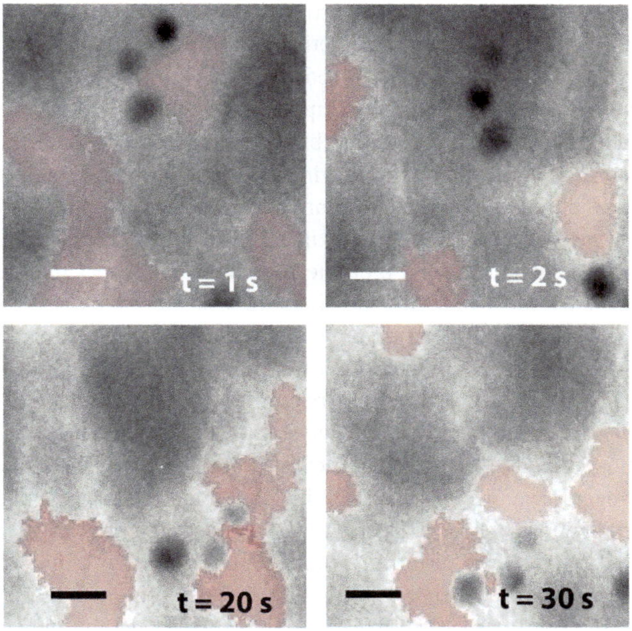

Fig. 5 Snapshots of a continuous TEM movie of a trimeric artificial molecule. Red shadows highlight areas of high intensity. Scale bar: 6 nm.

Interaction with the liquid medium. One unique aspect inherent to all liquid cell EM experiments is that the electron beam can also interact with the liquid medium in addition to the sample of interest.[19] Previous studies have been focused on possible radiation chemistries, where electron beam can produce reactive species to initiate chemical reactions *in situ*.

We have looked into another frequently observed phenomenon: the intensity fluctuation of the liquid environment during the liquid cell imaging. As shown in Fig. 5, artificial trimeric molecules rotate in the series of TEM images, together with a visible background fluctuation. One can clearly see the distinct lighter and darker islands in the surrounding liquid environment, with the lighter domains highlighted by red shadows in the TEM images. The positions of the red domains relative to those of the trimers randomize from frame to frame, exhibiting little correlation with the motion of the trimers. In other words, the motions of the trimers are independent of the dynamic fluctuation of the liquid environment, which indicates a lower possibility of the background fluctuation being of a convection nature. This lack of correlation, although the origin of the fluctuation is still not resolved, is important since the motions, and consequently the imaged configurations, of the nanocrystals are not then disturbed by the instantaneous flow pattern of the solution.

Conclusions and outlook

The development of bio-imaging techniques has been associated with important scientific understandings of structures and functions of living systems, and practical applications such as diagnosis and drug engineering. We discussed and demonstrated the unprecedented advantages brought by the recently developed graphene liquid cell EM as a new method of bio-imaging. Graphene liquid cells have a unique combination of properties: they provide both the native liquid environment enabled by optical microscopy and the nanometer resolution enabled by electron microscopy. This combination can potentially answer questions that were previously simply not approachable by other experimental techniques, such as the structures of membrane proteins in action in their working environment, the conformational docking of drugs at target proteins, and eventually the real nanoscale imaging of biomolecules functioning and in motion. In the paper, we used Au NP–DNA artificial molecules, which are frequently employed for the detection of a biomolecular environment, as model systems to illustrate how we can work with the challenges of bio-imaging using graphene liquid cells. We were able to generate 3D shapes of a certain artificial molecule by following its series of 2D projections through continuous rotations. Note that, previously, people have had to either rely on the monodispersity of millions of biomolecules (cryo EM) or on the bulk crystalline form of biomolecule crystals (X-ray crystallography) to do 3D reconstruction. In addition, we show how the small "staining" agents we used to make the invisible biomolecules visible are clearly imaged under graphene liquid cell EM, a unique capability of holding liquid samples in between two ultrathin graphene layers, rather than other thicker microfabricated chips. Furthermore, based upon these two capabilities, we followed the structural damage of artificial molecules under the imaging electron beam and found that the structural backbone of artificial molecules was kept stable under a higher electron dose than the previously determined threshold.

Although the stability of the artificial molecules may be different from the stability of their biomolecular linkers, more studies from other groups have also shown that graphene wraps can protect both dry and aqueous samples under an electron beam. This protection mechanism, if fully quantified and understood, can have a huge impact beyond the graphene liquid cell EM technique itself. For example, conventional cryo EM could potentially use graphene wraps to make samples more stable and allow better contrast under the imaging beam.

Acknowledgements

This research was supported in part by the Defense Threat Reduction Agency (DTRA) under award HDTRA1-13-1-0035, which provided for *in situ* TEM experiments, as well as DNA–Au nanoparticle sample preparation; by the National Science Foundation within the Center of Integrated Nanomechanical Systems, under Grant EEC-0832819, which provided for early development of graphene lamination methods. Q.C. was supported by a Miller fellowship from Miller Institute for Basic Research in Science at UC Berkeley. J.S. was supported by Agilent Technologies Applications and Core Technology University Research Grant.

References

1 L. F. Kourkoutis, J. M. Plitzko and W. Baumeister, *Annu. Rev. Mater. Res.*, 2012, **42**, 33.
2 J. N. Onuchic and P. G. Wolynes, *Curr. Opin. Struct. Biol.*, 2004, **14**, 70.
3 A. Pulk and J. H. D. Cate, *Science*, 2013, **340**, 1235970.
4 J. W. Miao, H. N. Chapman, J. Kirz, D. Sayre and K. O. Hodgson, *Annu. Rev. Biophys. Biomol. Struct.*, 2004, **33**, 157.
5 K. Xu, G. Zhong and X. Zhuang, *Science*, 2013, **339**, 452.
6 U. M. Mirsaidov, H. Zheng, C. Yosune and P. Matsudaira, *Biophys. J.*, 2012, **102**, L15.
7 F. Banhart, *ChemPhysChem*, 2011, **12**, 1637.
8 M. T. Proetto, *et al.*, *J. Am. Chem. Soc.*, 2014, **136**, 1162.
9 C. Wang, Q. Qiao, T. Shokuhfar and R. F. Klie, *Adv. Mater.*, 2014, **26**, 3410.
10 D. B. Peckys and N. de Jonge, *Nano Lett.*, 2011, **11**, 1733.
11 N. Mohanty, M. Fahrenholtz, A. Nagaraja, D. Boyle and V. Berry, *Nano Lett.*, 2011, **11**, 1270.
12 S. M. Hoppe, D. Y. Sasaki, A. N. Kinghorn and K. Hattar, *Langmuir*, 2013, **29**, 9958.
13 Q. Chen, J. M. Smith, J. Park, K. Kim, D. Ho, H. I. Rasool, A. Zettl and A. P. Alivisatos, *Nano Lett.*, 2013, **13**, 4556.
14 S. van de Linde, M. Heilemann and M. Sauer, *Annu. Rev. Phys. Chem.*, 2012, **63**, 519.
15 T. J. Gould, S. T. Hess and J. Bewersdorf, *Annu. Rev. Biomed. Eng.*, 2012, **14**, 231.
16 J. M. Yuk, *et al.*, *Science*, 2012, **336**, 61.
17 N. de Jonge and F. M. Ross, *Nat. Nanotechnol.*, 2011, **6**, 695.
18 J. Lu, Z. Aabdin, N. D. Loh, D. Bhattacharya and M. Utkur, *Nano Lett.*, 2014, **14**, 2111.

19 J. M. Grogan, N. M. Schneider, F. M. Ross and H. H. Bau, *Nano Lett.*, 2014, **14**, 359.

20 H.-G. Liao, L. Cui, S. Whitelam and H. Zheng, *Science*, 2012, **336**, 1011.

21 H. Zheng, S. A. Claridge, A. M. Minor, A. P. Alivisatos and U. Dahmen, *Nano Lett.*, 2009, **9**, 2460.

22 R. Zan, Q. M. Ramasse, R. Jalil, T. Georgiou, U. Bangert and K. S. Novoselov, *ACS Nano*, 2013, **7**, 10167.

23 G. Algara-Siller, S. Kurasch, M. Sedighi, O. Lehtinen and U. Kaiser, *Appl. Phys. Lett.*, 2013, **103**, 203107.

24 S. E. Lee, A. P. Alivisatos and M. J. Bissell, *Systems Biomedicine*, 2013, **1**, 12.

25 A. J. Mastroianni, S. A. Claridge and A. P. Alivisatos, *J. Am. Chem. Soc.*, 2009, **131**, 8455.

26 A. J. Mastroianni, D. A. Sivak, P. L. Geissler and A. P. Alivisatos, *Biophys. J.*, 2009, **97**, 1408.

Faraday Discussions

DISCUSSIONS

Optical nanoparticles: general discussion

Peter Dobson, Lanry L. Yung, Liane Rossi, Zoe Pikramenou, Sara Carreira, Paresh Ray, Catherine Amiens, Katherine Brown, Maha Abdollah, Dejian Zhou, Mauro Prato, Sandhya Moise, Lucio Litti, Matthew Todd, Kristian Göeken, Stefan Borsley, Amelie Heuer-Jungemann, Oliver Reiser, Peter Harvey, Thomas Carter, Maya Thanou, Dalibor Soukup, Anna Lesniak, Hedi Mattoussi and Siti Fatimah Abdul Ghani

DOI: 10.1039/C4FD90079K

Stefan Borsley opened the discussion of the paper by Hedi Mattoussi : Are there any advantages to using this polymeric His binding system over traditional thiol-bound ligands?

Hedi Mattoussi responded: The use of a multi-coordinating polymer ligand having several histidine groups for surface coordination may be beneficial for gold nanorods (AuNRs). These NRs tend to have large surface areas, and a polymer structure provides stronger binding and better interactions with the hydrophilic surrounding medium. Multi-thiol polymer coating would also provide stronger affinity, nonetheless. I should also stress the importance of using a PEG-rich polymer for these systems.

Dalibor Soukup asked: In your presentation you have shown that your particles are very stable in media. However, I was wondering if it is pure medium without serum or a complete medium with serum, as serum proteins can spontaneously adsorb on the surface of highly charged nanoparticles? This would increase the hydrodynamic size of particles and in turn affect the stability of nanoparticles. Typically, in order to avoid this, PEG-functionalized nanoparticles have been introduced, however, are there any other chemical coatings known that could do the same?

Hedi Mattoussi answered: The polymer-coated nanoparticles we discussed are stable in complete growth media. I should add that protein corona formation, which stems from adsorbed serum proteins on the nanoparticles (often discussed these days), takes place with electrostatically-stabilized nanoparticles, such as the very commonly used citrate-stabilized gold nanoparticles (AuNPs). One of the most effective routes to reducing those problems is to use PEG or zwitterion coating. Such coating circumvents reliance on charge stabilization and eliminates protein adsorption on the nanoparticles.

Sandhya Moise queried: Is it advantageous to have a protein corona form on the nanoparticle surface and, if so, when?

Hedi Mattoussi answered: The presence of a lateral PEG shell on His–PIMA–PEG-capped nanoparticles prevents protein corona formation by simply eliminating nonspecific interactions. These are the primary promoter of protein adsorption on electrostatically-stabilized nanoparticles, such as citrate-capped AuNPs. Corona formation can stabilize charged nanoparticles, but in general it is a problem when applications require control over nanoparticle size and/or coupling to target molecules.

Oliver Reiser remarked: When you add ligands you showed that this should have potential for chelating metals to your iron nanoparticles – do you observe the formation of discrete iron complexes dissociating into solutions?

Hedi Mattoussi replied: We have not observed metal complexes dissociating from the nanoparticle surfaces. If such complex dissociation occurred, it would certainly be very weak, as the overall size of the nanoparticles before and after ligand exchange remained the same, as verified by TEM data collected from dispersions of these materials. Also, the measured hydrodynamic size of these nanoparticles stayed constant even after several months of storage, indicating no deterioration or etching of the inorganic cores by the ligands.

Dejian Zhou enquired: Are these poly-His–PEG ligand-capped nanoparticles still able to bind to His-tagged proteins?

Hedi Mattoussi answered: The His–PIMA–PEG-capped AuNPs and AuNRs do not bind to His-tagged proteins, a problem primarily attributed to the steric hindrance imposed by the PEG shell on a full size protein. Nonetheless, if the protein is engineered with a longer linker between the polyhistidine tag and the protein itself, self-assembly can take place. This mode of coupling requires direct interactions between the histidine groups and the metal surface of the nanoparticles.

Maya Thanou said: To what extent does the polymer added affect the linkage – is the polymer (molecular weight) critical?

Hedi Mattoussi replied: No, the polymer molecular weight is not critical in itself. What matters most is the presence of several metal-coordinating groups to allow strong anchoring on the nanoparticle surface. Introducing specific hydrophilic moieties (*e.g.* PEG) enhances its affinity to water, while maintaining strong binding on the metal surface. It is, however, better to use a small molecular weight polymer to avoid a large hydrodynamic size for the hydrophilic nanoparticles.

Lucio Litti asked: Did you perform some experiments on the yield of ligand exchange between your multi-His polymers and mono-thiolated molecules on the surface of gold nanoparticles?

In your opinion, which one gives better and more stable coverage of the surface: a molecule with several weak ligands (such as your polymers) or one with only one stronger bond, such as thiolated molecules?

Hedi Mattoussi responded: We have not explored this idea of cap exchanging AuNPs capped with mono-thiol-ligands with the His–PIMA–PEG. We do believe that such polyhistidine-modified polymers work better with hydrophobic ligands or citrate- and CTAB-capped NPs/NRs. Conceptually, a polymer presenting 10-20 His groups along its backbone will bind tightly onto the NP/NR and should provide better long term colloidal stability than small molecules (mostly for AuNRs), due to the polymer nature of the coating. The latter provides higher affinity to water media, due to favorable Flory–Huggins interactions. We thus think that a polymer coating should provide better colloidal stability than a mono-dentate ligand, even if the latter presents a strong coordination group.

Sara Carreira opened the discussion of the paper by Siti Fatimah Abdul Ghani : Do you know the mechanism of the toxicity of the naked quantum dots and what the coating does to prevent that toxicity mechanism?

Siti Fatimah Abdul Ghani replied: Trioctylphosphine oxide (TOPO)-modified quantum dots (QDs) are poorly water-soluble and may contain solvent residues. As aggregates interact with cell membranes and compartments, this leads further to the rapid destabilisation of the core and metal leakage, which contributes to rapid cell death.

Water-soluble QDs modified with hydrophilic ligands, such as biphosphonates, enhance colloidal stability and aqueous dispersion, preventing aggregation and minimizing leakage within the cells.

Catherine Amiens asked: How do you think that the biphosphonate ligands are linked to the nanoparticles surface? Are both phosphonate end groups anchored onto the surface as suggested by the low zeta values of the solutions? In this case, what is the reason for the observed colloidal stability? No neat surface charge, hence no electrostatic stabilisation; no long chain ligands, hence no steric stabilisation: I do not understand the system.

Siti Fatimah Abdul Ghani responded: Although thorough studies need to be performed to identify the coordination of the biphosphonates on the quantum dots surface (ZnS), we believe that association of these molecules with the QD surface may be with either one or both phosphonate groups (or in an equilibrium). The zeta potential of the dispersions suggests that the particles appear neutral in solution and biphosphonates do not contribute to the nanoparticle charge. We hypothesize that stabilization is achieved because of the large number of biphosphonate ligands (as indicated by thermogravimetric analysis) associated with the surface of QDs, providing a thick hydrophilic "layer" that contributes to the colloidal stability.

Catherine Amiens queried: Did you try to use NMR spectroscopy to investigate the coordination of the biphosphonate ligands on the surface of the nanoparticles?

Siti Fatimah Abdul Ghani replied: We did not use NMR to investigate the coordination of the biphosphonate ligands on the surface of the nanoparticles. We have used ^{31}P-NMR to confirm ligand exchange reaction.

Amelie Heuer-Jungemann enquired: What is the advantage of your new ligand compared to other conventional ligands used?

Siti Fatimah Abdul Ghani replied: The biphosphonate ligands are safe (non-toxic) and efficient in stabilizing QDs. They provide a colloidally-stable water dispersion and can maintain this stability for days.

Amelie Heuer-Jungemann said: Did you try your ligand on different sizes of QDs?

Siti Fatimah Abdul Ghani responded: We have tried our ligands on CdSe/ZnS and $CuInS_2$/ZnS, which have a small difference in size. However, we have not tried this ligand on larger dots, *e.g.* polymer dots (80 nm)

Stefan Borsley commented: The TEM image presented of the biphosphonate-stabilised QDs appeared to show the presence of aggregates. Are these genuine aggregates or is this a drying effect? Have you tried verifying the colloidal stability of the particles in solution by another technique, such as dynamic light scattering (DLS)? I agree that your stability studies appear to show no change in the sample, however, this does not mean that the particle is not initially aggregated, only that the aggregation of the sample is not changing.

Siti Fatimah Abdul Ghani replied: The apparent aggregation is because of the drying effect. We have verified colloidal stability with DLS and Nanosight.

Maya Thanou added: DLS is used to measure size. However, if the NPs have fluorescent properties , these may affect the readings.

Mauro Prato asked: To perform cytotoxicity studies, you have tested an ovarian cancer cell line as a model. However, a major issue in general in tumor-targeting chemotherapy is the side effects on surrounding healthy tissues. Have you performed any cytotoxicity studies on normal healthy cells as controls? This could provide much more informative data on the potential toxicity of your molecules.

Siti Fatimah Abdul Ghani answered: This suggestion is correct. All nanomaterials should be tested on both cancer cell line and /or primary cell lines.

Lanry L. Yung commented: As a materials scientist, we usually test the toxicity of our nanoparticles using secondary and cancer cell lines (e.g. 3T3, HeLa, *etc.*). However, in the eyes of a biologist, such toxicity tests should be conducted using primary cell lines, such as HUVEC, since secondary and cancer cell lines are not sensitive towards a toxic environment unless the toxicity dose becomes very high.

Mauro Prato added in response to Lanry L. Yung's remark: As a biologist, I definitely agree on that point. Nevertheless, I would like to stress that it is crucial to keep in mind that the physiology (and phenotype) of cancer cells is far different to that of normal healthy cells. If the new nanomolecule is aimed at targeting cancer cells, a comparison with normal tissues should always be deemed necessary. Of course, if the target is normal cells (such as in wound healing, for instance), no further comparisons are needed. Please also note that the main disadvantage of primary cultures lays on the high donor's variability. Therefore, depending on the context, using immortalised cell lines might also do the job. For instance, if I am studying normal (noncancer) skin cells for helping wound healing, HaCaT cell line might be sufficient instead of using highly variable primary keratinocytes.

Maya Thanou agreed: The point and suggestion is correct. HUVEC cells mimic the vein endothelium and indeed such tests would provide an indication of the potential toxicity of nanomaterials to the blood vessels. There is a list of suitable assays for new nanomaterials[1] by the Nanotechnology Characterisation Lab (NCL) that suggests proper immunological and toxicological evaluation of novel nanomaterials. The only primary cell line assay found in that list is the cell line of primary hepatocytes (for identification of reactive oxygen species using novel nanomaterials). We have performed MTT, LDH and glutathione assays using cancer cell lines to identify the toxicity of previously widely studied QDs, with the addition of surface modification with biphosphonates.

1 Assay Cascade Protocols, Nanotechnology Characterisation Laboratory, http://ncl.cancer.gov/working_assay-cascade.asp

Paresh Ray remarked: Why do the fluorescence spectra show a blue shift of about 100 nm after coating?

Siti Fatimah Abdul Ghani replied: We have observed this effect with methylene diphosphonate capping (coating). We are not aware of any similar blue shift upon capping with (other) water-soluble molecules. This may require thorough investigation with structurally different capping agents.

Kristian Göeken asked: In the paper, you state that QDs coated with ethylene diphosphonate (EDP) are found in the cytosol, while methylene diphosphonate (MDP) QDs are located around the nucleus. Do you have an explanation for the possible mechanism behind this "targeting"?

Siti Fatimah Abdul Ghani responded: We would not like to call it targeting. It is an observation that shows that the structure of a small molecule used as capping agent on the quantum dots may have an effect on its cellular distribution.

Peter Harvey opened the discussion of the paper by Matthew Todd : When observing the signal corresponding to the nanoparticle in the dye-sensitised example, is the lifetime multi-component? If so, do the components correspond to lifetimes associated with both the dye and the dot?

Matthew Todd responded: The data related to this question were presented at the meeting, but not in the relevant discussion paper, so we would prefer to address this question after the relevant data have been through formal peer review.

Peter Harvey asked: Does the normalisation of the spectra take into account the fact that the increase in the QD emission intensity could be due to a contribution from a "shoulder" of the dye emission?

Matthew Todd replied: Yes. The spectra were deconvoluted such that contributions from the dye were removed.†

Zoe Pikramenou queried: Can you exclude photo-induced electron transfer taking place between the QD and the organic dye?

Matthew Todd answered: We can exclude electron transfer from dot to dye, since the 4-aminonaphthalimide is linked to the QD *via* an ethylene bridge. The imide functional group, being electron rich, is a very poor electron acceptor and thus not prone to PET. Studies on the direction of PET in structurally very similar molecules have been reported by A. P. de Silva and give proof of a photo-induced electron transfer happening only in one direction, which is from the dimethylamine to the 4-aminonaphthtalimide core.[1] The reversed case of electron transfer from the QD to the dye can reasonably be excluded, since the QD lacks appropriate amines to trigger the electron transfer.†

1 A. P. de Silva *et al.*, *Angewandte Chemie*, 1995, **34**, 1728–1731.

Zoe Pikramenou asked: How does the pH dependent fluorescence response of the dye on its own compare to the dye on the QD?

Matthew Todd replied: The pH response of the dye in conjugation with the QD is very similar to its response as "free" dye in its monomer form (in the range of pH 7.5 to 9.6), which we can conclude by comparison with the reference compound 2. Studies on compound 2 have been published in our related paper.[1] What differs is the change in intensity, which is large for compound 2 (7-fold), but small for the dye–QD conjugate (2-fold). This loss in intensity could be due to the energy transfer to the QD.†

1 S. Ast *et al.*, *Phys. Chem. Chem. Phys.*, 2014, **16**, 25255–25257.

Sara Carreira asked: If your pH responsive dye works best at pH 7–10, would you expect it to be sensitive enough to work in the intracellular space, where the pH can be much more acidic? If not, could you use a dye that has a better response to acidic pH?

Matthew Todd answered: Yes, this prototype has a pH range that makes it useful for measurements of physiological conditions, but, if we were to want to image more acidic environments, we would need to adapt the system. Note that the dots

† Both Matthew Todd and Sandra Ast collaborated on the response after the meeting.

themselves were not particularly stable at acidic pH, so that would be a component of the system that would need to be modified, as well as the appended dye.

Thomas Carter added: Extracellular physiological pH varies within a very sensitive range, and tumours may affect local extracellular pH, making the pH slightly lower than this. Would the pH-responsive nanoparticles be sensitive enough within the appropriate pH range to provide diagnostic information, or is the response range something which can be modified?

Matthew Todd responded: For use in realistic sensing environments (*i.e.* for diagnostics), we would need to show that the particle is bright enough (*e.g.* for cell culture experiments) and is stable at relevant physiological conditions – note that we observed significant instability of the QD itself at low pH values *in vitro*. While we certainly see clear responsiveness in model cases, it is likely we would want to try to increase the responsiveness for what will presumably be "noisier" real-world applications. These are the general areas it would be interesting to target next.

Dejian Zhou asked: Have you tried the QDs without the dye – are they still pH sensitive?

Matthew Todd replied: Yes, that has been tested and is reported in the paper (*Faraday Discuss.*, 2014, DOI: 10.1039/C4FD00110A) for the case of QD 570. The relevant data are shown in Fig. 7 (acidic sweep) and Fig. 8a and b (alkaline sweep). The QD signal remains stable from 7.5–9.6, so that the pH-induced changes in the QD emission of the conjugate can only be attributed to energy transfer from the pH-responsive dye.†

Paresh Ray asked: What is the mechanism of transfer – is it fluorescence resonance energy transfer (FRET)?

Matthew Todd replied: We assume so, based on voluminous literature of related systems, but we have not explicitly shown the transfer to be FRET here. Recent work by others has shown the distance dependence of dot to dye transfer follows the expected FRET model.[1]

1 G. Beane, K. Boldt, N. Kirkwood and P. Mulvaney, *J. Phys Chem. C*, 2014, **118**, 18079–18086.

Peter Dobson asked: Have you got information about all the HOMO and LUMO levels – how do these line up with the QD levels?

Should this information be reported more widely in any discussion involving energy transfer between light, quantum dots and dyes?

Matthew Todd responded: This is an interesting question. While we do this for small molecule sensing applications of related molecules,[1] we have not analysed the HOMO and LUMO levels for a dot. I agree that there is an underlying assumption in the literature here that we could explicitly address in the future. We did match the energy levels according to their photophysical data: the basic requirement for FRET to occur is obviously a match between the absorption of the QD (at the first exciton peak) with the emission of the dye.†

1 S. Ast, P. J. Rutledge and M. H. Todd, *Eur. J. Inorg. Chem.*, 2012, 5611–5615.

Dejian Zhou asked: Have you checked the hydrodynamic sizes of the QDs after the addition of the dye molecules?

Matthew Todd answered: The dots are covered in a polymer shell, to aid water solubility. The dye attachment occurs beneath that. We do not expect the hydrodynamic radius of the dye to change appreciably after the addition of the dye, particularly as the attachment method is a substitution for a surface-bound molecule of similar size.

Katherine Brown remarked: Does your QD–dye have enhanced photophysical properties appropriate for studying dynamic biological processes, considering the limitations of organic dyes (bleaching) and QDs (blinking)?

Matthew Todd responded: We have not seen either bleaching or blinking in these systems to date. One of the reasons for our wanting to look at the dot, rather than the dye, in a sensing environment is because the dot ought to be bright, and dots have been used for this reason in biomedical imaging applications. Whether our conjugate behaves well in a real biological setting (*i.e.* displays the same mode of energy transfer and is clearly detectable in cell culture) is a subject of future work.

Matthew Todd opened a general discussion of Hedi Mattoussi's, Siti Fatimah Abdul Ghani's and his own papers: One of the features of Hedi's work[1] is the highlighting of lifetime as a key determinant of whether it is sensible to expect a dye to be able to excite a dot. The lifetime of a dye is typically much shorter than that of a dot, so it makes little sense to expect energy transfer towards the dot when it is likely that the dot is in an excited state. We wondered whether in our case such transfer might be possible, because there are so many dyes per dot (around 100), meaning there is a statistical chance that the dot could be receptive to transfer from some dyes (essentially the "long tail" of the decay profile).

1 A. R. Clapp, I. L. Medintz, B. R. Fisher, G. P. Anderson and H. Mattoussi, *J. Am. Chem. Soc.*, 2005, **127**, 1242–1250.

Hedi Mattoussi responded: The data by Todd and co-workers confirms that FRET between organic dyes and QDs is much higher and better controlled when the QDs are used as energy donors. This produces a FRET quenching that is commensurate with the number of dyes interacting with the same QDs. The reverse configuration, energy transfer from a dye to a QD, is very weak even when a large number of dyes are used. This is caused by the fact that direct excitation of the QDs is highly efficient; these systems have very large extinction coefficients.

Sandhya Moise asked: How stable are the surface coatings on the quantum dots and cobalt particles in biological systems? How much knowledge is there about *in vivo* retention times and modes of excretion, and are the surface coatings stable enough to protect the cores sufficiently throughout the time scales and conditions to make them safe for *in vivo* applications?

Oliver Reiser replied: We do not have any *in vivo* data yet. From the chemical data we have, the carbon coated particles are extremely robust over a high temperature (>100 °C) and pH range (pH 4–10). For biological applications, we will use the analogous iron particles, which have the same properties, but the metal core obviously less toxic. The reason we have conducted the study shown here with cobalt particles is that they were available first through our collaborators (W. Stark, ETH Zurich and Turbobeeds Inc.). The iron particles are a more recent development by them.

Hedi Mattoussi responded: Knowledge about the *in vivo* retention of nanoparticles is still rather scattered, though several studies have recently looked into these issues. The stability of the surface coating on nanoparticles (such as QDs, metallic and magnetic nanoparticles) in biological media strongly depends on the strength of the ligand interactions with the metal-rich surface of these nanoparticles. These interactions are driven by metal-coordination. That is why ligands presenting multiple strongly coordinating groups (anchors) can bind more tightly onto the nanoparticles. For example, we have shown that iron oxide nanoparticles coated with multi-dopamine polymer ligands exhibit long-term colloidal stability (for more than 1 year) using *in vitro* test conditions, including acidic and basic pH, 1 M NaCl and in RPMI growth media.

1 W. Wang, X. Ji, H. B. Na, M. Safi, A. Smith, G. Palui, J. M. Perez and H. Mattoussi, *Langmuir*, 2014, **30**, 6197–6208.

Peter Dobson said: When dealing with complex systems such as QDs with a core and a shell and ligands attached to the surface, few researchers are reporting on the energy levels of the valence and conduction bands and HOMO/LUMO levels relative to the surroundings. This situation is not helpful, because to ascertain the mechanisms of detection or the mechanisms of destroying cells or pathogens it is necessary to know these facts.

Hedi Mattoussi added: There actually have been studies where groups looked into characterizing the energy levels of semiconductor QDs and their dependence on the nanocrystal size. Nonetheless, often only the lowest level in conduction and the highest level in the valence band are reported for simplicity. The ligand shell has little to no contribution to these levels for common alkyl type molecules, though recently groups have probed the effects of using benzenethiol ligands on the energy levels of PbS QDs. The shell can, however, strongly affect the rate of electron–hole radiative recombination, which manifests itself in the measured photoluminescence quantum yields of the nanocrystals.

The issue of QD interactions with cells is not necessarily associated with the energy levels *per se*. Often this problem is discussed in terms of generation of reactive oxygen species in the medium following UV excitation, or possible leaching of the metal ions into the surrounding medium. The latter can be controlled by the nature of the ligand used (stronger binding ligands provide better shielding).

Liane Rossi addressed Hedi Mattoussi and Matthew Todd :The purification of nanoparticles after ligand-exchange can be challenging. Can you please discuss the importance of removing the excess ligands and how to achieve this?

Hedi Mattoussi replied: Removing excess ligands from the nanoparticle dispersions is crucially important for using these systems to develop applications, such as sensor design and cell and tissue imaging. Excess free ligands can interfere with simple procedures, such as coupling to target biomolecules. As an example, we address ligand exchange as applied to hydrophobic AuNPs. The nanoparticles are purified in two steps. First, we apply two rounds of precipitation by adding hexane to the dispersion after ligand exchange; this primarily removes oleylamine (the native ligand). Second, the nanoparticles are dispersed in deionised (DI) water, sonicated for 2–3 min. and filtered through a 0.45 μm disposable syringe filter; this step is carried out without loss of materials. Removal of excess free hydrophilic ligands is achieved by applying 3 to 4 rounds of concentration/dilution using a centrifugal filtration device (Millipore, M_w cutoff = 50 kDa). To prepare the samples used to collect the NMR spectra, the solvent is switched from hydrogenated DI water to deuterated water by applying two rounds of concentration/dilution using 2 mL D_2O each. We also characterize the filtrate collected in the bottom of the device using ^{1}H-NMR spectroscopy. We have found no sign of free ligands in the filtrates.

Matthew Todd responded: We were unable to isolate the conjugate in a stable form due to non-specific interactions with the carboxylic acid groups. Instead, as described in the paper (*Faraday Discuss.*, 2014, DOI: 10.1039/C4FD00110A), we studied the ligand uptake by the dot and worked with a concentration at which we were certain that all of the dye was conjugated to the QD.

Peter Dobson opened the discussion of the paper by Gil Lee‡: I would just like to check that your image is really convincing visual proof that the nanorods and the target molecule are sticking on the cell surface?

Anna Lesniak replied: Movies showing the active targeting of MCF-7 cells by HRG-functionalized rods and their subsequent mechanical stimulation can be found in the supplementary material of our most recent publication.[1]

1 D. Kilinc, A. Lesniak, S. A. Rashdan, D. Gandhi, A. Blasiak, P. C. Fannin, A. von Kriegshelm, W. Kolch and G. U. Lee, Mechanochemical stimulation of MCF7 cells with rod-shaped Fe–Au Janus particles induces cell death through paradoxical hyperactivation of ERK, *Adv. Healthc. Mater.*, DOI: 10.1002/adhm.201400391.

Peter Dobson asked: If you remove the nanorods using your magnetic tweezers and try again, do the nanorods attach in the same places?

Anna Lesniak replied: When higher forces were applied, nanorods resulted in membrane tether formation, possibly through the disruption of the bonds between the ErbB3 receptors and the underlying cytoskeleton. We have not tried to target stretched cells one more time.

Maha Abdollah asked: Referring to Fig. 4 of your paper (*Faraday Discuss.*, 2014, DOI: 10.1039/C4FD00115J), can you explain how you can differentiate between bound and unbound (passing) nanorods?

‡ Gil Lee's paper was presented by Anna Lesniak, *Bionanotechnology Group, Conway Institute and School of Chemistry, UCD, Dublin, Ireland.*

Anna Lesniak replied: The experiment was recorded at a rate of 1 frame per second and nanorods with a velocity lower than 15 μm s^{-1} were considered for cell targeting analysis. This time resolution is sufficient to differentiate between bound and passing rods. Corresponding movies are provided as supplementary material for our most recent publication.[1]

1 D. Kilinc, A. Lesniak, S. A. Rashdan, D. Gandhi, A. Blasiak, P. C. Fannin, A. von Kriegshelm, W. Kolch and G. U. Lee, Mechanochemical stimulation of MCF7 cells with rod-shaped Fe–Au Janus particles induces cell death through paradoxical hyperactivation of ERK, *Adv. Healthc. Mater.*, DOI: 10.1002/adhm.201400391.

Maha Abdollah queried: When heregulin (HRG)-nanorods bind to the cells, do they stimulate any downstream signalling, or cause any cell death?

Anna Lesniak answered: They indeed activate ERK signalling pathway in active zones, but fail to induce a global cellular ERK activation. This was only possible with subsequent mechanical force application.

Paresh Ray asked: Since the size of a cell is usually 10 microns and the nanorod you have developed is more than one micron, why are the nanorods not visible in TEM images, like the cancer cells?

Anna Lesniak replied: Figure 4A in our paper (*Faraday Discuss.*, 2014, DOI: 10.1039/C4FD00115J) is a phase contrast image of a cluster of cells being targeted by an individual rod. Once adhered, the cells we used in this study assume a diameter of 20–25 μm and, at the magnification we used, both the cells and the rods are visible.

Mauro Prato commented: Your study was performed by culturing cells in serum-free medium. Do you have any data showing binding to plasma proteins?

Anna Lesniak replied: Cells were starved for only 1 hour prior to targeting experiments, which is not enough to remove all the proteins from the cell membrane. Targeting experiments were also performed in complete cell culture media (10% fetal bovine serum (FBS) and media with increasing amounts of FBS up to 80%). Targeting rates decreased with increasing FBS concentration (data not shown).

Mauro Prato asked: You have stated that, before performing your experiments, cells were starved for 1 h in serum-free medium. What are the effects of starving your breast cancer cell lines on the phosphorylation status of ERK1/2?

Anna Lesniak responded: We did not directly compare ERK phosphorylation levels in starved *vs.* non-starved cells; however, in serum-starved cells we have only observed an increase in the ERK phosphorylation when MCF-7 cells were targeted with HRG-rods. Control cells (MDA) and control particles (STR-nanorods) were used to confirm this finding.

Kristian Göeken asked: ErbB receptors are known to cluster upon activation. Would it be feasible to use this technique to follow clustering kinetics? For instance by determining an increase in interaction strength over time, or by following the binding of rods to specific locations over time.

Anna Lesniak responded: Indeed, specific binding of HRG-conjugated Fe–Au rods to ErbB receptors resulted in the clustering of these receptors (Fig. 4 in ref. 1). We believe that it would be possible to determine the change in the interaction strength by applying a constant force after different periods of nanorod–cell interaction. Regarding your second suggestion, our method cannot determine where on the cell the rods will bind. However, one could study different regions of the cell to estimate local differences in the receptor presence or clustering.

1 D. Kilinc, A. Lesniak, S. A. Rashdan, D. Gandhi, A. Blasiak, P. C. Fannin, A. von Kriegshelm, W. Kolch and G. U. Lee, Mechanochemical stimulation of MCF7 cells with rod-shaped Fe–Au Janus particles induces cell death through paradoxical hyperactivation of ERK, *Adv. Healthc. Mater.*, DOI: 10.1002/adhm.201400391.

Peter Dobson enquired: This allows you to study cells in tremendous detail – can you use magnetic tweezers to measure stiffness of cell membrane? Could you "shake cells to death"?

Anna Lesniak responded: Yes, since the applied force is known and the displacement (if any) in the cell membrane can be measured, the stiffness of the membrane can be estimated. No, in our hands, periodically stretching cells without any pharmaceutical intervention did not lead to cell damage. Please refer to our most recent paper[1] where we describe the local activation of the ERK pathway downstream of mechanical stimulation and how this could be used to induce cell death *in vitro*.

1 D. Kilinc, A. Lesniak, S. A. Rashdan, D. Gandhi, A. Blasiak, P. C. Fannin, A. von Kriegshelm, W. Kolch and G. U. Lee, Mechanochemical stimulation of MCF7 cells with rod-shaped Fe–Au Janus particles induces cell death through paradoxical hyperactivation of ERK, *Adv. Healthc. Mater.*, DOI: 10.1002/adhm.201400391.

Amelie Heuer-Jungemann queried: Is cell death induced by overstretching of the cells, or by creating lesions in the cell membrane upon nanorod removal?

Anna Lesniak responded: Periodically stretching cells at low forces, which do not cause nanorod separation, activates cell signalling pathways that could be used for killing the cells, *e.g.* through combination with other treatments. Although we did not test this possibility directly, overstretching the cell membrane may cause mechanoporation, potentially leading to cell death.

Lanry L. Yung asked: In cases where a more gentle mechanical stretching was used, do the cells die from apoptosis pathway? Have any apoptosis pathway biomarkers been investigated?

Anna Lesniak responded: Although markers of apoptotic pathways were not directly assessed, we did not observe any noticeable cell damage following mechanical stretching at low forces (in the range of 10 pN). Our more recent results show that combining mechanical stimulation with B-Raf inhibitors, which activate the ERK pathway, results in cell death, likely through the hyperstimulation of the ERK.[1]

1 D. Kilinc, A. Lesniak, S. A. Rashdan, D. Gandhi, A. Blasiak, P. C. Fannin, A. von Kriegshelm, W. Kolch and G. U. Lee, Mechanochemical stimulation of MCF7 cells with rod-shaped Fe-Au Janus particles induces cell death through paradoxical hyperactivation of ERK, *Adv. Healthc. Mater.*, DOI: 10.1002/adhm.201400391.

Dejian Zhou asked: Have you done any statistical analysis? How many nanorods are attached to each cell? This may allow you to identify the cell targeting specificity of the nanorod.

Anna Lesniak replied: For each condition, a number of independent experiments were performed and on average 45 particle–cell interactions were analyzed per experiment. We calculated the targeting rate by dividing the number of bound rods to the number of all passing rods (also normalized by the cell surface area). The targeting rates of MCF-7 cells by HRG-nanorods were indeed statistically significant compared to all other experimental groups (One-way ANOVA, followed by t-test, $p < 0.01$).

Lucio Litti enquired: In your opinion, on applying a magnetic field, could your nanorods self-assemble into an ordered surface, suitable for cell trapping and recovery after the removal of the magnetic field?

Anna Lesniak responded: We did not try to self-assemble rods into an ordered surface but similar rods, composed of Ni–Au, were previously used to capture different types of analytes.[1] Depending on the ligand or antibody functionalization at the tip of the rods it would be conceivable to trap specific cells in solution.

1 M. Platt, G. R. Willmott and G. U. Lee, Resistive pulse sensing of analyte-induced multicomponent rod aggregation using tunable pores, *Small*, 2012, **8**, 2436–2444.

Mauro Prato asked: The cell lines that you employed as a breast cancer *in vitro* model display prolonged ERK1/2 MAPK phosphorylation throughout the observational period. What is the phosphorylation status of other MAPKs, such as p38 and JNK?

Anna Lesniak responded: Only ERK1/2 phosphorylation was assessed in this study. We appreciate the suggestion of other pathways and will consider this for our future experiments.

Faraday Discussions

PAPER

Precious metal carborane polymer nanoparticles: characterisation of micellar formulations and anticancer activity

Nicolas P. E. Barry,*[a] Anaïs Pitto-Barry,[a] Isolda Romero-Canelón,[a] Johanna Tran,[a] Joan J. Soldevila-Barreda,[a] Ian Hands-Portman,[b] Corinne J. Smith,[b] Nigel Kirby,[c] Andrew P. Dove,[a] Rachel K. O'Reilly[a] and Peter J. Sadler*[a]

Received 7th May 2014, Accepted 16th June 2014
DOI: 10.1039/c4fd00098f

We report the encapsulation of highly hydrophobic 16-electron organometallic ruthenium and osmium carborane complexes [Ru/Os(*p*-cymene)(1,2-dicarba-*closo*-dodecarborane-1,2-dithiolate)] (**1** and **2**) in Pluronic® triblock copolymer P123 core–shell micelles. The spherical nanoparticles **RuMs** and **OsMs**, dispersed in water, were characterized by dynamic light scattering (DLS), cryogenic transmission electron microscopy (cryo-TEM), and synchrotron small-angle X-ray scattering (SAXS; diameter *ca.* 15 and 19 nm, respectively). Complexes **1** and **2** were highly active towards A2780 human ovarian cancer cells (IC_{50} 0.17 and 2.50 μM, respectively) and the encapsulated complexes, as **RuMs** and **OsMs** nanoparticles, were less potent (IC_{50} 6.69 μM and 117.5 μM, respectively), but more selective towards cancer cells compared to normal cells.

1 Introduction

Dicarba-*closo*-dodecarboranes are a class of boron-rich compounds with globular structures and diameters of *ca.* 1 nm (diameter of a rotating phenyl) that possess unusual properties, including high symmetry and remarkable stability.[1] These characteristics have given rise to numerous applications, and carboranes have been used as building blocks in various systems, such as dendrimers,[2] polymers[3] and nanoparticles.[4,5] They also have been extensively studied as potential boron neutron capture (BNCT) therapeutic agents,[6,7] and as bioisosters of phenyl groups and pharmacophores for targeted drug development.[8] Nevertheless, selective and effective delivery of boron agents is still a key issue that hinders their further clinical development.[9] On the other hand, half-sandwich complexes of ruthenium

[a]*Department of Chemistry, University of Warwick, Gibbet Hill Road, Coventry CV4 7AL, UK. E-mail: N.Barry@warwick.ac.uk; P.J.Sadler@warwick.ac.uk*
[b]*School of Life Sciences, University of Warwick, Gibbet Hill Road, Coventry CV4 7AL, UK*
[c]*Australian Synchrotron, 800 Blackburn Road, Clayton, Victoria 3168, Australia*

and osmium are a versatile class of organometallic compounds. Their biological properties have raised considerable expectations for their use in the treatment of cancer since the early 2000s, and they are considered a promising alternative to platinum-based chemotherapeutics.[10,11] We have recently discussed how the combination of arene ruthenium(II)/osmium(II) complexes and carboranes has unexplored potential in medicine.[12] Furthermore, such complexes exhibit unusual chemistry: coordination of the bulky, electron-deficient carborane ligand 1,2-dicarba-*closo*-dodecarborane-1,2-dithiolato to an arene-Ru or Os metal centre leads to the isolation of rare stable 16-electron complexes, such as [Ru/Os(*p*-cymene)(1,2-dicarba-*closo*-dodecarborane-1,2-dithiolate)] (**1** and **2**). However, these complexes are highly hydrophobic, and their biological applications are impaired by the lack of solubility in water. We have recently discussed how nanotechnology may help to overcome such challenges in medicinal inorganic chemistry.[13] Here, we report the encapsulation of these hydrophobic complexes in water-soluble polymer micelles.

The class of ABA triblock copolymers, where A = hydrophilic block poly(ethylene oxide) (PEO) and B = hydrophobic block poly(propylene oxide) (PPO), are commercially available as Pluronic® (non-proprietary name "poloxamers") and offer a pool of more than 50 amphiphilic,[14] water-soluble and polymorphic materials. The physical and chemical properties of Pluronic® copolymers can be finely tuned by modifying the molar mass ratio between the PEO and PPO blocks (from 1 : 9 to 8 : 2), which directly modifies their *in vivo* properties and interactions with cells and cell membranes, thus providing high potential for the design of innovative nanomedicines and new biomaterials.[15] To exploit the chemistry of carborane-containing arene ruthenium complexes in aqueous solution, and to take advantage of their unique properties, we have encapsulated the 16-electron complexes **1** and **2** in Pluronic® triblock copolymer P123 micelles (Fig. 1). The resulting polymeric micelles **RuMs** and **OsMs**, dispersed in water, have been characterized by UV-visible spectroscopy, DLS, cryo-TEM, and SAXS studies.

Complexes **1** and **2** and micelles **RuMs** and **OsMs** were tested against the A2780 human ovarian cancer cell line. The ruthenium-based complex and micelles were also tested against MRC5 fibroblasts, as an example of non-cancerous, but fast dividing, cells to determine their selectivity and potential therapeutic window. Such a nanotechnology-based strategy not only allows the utilisation of precious

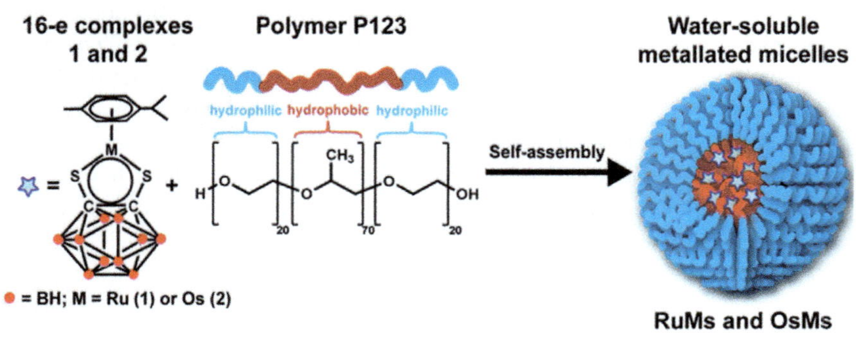

Fig. 1 Self-assembly formation of **RuMs** and **OsMs** (the red dots in **1** and **2** are boron–hydrogen vertices).

metal complexes containing carborane ligands in aqueous solution, but also offers the possibility of formulating and delivering a large number of these complexes (*ca.* 60 per particle) to cancer cells, thus generating a high intracellular boron concentration (*ca.* 600 per particle). There is currently much interest in the therapeutic use of boron-based pharmaceuticals, owing to their chemico-biological properties, to expand potential medical applications in prevention, diagnosis and therapy.[16]

2 Characterisation of the micellar formulations RuMs and OsMs

The organometallic half-sandwich Ru^{II} and Os^{II} arene complexes [Ru/Os(*p*-cymene)(1,2-dicarba-*closo*-dodecarborane-1,2-dithiolate)] (**1/2**) were synthesized as previously reported.[17] These complexes have a *pseudo*-octahedral structure, with a π-bonded arene occupying 3 coordination sites, a S-bound chelated dithiolato dicarba-*closo*-dodecarborane ligand, and a vacant 6th site (Fig. 1). They are 16-electron complexes and therefore electron-deficient at the metal.[18] Complexes **1** and **2** are highly hydrophobic and insoluble in water.[8] To achieve dispersion in water,[13] we encapsulated **1** and **2** in the water-soluble amphiphilic triblock copolymer P123 (poly(ethylene glycol)-*block*-poly(propylene glycol)-*block*-poly(ethylene glycol)) (PEO-PPO-PEO), by mixing a tetrahydrofuran (THF) solution of **1** or **2** with an aqueous solution of polymer P123 (THF–H_2O: 1/10 v/v; [complex] = 5 mg mL^{-1}), at ambient temperature for 4 h (Fig. 1). The encapsulation of the metal complexes in P123 micelles (P123Ms) was accompanied by a dramatic colour change from transparent to blue and purple for **RuMs** and **OsMs**, respectively, indicative of the presence of the 16-electron complexes **1** and **2** in the micelles. We have recently reported the full characterisation of **OsMs** micelles,[19] and now present the characterisation of micellar formulation **RuMs**, in comparison with previous data obtained on **OsMs** micelles.

Dynamic light scattering (DLS) experiments (Fig. 2) unambiguously demonstrated that polymer P123 and complex **1** and **2** self-assemble in aqueous solution. Encapsulation decreased the micellar size (hydrodynamic diameter) of P123 from

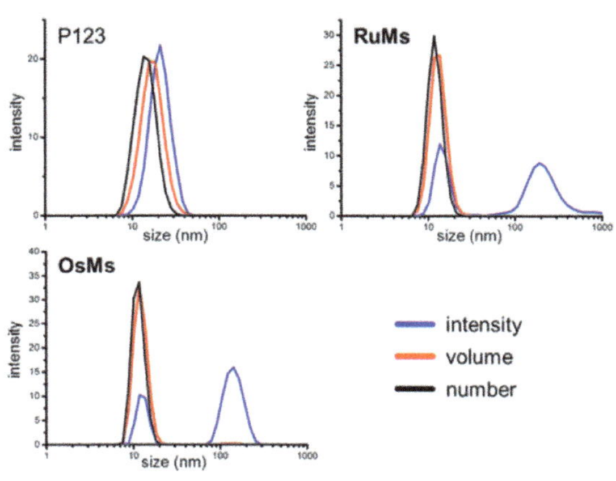

Fig. 2 DLS spectra of P123Ms, **RuMs**, and **OsMs** (1 mg mL^{-1}, H_2O).

19.6 ± 1.8 nm to 11.9 ± 4.1 nm for **RuMs** and 11.5 ± 2.3 nm for **OsMs** with dispersities (*Đ*) of 0.04 and 0.03, respectively. Although micellar size usually increases after encapsulation of organic molecules, incorporation of hydrophobic molecules can result in expulsion of water from micelles, causing a contraction.[20] The hydrophobicity of **1** and **2** probably results in a stronger folding of the PPO chains around the complexes through hydrophobic interactions, with concomitant expulsion of water from the core. A small-in-number (<0.01% in number) second population of **RuMs** and **OsMs** particles is found at $D_h \sim 220$ nm, which exhibits a strong intensity in DLS and is due to aggregation of some particles.

Cryo-TEM analysis without staining was performed on Quantifoil® carbon-coated grids, in order to observe the morphology of the nanoparticles in solution. The high contrast provided by the presence of the heavy ruthenium and osmium centres allowed facile imaging without staining (Fig. 3). From these analyses, it was clear that spherical micellar morphologies are formed when polymer P123 encapsulates complexes **1** or **2**. The observed diameters of these nano-spheres are 10.45 ± 3.43 nm for **RuMs** and 7.85 ± 1.97 nm for **OsMs** with very low dispersities (<1.12). These data are in accordance with the hydrodynamic diameters determined by DLS, within experimental error.

To gain further insight into their structures in aqueous solution, and to confirm cryo-TEM and DLS results, **RuMs**, **OsMs** and P123Ms were analyzed by SAXS (Fig. 4). The experimental profiles were fitted using IgorPro software[21] to three model functions for spherical micelles: SphereForm,[22] CoreShellSphere,[22] and PolyCoreShellRatio[23] (PCR). The PCR model fitted excellently for all micelles with very low dispersity parameters (0.124 for **RuMs**, 0.161 for **OsMs** and 0.146 for P123Ms, 0 being an ideal mono-disperse system, Table 1). These analyses demonstrated that **RuMs** and **OsMs** self-assembly leads to core–shell micelles with a core diameter of 9.88 ± 0.25 nm and 9.06 ± 0.12 nm, and a shell diameter of 9.09 ± 0.29 nm and 6.50 ± 0.15 nm, respectively (Table 1). The diameters of

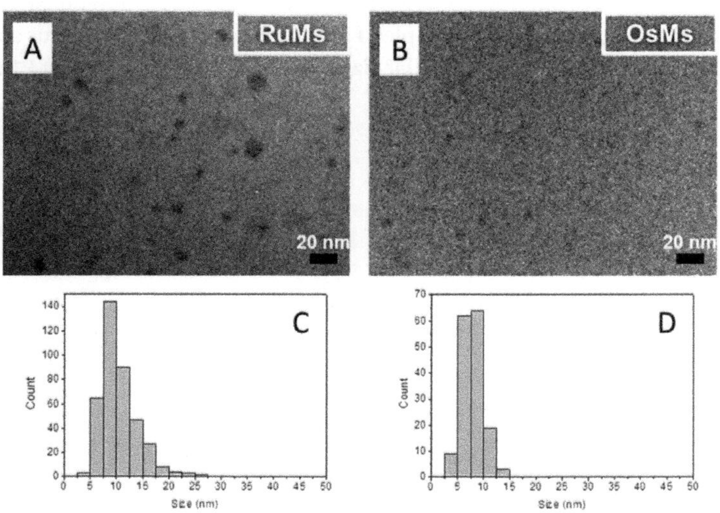

Fig. 3 Cryo-TEM of the self-assembled block copolymer micelles containing encapsulated ruthenium and osmium carborane complexes. (A) and (B) are images and (C) and (D) are particle counting/histogram analyses for **RuMs** and **OsMs**, respectively.

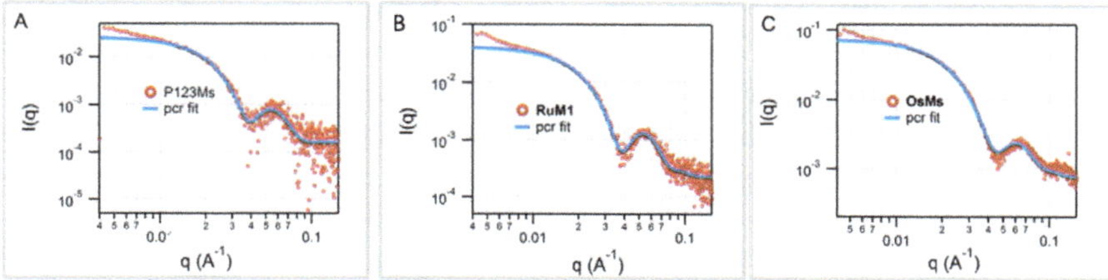

Fig. 4 Small-angle X-ray scattering (SAXS) experimental profiles and fitting with the PolyCoreShellRatio model of micelles (A) P123Ms, (B) **RuMs**, (C) **OsMs**: 1 mg mL^{-1} aqueous solutions.

Table 1 Physical characteristics of **RuMs** and P123Ms micelles determined by DLS, cryo-TEM and SAXS at 1 mg mL^{-1}

	Micelles		
Parameter	**RuMs**	**OsMs**[a]	P123Ms
Aggregation number	66 ± 4	52 ± 6	20 ± 2
Ru/Os complexes per micelle	59 ± 14	52 ± 11	0
DLS diameter (nm)	11.9 ± 4.1	11.5 ± 2.3	19.6 ± 1.8
DLS dispersity	0.04	0.03	0.03
Cryo-TEM diameter (nm)	10.45 ± 3.43	7.85 ± 1.97	nd[b]
Cryo-TEM dispersity	1.11	1.06	nd[b]
SAXS total diameter (nm)	18.97 ± 0.54	15.56 ± 0.27	18.96 ± 0.23
SAXS core diameter (nm)	9.88 ± 0.25	9.06 ± 0.12	6.74 ± 0.06
SAXS shell diameter (nm)	9.09 ± 0.29	6.50 ± 0.15	12.22 ± 0.17
SAXS dispersity	0.124	0.161	0.146

[a] Data from ref. 19. [b] Not determined.

RuMs and **OsMs** micelles determined by DLS and cryo-TEM are in accordance with the core diameter determined by SAXS within experimental error. From these data (scattering length density calculations, degrees of polymerization of Pluronic® P123, and the molecular formulae of the polymer and of complex **1**/**2**), aggregation numbers for **RuMs, OsMs** and P123Ms micelles were determined as 20 ± 2 monomer chains per P123Ms micelle, 66 ± 4 monomer chains per **RuMs** micelles and 52 ± 6 monomer chains per **OsMs** micelles. Determinations of ruthenium and osmium by inductively coupled plasma mass spectrometry (ICP-MS) gave a polymer/complex **1** ratio of 1/0.9 ± 0.11 for **RuMs** and 1/1 ± 0.09 for **OsMs** showing that the 66 polymer chains self-assembled with 59 ± 14 complexes **1** in **RuMs** and 52 monomer chains self-assembled with 52 ± 11 complexes **2** (see Table 1).

3 Anticancer activity of complexes 1 and 2 and micellar formulations RuMs and OsMs

Synthetic polymer therapeutics are of particular interest in medicine, due to their synthetic versatility, as well as their tunable properties.[24] A number of biologically active polymer–drug conjugates and polymeric formulations, such as micelles, hydrogels and polymer-coated nanoparticles, are currently in clinical development.[25] Among the most commonly used polymers for applications in medicine are the ABA Pluronic® triblock block copolymers, which are particularly suitable for the design of bio-inspired, bioengineered and biomimetic polymer nanoparticles. The utilisation of Pluronic® block copolymers as drug delivery systems,[26–31] biological response modifiers,[32–36] pharmaceutical ingredients,[27,37,38] and steric stabilizers to lyotropic liquid crystalline particles[39–41] has led to recent clinical advances.[15]

Inorganic compounds offer different mechanisms of drug action depending on the metal used, their structures and their redox properties.[11,42–85] They can this be utilized for the design of novel drugs in the treatment of a broad range of diseases,[86] and their combination with nanotechnology tools, such as polymer nanoparticles, may provide opportunities for tackling medical challenges in the near future.[13] Surprisingly, only a few studies have been devoted to the encapsulation of inorganic compounds (mainly cisplatin) in Pluronic® block copolymer-based nanostructures.[87,88]

We studied the antiproliferative activity of complexes **1**, **2** and micelles **RuMs** and **OsMs** in A2780 human ovarian cancer cells (Table 2). Cells were exposed for 24 h to the complexes (dissolved in 5% dimethyl sulfoxide (DMSO)/95% saline and further diluted in cell culture medium until working concentrations in the range of 100–0.01 μM were achieved with a maximum DMSO concentration of 1%) or micelles (100% saline, further diluted with cell culture medium to a range of concentration 200–0.10 μM). After this, drugs were removed and cells were washed and placed in fresh growth medium for a further 72 h as a recovery period. Cell viability was then assessed using the sulforhodamine B (SRB) colorimetric assay. Complex **1** is highly potent (IC_{50} 170 nM), about 7× more potent than the clinical drug cisplatin. Complex **1** is also 39× more potent than **RuMs** micelles, which still exhibit good (micromolar) activity towards cancer cells and are as potent as the clinical drug carboplatin. Interestingly, **RuMs** micelles are 4× more selective than **1**, showing only moderate toxicity towards MRC5-fibroblast cells. Finally, the formulation of complexes **1** and **2** in polymer micelles has allowed dispersal of hydrophobic ruthenium and osmium arene complexes in water in a

Table 2 IC_{50} values (μM) of complexes **1** and **2** and micelles P123Ms, **RuMs** and **OsMs** towards A2780 human ovarian cancer cells, IC_{50} values (μM) of complex **1** and micelles **RuMs** for MRC5-fibroblast cells, and selectivity factors ($IC_{50}(MRC_5)/IC_{50}(A2780)$)

	IC_{50} (μM)		
Compound	A2780	MRC5-fibroblast	Selectivity factor
P123Ms	>100	nd	—
1	0.17 ± 0.02	0.31 ± 0.03	1.82
2	2.50 ± 0.09	nd	—
RuMs	6.69 ± 0.33	51.6 ± 0.9	7.71
OsMs	117.50 ± 0.18	nd	—
Cisplatin	1.20 ± 0.10	9.5 (ref. 89)	7.92
Carboplatin	6.0 ± 0.7 (ref. 90)	nd	—

manner suitable for administration to cancer cells (administration to cancer cells in water, without the need to add DMSO).

4 Conclusion

We have encapsulated highly hydrophobic carborane-containing precious metal complexes in triblock copolymer micelles. This has allowed dispersal of hydrophobic ruthenium and osmium arene complexes in water in a manner suitable for administration to cancer cells. Although entrapment of the 16-electron complexes **1** and **2** in Pluronic® micelles leads to a reduction in their anticancer potency, the micelles exhibit enhanced selectivity towards cancer cells compared to normal cells. Polymer encapsulation of metal carborane complexes also provides the potential for delivering high amounts of boron to cells, which is of interest for boron neutron capture therapy (BNCT). The neutron capture ability of such micelles is currently under investigation.

5 Experimental section

Materials

The preparation of the complexes [Ru/Os(*p*-cym)(1,2-dicarba-*closo*-dodecaborane-1,2-dithiolato)] (**1** and **2**) was based on a previous report.[17] The preparation of the **OsMs** micelles was based on a previous report.[19] The triblock copolymer P123 [poly(ethylene glycol)-*block*-poly(propylene glycol)-*block*-poly(ethylene glycol)] was purchased from Sigma-Aldrich and used as received. Anhydrous tetrahydrofuran (Aldrich) was used. 18.2 mega-ohm purity water was collected from a Purelab® UHQ USF Elga system. Holey carbon grids with 200 mesh and lacey carbon grids were purchased from Quantifoil Micro Tools Gmbh and Elektron Technology UK Ltd, respectively, and used as received.

Synthesis of RuMs

A tetrahydrofuran (THF) solution (1 mL) of complex **1** (5 mg mL^{-1}) was added to an aqueous solution (10 mL) of polymer P123 (5 mg mL^{-1}) and the resultant mixture was stirred at ambient temperature for 4 h. The solution was then dialyzed to remove the THF (MWCO = 1000 Da), for 48 h, and then freeze-dried to give **RuMs.** A similar procedure was used for synthesizing **OsMs** with 1 mol equiv. of the Os analogue and 1 mol equiv. of polymer P123.

Instrumentation

UV-visible spectroscopy. UV-visible absorption spectra were recorded on a temperature-controlled Varian CARY 300 Biospectrophotometer using 1 cm pathlength quartz cuvettes (0.5 mL).

Inductively coupled plasma mass spectrometry (ICP-MS). Osmium or ruthenium content was determined using an ICP-MS Agilent technologies 7500 series instrument. The standard for osmium was purchased from Aldrich. Calibration curves were prepared using Ru/Os standard solutions in double deionised water (ddw) with 3% nitric acid, ranging between 50 and 0.5 ppb (9 points). Samples were freshly prepared in ddw with 3% nitric acid. Readings were made in no-gas mode with a detection limit of 1 ppt.

Dynamic light scattering (DLS). The hydrodynamic diameter (D_h) of nanoparticles was determined by DLS. Typically, an aqueous nanoparticle solution was measured with a Malvern Zetasizer NanoS instrument equipped with a 4 mW He–Ne 633 nm laser module at 25 °C. Measurements were carried out at a detector angle of 173° (back scattering). Data were analyzed by the Malvern DTS 6.20 software. D_h was calculated by fitting the apparent diffusion coefficient in the Stokes–Einstein equation $D_h = kT/(3\pi\eta D_{app})$, where k is the Boltzmann constant, T is the temperature and η is the viscosity of the solvent. D_h coincides with the hydrodynamic diameter when the sample is made of monodispersed spherical particles (D_{app} equals the translational diffusion D_t).

Cryogenic electron microscopy (cryo-TEM). A JEOL 2010F TEM was operated at 200 keV and images were recorded on a Gatan UltraScan 4000 camera for cryo-TEM and glow discharge. The samples were prepared at ambient temperature by placing a droplet on a TEM grid. The extra liquid was then blotted with a filter paper and the grid was inserted into liquid ethane at its freezing point. The frozen samples were subsequently kept under liquid nitrogen.

Small-angle X-ray scattering (SAXS). Measurements were carried out on the SAXS beamline at the Australian Synchrotron facility at a photon energy of 11 keV. The samples in solution were in 1.5 mm diameter quartz capillaries. The measurements were collected at a sample to detector distance of 3.252 m to give a q range of 0.004 to 0.2 Å^{-1}, where q is the scattering vector and is related to the scattering angle (θ) and the photon wavelength (λ) by the following eqn (1):

$$q = \frac{4\pi \sin(\theta)}{\lambda} \quad (1)$$

The scattering from a blank (H_2O) was measured in the same location as sample collection and was subtracted for each measurement. Data were normalized for total transmitted flux using a quantitative beamstop detector and absolute scaled using water as an absolute intensity standard. The two-dimensional SAXS images were converted in one-dimensional SAXS profiles ($I(q)$ *versus* q) by circular averaging, where $I(q)$ is the scattering intensity. Functions were used from the NCNR package. Scattering length densities were calculated using the "Scattering Length Density Calculator" provided by NIST Center for Neutron Research.

Cell culture. A2780 human ovarian carcinoma and MRC5 human fetal lung fibroblasts were obtained from the European Collection of Cell Cultures (ECACC). Both cell lines were grown in Roswell Park Memorial Institute medium (RPMI-1640) supplemented with 10% of fetal calf serum, 1% of 2 mM glutamine and 1% penicillin/streptomycin. All cells were grown as adherent monolayers at 310 K in a 5% CO_2 humidified atmosphere and passaged at approximately 70–80% confluency.

***In vitro* growth inhibition assay.** Briefly, 96-well plates were used to seed 5000 cells per well. The plates were left to pre-incubate in drug-free media at 310 K for 48 h before adding different concentrations of the compounds to be tested. A drug exposure period of 24 h was allowed. After this, supernatants were removed by suction and each well was washed with PBS. A further 48 h was allowed for the cells to recover in drug-free medium at 310 K. The SRB assay was used to determine cell viability. IC_{50} values, as the concentration which caused 50% of cell death, were determined as duplicates of triplicates in two independent sets of experiments and their standard deviations were calculated.

Acknowledgements

We thank the Leverhulme Trust (Early Career Fellowship no. ECF-2013-414 to NPEB), the University of Warwick (Grant no. RDF 2013-14 to NPEB), the Swiss National Science Foundation (Grant no. PA00P2_145308 to NPEB and PBNEP2_142949 to APB), the ERC (Grant no. 247450 to PJS), EPSRC (EP/G004897/1 to APB, and EP/F034210/1 to PJS), Institute of Advanced Study (IAS) – University of Warwick (Fellowship to JJSB), and Science City (AWM/ERDF) for support. We thank the Wellcome Trust (055663/Z/98/Z) for funding to the Electron Microscopy Facility, School of Life Sciences, University of Warwick. We also thank COST Action CM1105 for stimulating discussions, Thomas Wilks for supplying the micelle image for Fig. 1, Dr Magdalena Moss for technical asssistance with the cell culture, the Australian Synchrotron and the University of Monash for allocation of beamtime on the SAXS/WAXS beamline and funding.

Notes and references

1 Z. J. Lesnikowski, *Collect. Czech. Chem. Commun.*, 2007, **72**, 1646–1658.
2 R. Djeda, J. Ruiz, D. Astruc, R. Satapathy, B. P. Dash and N. S. Hosmane, *Inorg. Chem.*, 2010, **49**, 10702–10709.
3 K. Kokado, M. Tominaga and Y. Chujo, *Macromol. Rapid Commun.*, 2010, **31**, 1389–1394.
4 D. C. Kennedy, D. R. Duguay, L.-L. Tay, D. S. Richeson and J. P. Pezacki, *Chem. Commun.*, 2009, 6750–6752.
5 J. F. Valliant, K. J. Guenther, A. S. King, P. Morel, P. Schaffer, O. O. Sogbein and K. A. Stephenson, *Coord. Chem. Rev.*, 2002, **232**, 173–230.
6 W. Tjarks, R. Tiwari, Y. Byun, S. Narayanasamy and R. F. Barth, *Chem. Commun.*, 2007, 4978–4991.
7 I. B. Sivaev and V. V. Bregadze, *Eur. J. Inorg. Chem.*, 2009, **2009**, 1433–1450.
8 M. Scholz and E. Hey-Hawkins, *Chem. Rev.*, 2011, **111**, 7035–7062.
9 S. Korbe, P. J. Schreiber and J. Michl, *Chem. Rev.*, 2006, **106**, 5208–5249.
10 G. Süss-Fink, *Dalton Trans.*, 2010, **39**, 1673–1688.
11 A. M. Pizarro, N. P. E. Barry and P. J. Sadler, in *Comprehensive Inorganic Chemistry II*, ed. J. Reedijk, Elsevier, 2013, vol. 3, pp. 752–784.
12 N. P. E. Barry and P. J. Sadler, *Chem. Soc. Rev.*, 2012, **41**, 3264–3279.
13 N. P. E. Barry and P. J. Sadler, *ACS Nano*, 2013, **7**, 5654–5659.
14 http://worldaccount.basf.com/wa/NAFTA/Catalog/ChemicalsNAFTA/pi/BASF/Brand/pluronic.
15 A. Pitto-Barry and N. P. E. Barry, *Polym. Chem.*, 2014, **5**, 3291–3297.
16 M. A. Soriano-Ursúa, B. C. Das and J. G. Trujillo-Ferrara, *Expert Opin. Ther. Pat.*, 2014, **24**, 485–500.
17 M. Herberhold, H. Yan and W. Milius, *J. Organomet. Chem.*, 2000, **598**, 142–149.
18 N. P. E. Barry, T. F. Kemp, P. J. Sadler and J. V. Hanna, *Dalton Trans.*, 2014, **43**, 4945–4949.

19 N. P. E. Barry, A. Pitto-Barry, A. M. Sanchez, A. P. Dove, R. J. Procter, J. J. Soldevila-Barreda, N. Kirby, I. Hands-Portman, C. J. Smith, R. K. O'Reilly, R. Beanland and P. J. Sadler, *Nat. Commun.*, 2014, **5**, 3851.
20 A. Parmar, V. K. Aswal and P. Bahadur, *Spectrochim. Acta, Part A*, 2012, **97**, 137–143.
21 S. Kline, *J. Appl. Crystallogr.*, 2006, **39**, 895–900.
22 A. Guinier and G. Fournet, *Small-angle scattering of X-rays*, Wiley, New York, 1955.
23 J. B. Hayter, in *Physics of Amphiphiles–Micelles, Vesicles, and Microemulsions*, ed. V. DeGiorgio and M. Corti, North-Holland Publishing Company, 1983, pp. 59–93.
24 A. R. Kirtane and J. Panyam, *Nat. Nanotechnol.*, 2013, **8**, 805–806.
25 O. C. Steinbach, *Ther. Delivery*, 2014, **5**, 113–118.
26 D. Y. Alakhova, Y. Zhao, S. Li and A. V. Kabanov, *PLoS One*, 2013, **8**, e72238.
27 E. V. Batrakova, S. Li, A. M. Brynskikh, A. K. Sharma, Y. Li, M. Boska, N. Gong, R. L. Mosley, V. Y. Alakhov, H. E. Gendelman and A. V. Kabanov, *J. Controlled Release*, 2010, **143**, 290–301.
28 E. V. Batrakova and A. V. Kabanov, *J. Controlled Release*, 2008, **130**, 98–106.
29 T. Minko, E. V. Batrakova, S. Li, Y. Li, R. I. Pakunlu, V. Y. Alakhov and A. V. Kabanov, *J. Controlled Release*, 2005, **105**, 269–278.
30 A. V. Kabanov, E. V. Batrakova and V. Y. Alakhov, *J. Controlled Release*, 2002, **82**, 189–212.
31 E. V. Batrakova, S. Li, W. F. Elmquist, D. W. Miller, V. Y. Alakhov and A. V. Kabanov, *Br. J. Cancer*, 2001, **85**, 1987–1997.
32 D. Y. Alakhova, N. Y. Rapoport, E. V. Batrakova, A. A. Timoshin, S. Li, D. Nicholls, V. Y. Alakhov and A. V. Kabanov, *J. Controlled Release*, 2010, **142**, 89–100.
33 E. V. Batrakova, S. Li, V. Y. Alakhov, W. F. Elmquist, D. W. Miller and A. V. Kabanov, *Pharm. Res.*, 2003, **20**, 1581–1590.
34 E. Batrakova, S. Li, V. Alakhov, D. Miller and A. Kabanov, *J. Pharmacol. Exp. Ther.*, 2003, **304**, 845–854.
35 A. Kabanov, E. Batrakova and V. Alakhov, *Adv. Drug Delivery Rev.*, 2002, **54**, 759–779.
36 A. Venne, S. Li, R. Mandeville, A. Kabanov and V. Alakhov, *Cancer Res.*, 1996, **56**, 3626–3629.
37 E. R. Gariepy and J. C. Leroux, *Eur. J. Pharm. Biopharm.*, 2004, **58**, 409–426.
38 I. R. Schmolka, *Poloxamers in the Pharmaceutical Industry*, CRC Press, Boca Ratin FL, 1991.
39 J. Y. T. Chong, X. Mulet, L. J. Waddington, B. J. Boyd and C. J. Drummond, *Soft Matter*, 2011, **7**, 4768–4777.
40 K. W. Lee, T. H. Nguyen, T. Hanley and B. J. Boyd, *Int. J. Pharm.*, 2009, **365**, 190–199.
41 B. J. Boyd, Y. D. Dong and T. Rades, *J. Liposome Res.*, 2009, **19**, 12–28.
42 Z. Liu and P. J. Sadler, *Acc. Chem. Res.*, 2014, **47**, 1174–1185.
43 Y. Zhao, J. A. Woods, N. J. Farrer, K. S. Robinson, J. Pracharova, J. Kasparkova, O. Novakova, H. Li, L. Salassa, A. M. Pizarro, G. J. Clarkson, L. Song, V. Brabec and P. J. Sadler, *Chem.–Eur. J.*, 2013, **19**, 9578–9591.
44 N. A. Smith and P. J. Sadler, *Philos. Trans. R. Soc., A*, 2013, **371**, 1995.

45 I. Romero-Canelón, L. Salassa and P. J. Sadler, *J. Med. Chem.*, 2013, **56**, 1291–1300.
46 I. Romero-Canelón and P. J. Sadler, *Inorg. Chem.*, 2013.
47 Z. Liu, I. Romero-Canelón, B. Qamar, J. M. Hearn, A. Habtemariam, N. P. E. Barry, A. M. Pizarro, G. J. Clarkson and P. J. Sadler, *Angew. Chem., Int. Ed.*, 2014, **53**, 3941–3946.
48 Y. Fu, R. Soni, M. J. Romero, A. M. Pizarro, L. Salassa, G. J. Clarkson, J. M. Hearn, A. Habtemariam, M. Wills and P. J. Sadler, *Chem.–Eur. J.*, 2013, **19**, 15199–15209.
49 J. Schulz, J. Tauchman, I. Císařová, T. Riedel, P. J. Dyson and P. Štěpnička, *J. Organomet. Chem.*, 2014, **751**, 604–609.
50 A. A. Nazarov, C. G. Hartinger and P. J. Dyson, *J. Organomet. Chem.*, 2014, **751**, 251–260.
51 K. J. Kilpin, S. Crot, T. Riedel, J. A. Kitchen and P. J. Dyson, *Dalton Trans.*, 2014, **43**, 1443–1448.
52 C. M. Clavel, E. Paunescu, P. Nowak-Sliwinska and P. J. Dyson, *Chem. Sci.*, 2014, **5**, 1097–1101.
53 K. J. Kilpin and P. J. Dyson, *Chem. Sci.*, 2013, **4**, 1410–1419.
54 G. Gupta, A. Garci, B. S. Murray, P. J. Dyson, G. Fabre, P. Trouillas, F. Giannini, J. Furrer, G. Suss-Fink and B. Therrien, *Dalton Trans.*, 2013, **42**, 15457–15463.
55 F. Giannini, J. Furrer, G. Süss-Fink, C. M. Clavel and P. J. Dyson, *J. Organomet. Chem.*, 2013, **744**, 41–48.
56 A. Leonidova, V. Pierroz, R. Rubbiani, J. Heier, S. Ferrari and G. Gasser, *Dalton Trans.*, 2014, **43**, 4287–4294.
57 T. Joshi, V. Pierroz, C. Mari, L. Gemperle, S. Ferrari and G. Gasser, *Angew. Chem., Int. Ed.*, 2014, **53**, 2960–2963.
58 M. Patra, T. Joshi, V. Pierroz, K. Ingram, M. Kaiser, S. Ferrari, B. Spingler, J. Keiser and G. Gasser, *Chem.–Eur. J.*, 2013, **19**, 14768–14772.
59 M. Patra, K. Ingram, A. Leonidova, V. Pierroz, S. Ferrari, M. N. Robertson, M. H. Todd, J. Keiser and G. Gasser, *J. Med. Chem.*, 2013, **56**, 9192–9198.
60 M. Patra, J. Hess, S. Konatschnig, B. Spingler and G. Gasser, *Organometallics*, 2013, **32**, 6098–6105.
61 P. K. Sasmal, C. N. Streu and E. Meggers, *Chem. Commun.*, 2013, **49**, 1581–1587.
62 Y. Xiang, C. Fu, T. Breiding, P. K. Sasmal, H. Liu, Q. Shen, K. Harms, L. Zhang and E. Meggers, *Chem. Commun.*, 2012, **48**, 7131–7133.
63 A. L. Merkel, E. Meggers and M. Ocker, *Expert Opin. Ther. Pat.*, 2012, **21**, 425–436.
64 A. Kastl, A. Wilbuer, A. L. Merkel, L. Feng, P. Di Fazio, M. Ocker and E. Meggers, *Chem. Commun.*, 2012, **48**, 1863–1865.
65 E. Meggers, *Angew. Chem., Int. Ed.*, 2011, **50**, 2442–2448.
66 K. D. Mjos and C. Orvig, *Chem. Rev.*, 2014, **114**, 4540–4563.
67 M. A. Telpoukhovskaia and C. Orvig, *Chem. Soc. Rev.*, 2013, **42**, 1836–1846.
68 Y. L. K. Tan, P. Pigeon, S. Top, E. Labbe, O. Buriez, E. A. Hillard, A. Vessieres, C. Amatore, W. K. Leong and G. Jaouen, *Dalton Trans.*, 2012, **41**, 7537–7549.
69 G. Sava, G. Jaouen, E. A. Hillard and A. Bergamo, *Dalton Trans.*, 2012, **41**, 8226–8234.
70 A. Vessières, C. Corbet, J. M. Heldt, N. Lories, N. Jouy, I. Laïos, G. Leclercq, G. Jaouen and R.-A. Toillon, *J. Inorg. Biochem.*, 2010, **104**, 503–511.

71 J. W. Yi, N. P. E. Barry, M. A. Furrer, O. Zava, P. J. Dyson, B. Therrien and B. H. Kim, *Bioconjugate Chem.*, 2012, **23**, 461–471.
72 A. Pitto-Barry, O. Zava, P. J. Dyson, R. Deschenaux and B. Therrien, *Inorg. Chem.*, 2012, **51**, 7119–7124.
73 N. P. E. Barry, O. Zava, W. Wu, J. Zhao and B. Therrien, *Inorg. Chem. Commun.*, 2012, **18**, 25–28.
74 N. P. E. Barry, O. Zava, P. J. Dyson and B. Therrien, *J. Organomet. Chem.*, 2012, **705**, 1–6.
75 G. S. Smith and B. Therrien, *Dalton Trans.*, 2011, **40**, 10793–10800.
76 A. Pitto-Barry, N. P. E. Barry, O. Zava, R. Deschenaux and B. Therrien, *Chem.–Asian J.*, 2011, **6**, 1595–1603.
77 A. Pitto-Barry, N. P. E. Barry, O. Zava, R. Deschenaux, P. J. Dyson and B. Therrien, *Chem.–Eur. J.*, 2011, **17**, 1966–1971.
78 L. E. H. Paul, B. Therrien and J. Furrer, *Inorg. Chem.*, 2011, **51**, 1057–1067.
79 N. P. E. Barry, O. Zava, J. Furrer, P. J. Dyson and B. Therrien, *Dalton Trans.*, 2010, **39**, 5272–5277.
80 N. P. E. Barry, O. Zava, P. J. Dyson and B. Therrien, *Aust. J. Chem.*, 2010, **63**, 1529–1537.
81 N. P. E. Barry, F. Edafe, P. J. Dyson and B. Therrien, *Dalton Trans.*, 2010, **39**, 2816–2820.
82 G. Gupta, J. Mahesh Kumar, A. Garci, N. Rangaraj, N. Nagesh and B. Therrien, *ChemPlusChem*, 2014, **79**, 610–618.
83 M. Pernot, N. P. E. Barry, T. Bastogne, C. Frochot, M. Barberi-Heyob and B. Therrien, *Inorg. Chim. Acta*, 2014, **414**, 134–140.
84 G. Gupta, B. Murray, P. Dyson and B. Therrien, *Materials*, 2013, **6**, 5352–5366.
85 P. Chellan, K. M. Land, A. Shokar, A. Au, S. H. An, D. Taylor, P. J. Smith, T. Riedel, P. J. Dyson, K. Chibale and G. S. Smith, *Dalton Trans.*, 2014, **43**, 513–526.
86 N. P. E. Barry and P. J. Sadler, *Chem. Commun.*, 2013, **49**, 5106–5131.
87 A. Sonoda, N. Nitta, S. Ohta, A. Nitta-Seko, S. Morikawa, Y. Tabata, M. Takahashi and K. Murata, *CardioVasc. Interventional Radiol.*, 2010, **33**, 135–142.
88 J.-Y. Fang, S.-H. Hsu, Y.-L. Leu and J.-W. Hu, *J. Biomater. Sci., Polym. Ed.*, 2009, **20**, 1031–1047.
89 S. Dhar and S. J. Lippard, *Proc. Natl. Acad. Sci. U. S. A.*, 2009, **106**, 22199–22204.
90 R. E. Aird, J. Cummings, A. A. Ritchie, M. Muir, R. E. Morris, H. Chen, P. J. Sadler and D. I. Jodrell, *Br. J. Cancer*, 2002, **86**, 1652–1657.

Faraday Discussions

PAPER

Lanthanide nano-drums: a new class of molecular nanoparticles for potential biomedical applications†

Richard A. Jones,*[a] Annie J. Gnanam,[a] Jonathan F. Arambula,[a] Jessica N. Jones,[bc] Jagannath Swaminathan,[cd] Xiaoping Yang,[ae] Desmond Schipper,[a] Justin W. Hall,[a] Lauren J. DePue,[a] Yakhya Dieye,[f] Jamuna Vadivelu,[f] Don J. Chandler,[g] Edward M. Marcotte,[bcd] Jonathan L. Sessler,[a] Lauren I. R. Ehrlich[bc] and Katherine A. Brown[ah]

Received 23rd May 2014, Accepted 9th July 2014
DOI: 10.1039/c4fd00117f

We are developing a new class of lanthanide-based self-assembling molecular nanoparticles as potential reporter molecules for imaging, and as multi-functional nanoprobes or nanosensors in diagnostic systems. These lanthanide "nano-drums" are homogeneous 4d–4f clusters approximately 25 to 30 Å in diameter that can emit from the visible to near-infrared (NIR) wavelengths. Here, we present syntheses, crystal structures, photophysical properties, and comparative cytotoxicity data for six nano-drums containing either Eu, Tb, Lu, Er, Yb or Ho. Imaging capabilities of these nano-drums are demonstrated using epifluorescence, total internal reflection fluorescence (TIRF), and two-photon microscopy. We discuss how these molecular nanoparticles can to be adapted for a range of assays, particularly by taking advantage of functionalization strategies with chemical moieties to enable conjugation to protein or nucleic acids.

1. Introduction

"Nanodiagnostics" involves the use of engineered nanomaterials that can be biofunctionalized with target-specific molecules of interest, allowing ultra-

[a]*Department of Chemistry, The University of Texas at Austin, 105 E. 24th St. Stop A5300, Austin, Texas 78712, USA. E-mail: rajones@cm.utexas.edu; Tel: +1 512 471 1706*
[b]*Department of Molecular Biosciences, The University of Texas at Austin, Austin, Texas 78712, USA*
[c]*Institute for Cellular and Molecular Biology, The University of Texas at Austin, Austin, Texas 78712, USA*
[d]*Center for Systems and Synthetic Biology, The University of Texas at Austin, Austin, Texas 78712, USA*
[e]*College of Chemistry and Materials Engineering, Wenzhou University, Wenzhou 325035, China*
[f]*Department of Medical Microbiology, University of Malaya, Kuala Lumpur 50603, Malaysia*
[g]*Luminex, Austin, Texas, 78727, USA*
[h]*Cavendish Laboratory, Department of Physics, University of Cambridge, Cambridge CB3 0HE, UK*

† Electronic supplementary information (ESI) available. CCDC 921361–921363 and 1004328. For ESI and crystallographic data in CIF or other electronic format see DOI: 10.1039/c4fd00117f

sensitive detection. These materials include metal nanoparticles, DNA-based structures and circuits, nanowires and carbon nanotubes, quantum dots, and nanoporous materials.[1–3] Nanodiagnostic approaches are particularly prevalent in biomarker detection and imaging, and hold promise for facilitating the development of rapid, cost-effective, point-of-care diagnostics, which would be particularly attractive for implementations in both developing and developed countries.[3,4]

Many of these imaging and diagnostics applications rely upon the use of fluorescent dyes or labels as reporter molecules, because of their exceptional measurement contrast properties. However, both fluorescent dyes and protein-based fluorescent probes have technical limitations, including susceptibility to severe photo-bleaching (which limits long-term analysis), and small Stokes shifts. In comparison, lanthanide-based bioprobes have several advantages over conventional organic dyes or semiconductor quantum dots. These advantages include a very large pseudo-Stokes shift between the excitation and emission wavelengths, absence of photo-bleaching, long lived excited states, and narrow emission bands which enable selective detection.[5] The long emission lifetimes of lanthanide(III) complexes (1 μs to 5 ms) make them attractive bioprobes, because these lifetimes offer the possibility of using time-gated acquisition methods to enhance signal/noise by minimizing interference from light scattering or autofluorescence.[6–8]

As part of our research focused on high nuclearity lanthanide moieties, we recently reported a new class of luminescent high nuclearity self-assembled cadmium-lanthanide cluster molecules ("molecular nanoparticles") with unusual drum-like architectures.[9,10] In the current study, we have evaluated the potential of these molecules to serve as probes in biological applications. Our original report described nano-drum NIR emitters that featured Nd, Yb, Er, as well as Gd.[10] Since many lanthanide-based probes systems use Eu or Tb, which emit in the visible spectrum, we have also investigated the synthesis and photophysical properties of nano-drums based on these two metals. We describe here the synthesis of four new examples of molecular nano-drums, $[Ln_6Cd_{18}L_9Cl_8(OAc)_{28}]$ (Ln = Eu (**1**) and Tb (**2**)), $[Lu_6Cd_{18}L_9Cl_{10}(OAc)_{26}]$ (**3**), and $[Ho_8Cd_{24}L_{12}(OAc)_{48}]$ (**4**). We have structurally characterized each of these compounds by single crystal X-ray diffraction studies. We explore the potential of this class of compounds for biological applications by assessing their cytotoxicity using two cancer cell lines. We also investigate the imaging properties of nano-drums containing Er and Yb, which are analogues of **4**, using three microscopy platforms commonly used for visualization of cells and tissues: epifluorescence, TIRF, and two-photon microscopy.

2. Experimental

2.1 Synthesis of $[Ln_6Cd_{18}L_9Cl_8(OAc)_{28}]$ (Ln = Eu (1) and Tb (2)) and $[Lu_6Cd_{18}L_9Cl_{10}(OAc)_{26}]$ (3) and $[Ho_8Cd_{24}L_{12}(OAc)_{48}]$ (4)

All reactions were performed under dry oxygen-free dinitrogen atmospheres using standard Schlenk techniques. Metal salts and other solvents were purchased from Aldrich and used directly without further purification. The Schiff-base ligand H_2L (*N*,*N*′-bis(3-methoxysalicylidene)hexane-1,6-diamine) was prepared according to well-established procedures.[11] Melting points were obtained in sealed glass capillaries under dinitrogen and are uncorrected. Complexes $[Er_8Cd_{24}L_{12}(OAc)_{48}]$ (5) or $[Yb_8Cd_{24}L_{12}(OAc)_{48}]$ (6) were prepared as previously described.[10]

$[Eu_6Cd_{18}L_9Cl_8(OAc)_{28}]$ (1). 0.52 mmol (0.1382 g) of $Cd(OAc)_2 \cdot 2H_2O$ and 0.12 mmol (0.0439 g) of $EuCl_3 \cdot 6H_2O$ were dissolved in 60 mL MeOH at room temperature, and 0.26 mmol (0.1001 g) H_2L and 0.30 mmol Et_3N in 10 mL EtOH were then added. The resulting solution was stirred and heated under reflux for 30 min, allowed to cool, and then filtered. Diethyl ether was allowed to diffuse slowly into the filtrate at room temperature and pale yellow crystals were obtained after one week. The crystals were filtered off and washed with EtOH (5 mL). Yield (based on $EuCl_3 \cdot 6H_2O$): 0.0926 g (52%). m. p. > 175 °C (dec.). ^1H NMR (600 MHz, CD_3OD): δ (ppm) −19.644 (2H), −19.349 (2H), −13.287 (2H), −9.285 (4H), −2.337 (4H), −2.245 (12H), −1.767 (4H), −1.546 (2H), −1.417 (18H), −0.805 (6H), −0.636 (4H), 0.337 (2H), 0.479 (12H), 0.660 (6H), 0.863 (4H), 0.991 (8H), 1.075 (20H), 1.140 (10H), 1.201 (12H), 1.261 (6H), 1.449 (12H), 1.903 (12H), 2.023 (2H), 2.118 (2H), 3.592 (6H), 3.657 (6H), 3.856 (6H), 4.231 (12H), 4.495 (4H), 4.634 (4H), 4.890 (12H), 5.168 (6H), 5.438 (4H), 5.524 (8H), 6.184 (4H), 6.426 (4H), 6.780 (2H), 6.879 (4H), 6.912 (2H), 8.321 (2H), 9.642 (4H), 10.372 (4H), 11.252 (12H), 12.511 (12H), 12.929 (4H), 13.117 (8H), 13.344 (4H), 14.607 (12H), 16.249 (4H). IR (CH_3OH, cm^{-1}): 2930 (w), 1632 (m), 1572 (s), 1465 (m), 1408 (s), 1307 (m), 1212 (s), 1079 (w), 1017 (w), 850 (m), 734 (s).

$[Tb_6Cd_{18}L_9Cl_8(OAc)_{28}]$ (2). The procedure was the same as that for **1** using $TbCl_3 \cdot 6H_2O$ (0.0448 g, 0.12 mmol). Pale yellow single crystals of **2** were formed after one week. Yield: 0.062 g (35%). m. p. > 184 °C (dec.). ^1H NMR (600 MHz, CD_3OD. The numbers of protons cannot be determined.): δ (ppm) −57.144, −52.939, −44.543, −26.494, 1.164, 1.485, 1.932, 2.136, 3.317, 3.444, 3.620, 3.892, 4.085, 6.629, 6.808, 6.913, 7.212, 8.354, 9.939, 12.594, 13.139, 21.931, 23.010, 25.508, 26.425, 35.846, 37.827, 50.759, 64.747, 78.970, 81.387, 94.693, 133.359. IR (CH_3OH, cm^{-1}): 2933 (m), 2851 (m), 1631 (s), 1602 (w), 1548 (w), 1456 (s), 1412 (m), 1303 (m), 1237 (m), 1213 (s), 1074 (m), 1025 (w), 849 (m), 741 (s).

$[Lu_6Cd_{18}L_9Cl_{10}(OAc)_{26}]$ (3). The procedure was the same as that for **1** using $LuCl_3 \cdot 6H_2O$ (0.0467 g, 0.12 mmol). Pale yellow single crystals of **3** were formed after one week. Yield: 0.053 g (30%). m. p. > 174 °C (dec.). ^1H NMR (600 MHz, CD_3OD. The numbers of protons cannot be determined.): δ (ppm) 0.762, 1.010, 1.119, 1.153, 1.116, 1.247, 1.452, 1.860, 1.889, 1.908, 1.945, 1.975, 2.070, 2.116, 2.153, 2.272, 2.304, 2.365, 2.401, 2.848, 2.892, 2.923, 2.986, 3.152, 3.430, 3.439, 3.600, 3.660, 3.713, 3.791, 3.854, 3.891, 3.948, 6.454, 6.515, 6.685, 6.777, 6.876, 6.906, 6.977, 7.132, 7.198, 7.322, 7.440, 7.615, 7.675, 7.936, 8.184, 8.326, 9.919, 10.983, 13.850 (br). IR (CH_3OH, cm^{-1}): 2934 (w), 1635 (s), 1579 (s), 1549 (s), 1452 (s), 1406 (s), 1307 (m), 1238 (m), 1213 (s), 1080 (m), 742 (s).

$[Ho_8Cd_{24}L_{12}(OAc)_{48}]$ (4). $Ho(OAc)_3 \cdot 4H_2O$, (0.0497 g, 0.12 mmol), $Cd(OAc)_2 \cdot 2H_2O$, (0.1384 g, 0.52 mmol) and H_2L (0.1000 g, 0.26 mmol) were dissolved in MeOH (25 mL) at room temperature and Et_3N (1 mL) added. The resulting solution was stirred and heated under reflux for 30 minutes. After cooling to room temperature, the solution was filtered and the filtrate transferred to test tubes. Diethyl ether vapour was allowed to diffuse slowly into the filtrate at room temperature and pale yellow crystals were obtained after one week. The crystals were filtered off and washed with MeOH (2 mL). Yield (based on $Ho(OAc)_3 \cdot 4H_2O$): 0.0823 g (48%). m. p. > 178 °C (dec.). IR (CH_3OH, cm^{-1}): 2932 (w), 1889 (w), 1634 (m), 1574 (s), 1435 (s), 1307 (m), 1169 (w), 1213 (s), 1080 (m), 1044 (w), 962 (s), 873 (s), 818 (w), 742 (s), 669 (m), 641 (w).

2.2 X-ray structure determinations

Powder X-ray diffraction data were collected on a Rigaku R-AXIS RAPID II. Single crystal data were collected on a Rigaku Saturn Kappa CCD diffractometer with graphite monochromated Mo-Kα radiation ($\lambda = 0.71073$ Å) at 223 K. The data set was corrected for absorption based on multiple scans and reduced using standard methods. Data reduction was performed using DENZO-SMN.[12] The structures were solved by direct methods and refined anisotropically using full-matrix least-squares methods with the SHELX 97 program package.[13] Coordinates of the non-hydrogen atoms were refined anisotropically, while hydrogen atoms were included in the calculation isotropically but not refined. Neutral atom scattering factors were taken from Cromer and Waber.[14] Crystallographic data for all complexes are given in the ESI in Table S1† and selected bond lengths and angles are given in Tables S2–S4.†

2.3 Bead preparation

Polystyrene beads (6 μm, crosslinked) were suspended in a 50 : 50 (v/v) solution of methanol and chloroform in a 1.5 mL microfuge tubes. Typically 0.5–1 mg of compound was added and the suspension was then rotated from 2–4 days to permit bead loading. Suspensions were then dried under vacuum overnight and beads removed and then washed and re-suspended in phosphate buffered saline (PBS) prior to imaging.

2.4 Spectroscopic characterization

Physical measurements were made using the following instrumentation: NMR: VARIAN UNITY-plus. 600 spectrometer (^{1}H, 600 MHz) at 298 K in CD_3OD; IR: Nicolet IR 200 FTIR spectrometer. Absorption spectra were obtained on a BECKMAN DU 640 spectrophotometer, excitation and visible emission spectra on a QuantaMaster PTI fluorimeter. Fluorescence quantum yields were determined by using quinine sulfate ($\Phi_{em} = 0.546$ in 0.5 M H_2SO_4) and $[Ru(bipy)_3]Cl_2$ (bipy = 2,2′-bipyridine; $\Phi_{em} = 0.028$ in water) as standard for the Tb^{3+} and Eu^{3+} complexes, respectively.[15]

X-ray photoelectron spectra (XPS) were obtained using a Kratos AXIS Ultra spectrometer equipped with a monochromatized Al Kα source, hybrid optics, and a delay line detector coupled to a hemispherical analyzer. The analysis chamber base pressure was typically 2×10^{-9} Torr. All spectra were recorded using a single sweep with a spot size of 300 μm × 700 μm. Survey scans were collected from 0–1200 eV with a pass energy of 80 eV, step size of 1 eV, and a dwell time of 250 ms. The high resolution component spectra were collected with a pass energy of 20 eV, step size of 0.1 eV, and a dwell time of 3600 ms. Binding energies were referenced to the adventitious carbon line (C 1s, 284.8 eV) and a charge neutralization was applied during all acquisitions. Casa XPS analysis software was used for stoichiometry determination of the samples and Kratos sensitivity factors used for each element of interest. All samples were washed with methanol three times and dried under vacuum for 18 hours before placing in an Al cup within the XPS instrument.

Samples were carefully spread evenly within the cup using a blunt tool covered with Al foil between each procedure to avoid contamination.

2.5 Cytotoxicity assays

The proliferation of exponential phase cultures was carried out using either an A549 or an AGS cancer cell line. A549 cells were seeded in 96-well microliter plates at 1000 cells per well and allowed to adhere overnight in 100 μL RPMI 1640 medium supplemented with 2 mM L-glutamine, 10% heat inactivated fetal bovine serum, and antibiotics (200 U cm^{-3} penicillin and 200 μg cm^{-3} streptomycin). AGS cells were seeded in 96-well microliter plates at 8000 cells per well, grown in F-12 medium supplemented with 10% heat inactivated fetal bovine serum, 100 U mL^{-1} penicillin, and 100 μg mL^{-1} streptomycin, and allowed to adhere for 1 day. Cell viability was assessed by tetrazolium salt reduction. Stock solutions of $[Ho_8Cd_{24}L_{12}(OAc)_{48}]$ (**4**), holmium acetate, or H_2L ligand (all 5 mM) in 50/50 methanol–water were formulated and then diluted in medium for secondary stocks of 20–200 μM depending on the complex being tested. Secondary stock solutions were serially diluted in medium and immediately added to wells, whereupon plates were incubated at 37 °C under a 5% CO_2/95% air atmosphere. After a total of 3 days, a 50 μL aliquot of 3 mg mL^{-1} tetrazolium dye, 3-(4,5-dimethylthiazol-2-yl)-2,5-diphenyltetrazolium bromide (MTT, Sigma Chemical), was added to each well, followed by a 4 hour incubation at 37 °C. The medium was then removed and the resulting formazan was dissolved in 50 μL DMSO and absorbances measured at 560–650 nm using a microplate reader (Molecular Devices, Sunnyvale, CA). Absorbances were corrected for background and the values normalized to wells containing untreated cells to allow plate-to-plate comparison. The data are shown as mean inhibition of proliferation or growth as a percentage of control cells' proliferation or growth from 2–3 replicate values.

2.6 Microscopy

A flow chamber was constructed from a pre-cleaned glass slide and coverslip joined by double-sided sticky tape. About 1 mg of Yb and Er loaded beads were suspended in 20 μL of water and loaded in the flow chamber, which was used for both the epifluorescent and TIRF imaging. For epifluorescent imaging, the slide was mounted on a Nikon Eclipse TE2000-E inverted microscope (Nikon Inc, Japan) and images of the beads acquired at 1 frame per second using a Cascade II 512 camera (Photometrics, Tucson, AZ) on a Nikon Apo 60×/NA 0.95 objective. A combination of excitation filters DAPI – AT350/50× (340–380 nm), FITC – ET490/20× (465–495 nm) or TRITC – ET555/25× (528–553 nm) and emission filters DAPI – ET460/50 m (435–485 nm), FITC – ET525/36 m (515–555 nm), TRITC – ET605/52 m (590–650 nm) or Cy5 – ET700/60× (640–730 nm) were used to determine the capability of imaging the lanthanide dyes. The Sutter Lambda 10-3 filter wheels (Sutter Instrument, Novato, CA), motorized stage (Prior Scientific Inc; Rockland, MA), and image analysis were driven by Nikon NIS Elements Imaging Software.

TIRF imaging was performed in the objective mode using a Nikon Apo 60×/NA 1.49 oil objective set on a Nikon Eclipse Ti inverted microscope, fitted with a ProScan II motorized stage (Prior Scientific Inc; Rockland, MA) and a monolithic laser launch unit (Agilent Technologies; Santa Clara, CA) that is capable of providing tunable power and wavelength. The beads were excited by a 405 nm laser at 2 mW power (at source of fiber optic) and the images acquired using an EMCCD iXon 3 camera (Andor; Belfast, UK) at 1 frame per second exposures before being filtered by a TIRF Quad Cube (Chroma Technology, Bellows Falls, VT). The Quad band filter is a setup of BandPass filters and dichroics that allow for 405/488/561/647 +/−10 nm excitation and emission between 417–476 nm, 500–547 nm, 574–620 nm, and 637–670 nm.

Two-photon fluorescence microscopy images were acquired with an Ultima IV microscope using PrairieView acquisition software and an Olympus Plan FL 20× water immersion objective (Prairie Technologies; Middleton, WI). Cells and beads were illuminated with a MaiTai Ti:sapphire laser (SpectralPhysics; Santa Clara, CA) tuned to 760 nm, or scanned from 690–1000 nm in 10 nm steps. Emitted light was passed through 400/50 nm, 480/40 nm, 535/50 nm, and 607/45 nm bandpass filters (Chroma Technology; Bellows Falls, VT) to separate PMTs for detection.

3. Results

3.1 Synthesis and X-ray structures of nano-drums

Reactions of the flexible Schiff base ligand H_2L (Fig. 1) with $Cd(OAc)_2 \cdot 4H_2O$ and $LnCl_3 \cdot 6H_2O$ (Ln = Eu, Tb and Lu) in refluxing MeOH/EtOH produced yellow solutions from which the Cd–Ln complexes **1–3** were isolated as pale yellow crystalline solids in 52%, 35% and 30% yields, respectively. The larger 32-metal holmium complex **4** is isomorphous with the analogous Nd, Yb, and Er nano-drums reported recently[10] and was prepared in a similar fashion using $Ho(OAc)_2 \cdot 4H_2O$. Complexes **1** and **2** are isomorphous, and reveal a 24-metal drum-like structure. A key factor affecting the nano-drum self-assembly process is the use of the flexible Schiff-base ligand H_2L, which facilitates construction of the nano-drum. The central six carbon chains of the ligands form the sides of the drum, while the deprotonated phenolic oxygen atoms, imine nitrogen and methoxy oxygen atoms bind the metals at the top and bottom of the drum.

Two views of the crystal structure of **1** are shown in Fig. 2. The complex is of nanoscale proportions (16 × 21 × 21 Å). The ends of the drum are created by two rings of 12 metals (3 Eu(III) and 9 Cd(II)) coordinated to half of the N, O binding groups of the 9 Schiff base ligands, plus 14 OAc^- and 4 Cl^- units. The sides of the drum are formed by the $-(CH_2)_6-$ linkers of the Schiff base ligands. Each Schiff base ligand is thus coordinated to metals at both ends of the drum. In each 12-metal ring, three Eu ions are eight-coordinate. Two Eu(III) ions (Eu(1) and Eu(2)) have similar coordinate environments and each is bound by two Schiff base ligands and four bridging OAc^- anions. A bridging Cl^- anion replaces an acetate OAc^- in the coordinate environment of the third Eu(III) ion, (Eu(3)). For the nine Cd^{2+} ions, four of them (Cd(2), Cd(4), Cd(6) and Cd(9)) have bi-pyramidal

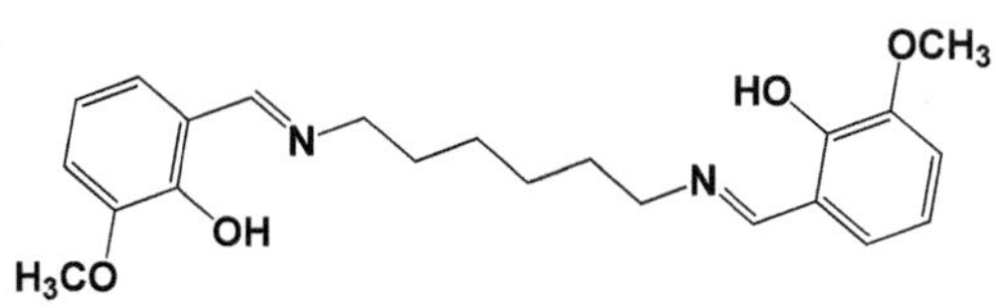

Fig. 1 Schiff base ligand H_2L.

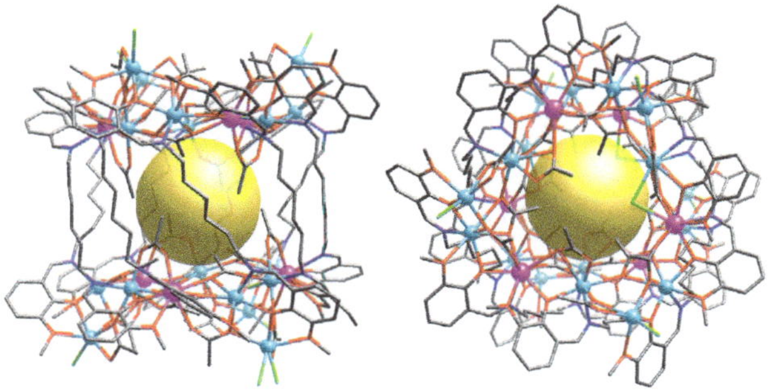

Fig. 2 A view of the crystal structure of **1**. Viewed along the *a*-axis (left) and *c*-axis (right) (Eu^{3+}: purple; Cd^{3+}: blue).

geometries. They are coordinated to one Schiff base ligand, one Cl^- anion, and two (or three) OAc^- anions. The other five Cd^{2+} ions are seven-coordinate: three of them (Cd(1), Cd(3) and Cd(7)) are bound by two Schiff base ligands and two OAc^- anions, while two of them (Cd(5) and Cd(8)) are coordinated to one Schiff base and four OAc^- anions (ESI†).

The drum-like architecture of complex **3** is *mm* symmetric, formed by two equivalent 12-metal rings ($Lu_3Cd_9Cl_5(OAc)_{13}$) linked by nine Schiff base ligands (Fig. S1, ESI†). The slight difference between **1** and **2** *vs.* **3** is that, in **3**, one bridged Cl^- anion replaces a tridentate OAc^- anion. In **1–3**, each Ln^{3+} ion and its closest two Cd^{2+} ions are linked by phenolic oxygen atoms of L^{2-}, OAc^- anions and Cl^- anions. The average distances between Ln^{3+} and Cd^{2+} ions are 3.675 Å, 3.671 Å and 3.669 Å for **1**, **2** and **3**, respectively. Powder XRD patterns of **1–3** are similar to their simulated patterns generated from single crystal X-ray data (Fig. S2†).

Two views of the crystal structure of the 32-metal Ho–Cd drum [$Ho_8Cd_{24}L_{12}(OAc)_{48}$] (**4**) are shown in Fig. 3. The ends of the drum are created by two rings of 16 metals (4 Ho(III) and 12 Cd(II)) coordinated to half of the N, O binding groups of the 12 Schiff base ligands plus 24 OAc^- anions. The sides of the drum are formed by the $-(CH_2)_6-$ linkers of the Schiff base ligands. Each Schiff base ligand is thus coordinated to metals at both ends of the drum. In each 16-metal ring, four Ho ions are eight-coordinate. For the 12 Cd^{2+} ions, four of them have bipyramidal geometries, and the other eight Cd^{2+} ions are seven-coordinate. Detailed descriptions of the structures of the isomorphous Er and Yb analogs of **4** are given in ref. 10.

3.2 Spectroscopic characterization

^1H NMR spectra of **1** and **2** in CD_2Cl_2 contain multiple broad peaks ranging from −20 to +18 ppm and −70 to +140 ppm, respectively, while **3** shows a narrower range of 0.7–14 ppm due to the diamagnetic Lu^{3+} ion. The ^1H NMR spectra of **1–3** remain unchanged for several weeks, indicating that they are stable in solution (Fig. S3–S5†). The photophysical properties of **1** and **2** were studied in CH_3CN. The free ligand H_2L exhibits absorption bands at 226, 260 and 327 nm which are all red-shifted upon coordination to metal ions in **1–3** (Fig. S6†). Complexes **1–3** exhibit similar solution absorption spectra. For the free ligand H_2L, excitation of the absorption band at 410 nm produces a broad emission band at $\lambda_{max} = 515$ nm

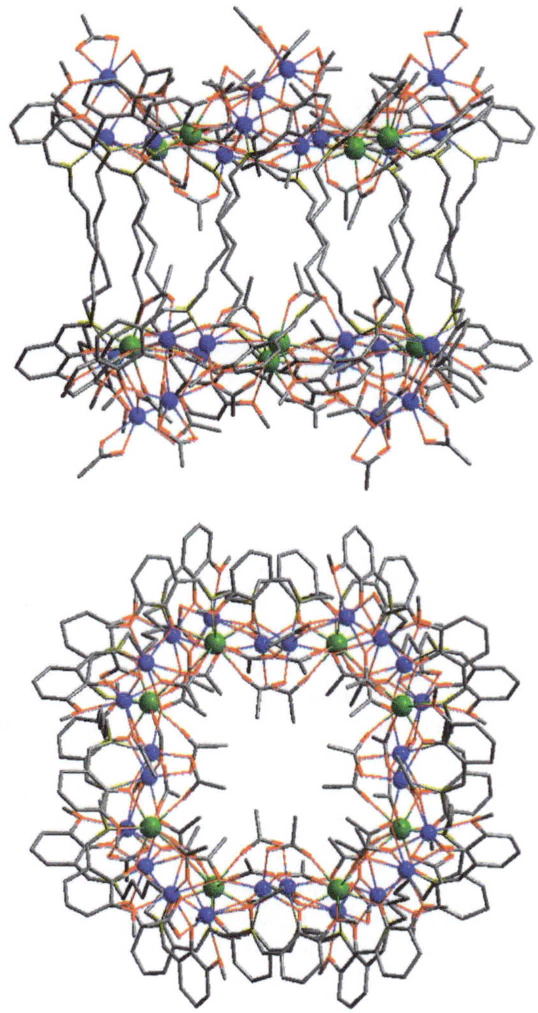

Fig. 3 A view of the crystal structure of 4. Viewed along the *a*-axis (top) and *c*-axis (lower). (Ho^{3+}: green; Cd^{3+}: blue).

(Fig. 4). For **3**, a broad blue-shifted emission from the Cd/L center can be detected at $\lambda_{max} = 440$ nm upon excitation of the ligand-centered absorption band at 380 nm. Complexes **1** and **2** show the visible emission bands of Eu^{3+} ($^5D_0 \rightarrow {}^7F_j$ transitions, $j = 0, 1, 2, 3$ and 4) and Tb^{3+} ($^5D_4 \rightarrow {}^7F_j$ transitions, $j = 6, 5, 4$ and 3) ions, respectively, while the Cd/L-centered emission is not detected (Fig. 4). The appearance of the symmetry-forbidden emission $^5D_0 \rightarrow {}^7F_0$ at 579 nm indicates that the Eu^{3+} ions in **1** occupy sites with low symmetry and have no inversion center.[17] This is in agreement with the result of the single-crystal X-ray analysis. It is also further confirmed by the intensity ratio of 3.8 for $I(^5D_0 \rightarrow {}^7F_2)/I(^5D_0 \rightarrow {}^7F_1)$, which is a good measure of the nature and symmetry of the first coordination sphere of Eu^{3+} ion.[18]

In **1** and **2**, there are no solvent molecules such as MeOH or H_2O coordinated to the lanthanide ions which could quench lanthanide luminescence. Also, the

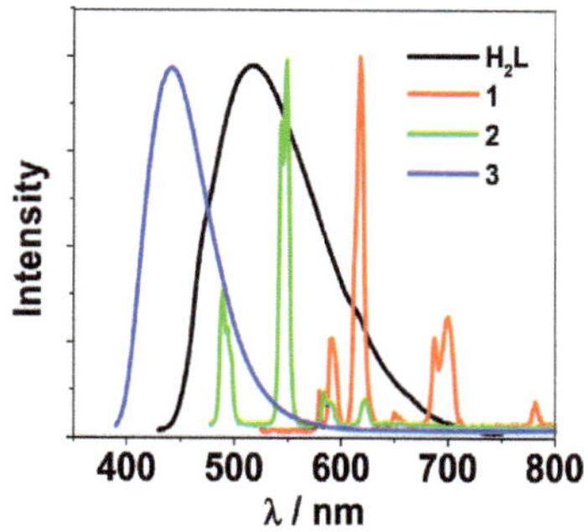

Fig. 4 Emission spectra of the free ligand H_2L and complexes **1–3** in CH_3CN at room temperature.

Ln(III) centers are shielded within the drum-like structures and therefore fairly well protected from the outside solvent environment. This shielding may help to improve their luminescent properties. The fluorescence quantum yields (Φ_{em}) of **1** and **2** are 0.277 and 0.093, respectively. These values are higher than those of polynuclear complexes and 1-D coordination polymers formed with the Schiff base ligands that feature shorter chain lengths between ends.[19] Interestingly, the Cd/L based emission is also observed as a broad band centered at about 540 nm for the larger 32-metal drums based on Er and Yb[10] while for the smaller drums based on Eu and Tb (**1** and **2**) the absence of Cd/L-centered luminescence in the emission spectra suggests that the L → Ln and Cd → Ln energy transfers take place efficiently.[20]

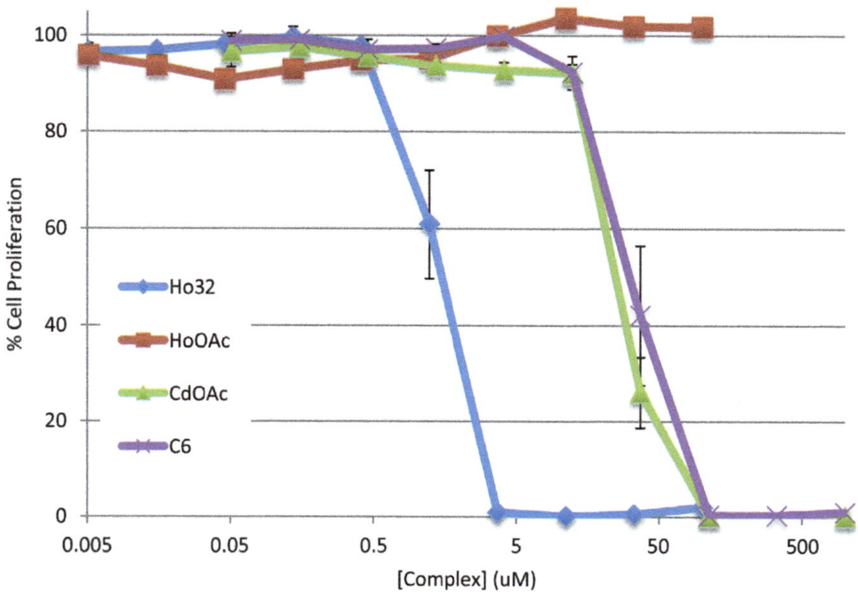

Fig. 5 Dose responsive cell proliferation curves of A549 lung cells treated with indicated complex. Error bars represent standard deviation.

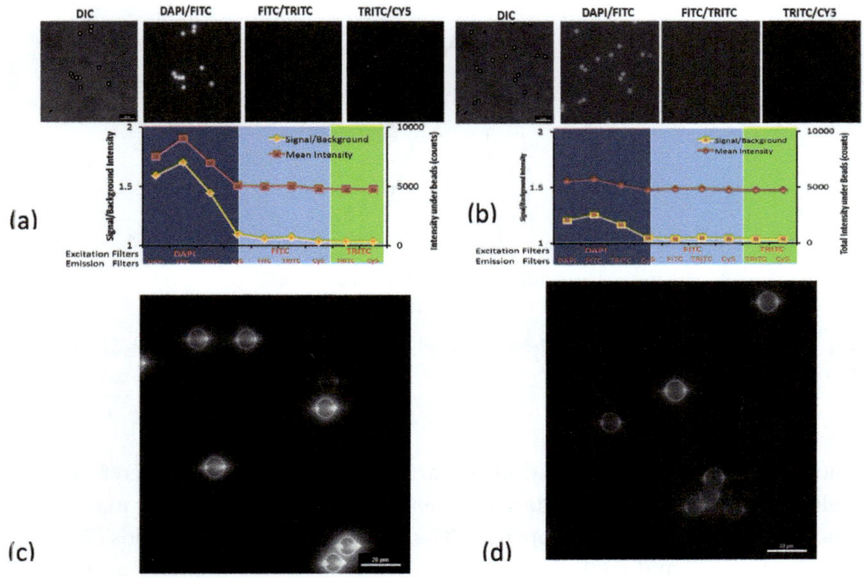

Fig. 6 Polystyrene beads doped with Er or Yb nano-drums show bright fluorescence under epifluorescent and total internal reflection fluorescence (TIRF) microscopy. Epifluorescence images of (a) Er or (b) Yb nano-drum-loaded polystyrene beads excited with an illumination source filtered by a DAPI filter. Images and associated signal/background intensities are displayed for the DAPI, FITC, TRITC and Cy5 emission filter channels. TIRF images of (c) Er or (d) Yb nano-drum-loaded polystyrene bead obtained by exciting at 405 nm.

3.3. Cytotoxicity screening

To determine the inherent cytotoxicity of the complexes, cell proliferation assays were conducted using A549 lung cancer cells and AGS gastric cancer cells using the holmium nano-drum, $[Ho_8Cd_{24}L_{12}(OAc)_{48}]$ (**4**), shown in Fig. 3. As shown in Fig. 5 for the A549 cells, **4** displayed an IC_{50} value of 1.32 ± 0.11 μM. A similar IC_{50} value was obtained for AGS cells, approximately 5 μM. Metal salt $Cd(OAc)_2 \cdot 2H_2O$ and the C6 ligand provided IC_{50} values of 31.0 ± 7.0 and 29.7 ± 1.2 μM, respectively, while $Ho(OAc)_3 \cdot 4H_2O$ displayed no inhibition of cell proliferation under the concentrations tested.

3.4. Microscopy studies

Crosslinked polystyrene beads (6 mm) containing either $[Er_8Cd_{24}L_{12}(OAc)_{48}]$ (**5**) or $[Yb_8Cd_{24}L_{12}(OAc)_{48}]$ (**6**) were visualized using epiflouresence, TIRF, and two-photon microscopy by observing emission in the visible spectrum. Using the epifluorescence platform, both bead types can be excited with an illumination source filtered with a DAPI filter. The bead images show high signal to background intensity and high mean intensity under the DAPI, FITC, TRITC and Cy5 emission filter channels (Fig. 6). The use of longer wavelength excitation filters did not result in significant emission from either bead type. The same beads exhibit high fluorescent signal with a 405 nm excitation laser with the TIRF imaging platform (Fig. 6). The two-photon fluorescence images shown in Fig. 7

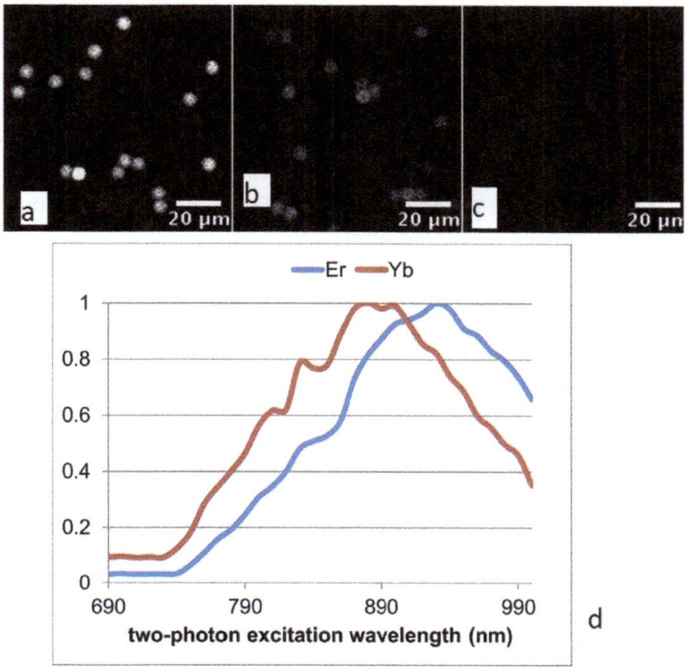

Fig. 7 Polystyrene beads loaded with Er and Yb nano-drums show bright fluorescence under two-photon fluorescence excitation microscopy. (a) Er- and (b) Yb-loaded polystyrene beads were excited at 760 nm with constant laser power and detector sensitivity, with the images shown obtained by PMT with 535/50 nm emission filter. As a control, unloaded beads lack fluorescence (c). Normalized fluorescence intensity in the 535/50 nm emission channel is shown for when the laser is scanned from 690 to 1000 nm in 10 nm steps (d).

were acquired at constant laser power and detector sensitivity. Beads were excited at 760 nm, and the images shown were obtained by PMT with a 535/50 nm emission filter. Within this range, Er beads emitted with greater brightness than Yb beads upon two-photon excitation. Relative normalized fluorescence intensity in the 535/50 nm emission channel is shown for laser scans from 690 to 1000 nm in 10 nm steps. The Er (**5**) and Yb (**6**) nano-drum-containing beads display an excitation peak at 930 nm and 880 nm, respectively. In addition, beads loaded with either **5** or **6** were shown by XPS to contain Ln : Cd ratios of 1 : 3 (Fig. S7–S9 and Table S5†), consistent with the ratio revealed by single crystal X-ray studies.[10] Control beads lacking nano-drums did not display significant fluorescence in any of the microscopy studies presented here.

4. Discussion

The studies presented here explore the potential of a novel class of molecular lanthanide-based nanoparticles, called nano-drums, as bioprobes for biological imaging applications. Desirable features of a bioprobe include aqueous stability, thermodynamic stability and favourable photophysical properties, including narrow band emissions and no photobleaching.[5,21,22] Their crystal structures reveal that these nano-drums are discrete multinuclear lanthanide clusters that self-assemble as homogeneous molecular species in a "drum" shape. The dimensions of these structures can be tuned at the molecular level, for example, by varying the anion, which changes the shape and composition of the nano-drum to yield a 24- or 32-metal cluster, but still yields a perfectly homogeneous particle size rather than a heterogeneous collection of different sized species. The changes that result from varying the anion also demonstrate the potential for enhancing emissions of this compound class in aqueous environments by caging or shielding the metal ions from solvents that could act as quenching agents. In addition, chemical modification of the flexible Schiff base linker with suitable functional groups may allow for further bioconjugation of the nano-drum.[23] Other modifications such as polyglutamation would also lead to enhanced aqueous solubility.[24]

Spectroscopic characterization of the nano-drums has also revealed that these molecules possess physical and chemical properties with potential for bioprobe development. ^{1}H NMR spectra remain unchanged for weeks, indicating that they are stable in solution, as well as in crystalline forms. In addition, polystyrene beads loaded with nano-drums could also be stored in aqueous buffer at room temperature for several months and still retain the imaging properties shown in Fig. 6 and 7. The photophysical properties of the nano-drums not only show the potential of these compounds as both visible and NIR-emitters,[10] but the presence of the Cd/L centre also allows their use as non-bleaching fluorescent labels that emit in the visible region of the spectrum and can be imaged using conventional, commercially available microscopy platforms. This added spectral flexibility cannot be attributed only to the photophysical properties of the Cd/L center; data indicate that the intensity of emission is significantly influenced by the presence of the lanthanide and that different lanthanides can enhance the emissive properties of these compounds. Of the two compounds imaged in beads, the Er nano-drum (**5**) displays a higher intensity visible emission spectrum compared to the Yb nano-drum (**6**) in solution (centered at 525 nm). This result is also consistent with the emissions from the beads observed by epifluorescence microscopy (Fig. 6). How the lanthanides modulate this emission is not well understood and demands further study.

To further explore the capacity of these nano-drums for applications in biological systems, we carried out cytotoxicity studies using two cancer cell lines. When corrected for the stoichiometries of the components within the assembled nano-drum (8 Ho, 24 Cd, 12 L), the IC_{50} of this compound lies in the range of 10–120 μM. This value is indicative of moderate to mild cytotoxicity and comparable to Eu(III) and Tb(III) molecular compounds that can be used as agents for cell imaging.[16,21] The mechanisms responsible for cellular uptake and cytotoxicity are not known at this point. Cellular uptake is likely to involve endocytosis and parameters influencing transport of these nano-drums into cells can include particle size, shape, surface chemistry, topology and mechanical properties.[25] Comparison of the cytotoxicity of compound **4** with components that form the assembled nano-drums suggest that the cytotoxic effects primarily originate from the Cd and ligand (L). These data are consistent with observations that the nature of the antenna (the Cd/L in the nano-drums) is the most important feature that controls toxicity in lanthanide bioprobes.[21]

Finally, the suitability of nano-drums for use in bioimaging applications was assessed using three microscopy platforms configured to detect fluorescence in the visible spectrum. As shown in Fig. 6 and 7, polystyrene beads loaded with these compounds could all be visualized by epifluorescence, TIRF or two-photon microscopy, and intensity scans of the beads showed spectral features consistent with previous data obtained on these compounds in solution.[10] In particular, the two-photon scan of the beads indicates a wide range for excitation, corresponding to a single photon excitation range maximum at 465 nm for the Er nano-drum (**5**) and 440 nm for the Yb nano-drum (**6**), This scan, along with other spectral data, explains why we can use the DAPI filter (340–380 nm) in the epifluorescence microscope and the 405 nm laser with the TIRF microscope to produce the fluorescence images presented here. These data also demonstrate that excitation in the visible region can be carried out with these compounds using conventional sources, in order to minimize damage to biological materials in situations where that is a genuine concern. However, excitation in the UV region (to induce NIR emission by the lanthanide) is also possible. Also, UV excitation is not a constraint when using the two-photon microscope, because excitation relies upon the absorption of two infrared photons simultaneously to permit imaging of living cells within intact tissues.

5. Conclusions

The studies presented here show that this new class of self-assembling molecular nano-drums have considerable potential for use as imaging bioprobes for a range of platforms that use fluorescent or luminescent detection. Compared to conventional organic dyes or quantum dots used in biomedical applications, these bi-metal molecular compounds do not photobleach or blink. In addition, our nano-drums self-assemble as homogeneous nanoparticles with structurally accessible chemical moieties that are amenable to synthetic modification. The low temperature, bottom-up self-assembly route to these materials should allow for improvements in thermodynamic stability and photophysical properties of the nano-drums, as well as other important properties such as lanthanide shielding, solubility and toxicity. Chemical modifications to the Schiff base ligand are a precise way to tailor the properties of the nano-drum at the molecular level. Our future studies are focused on creating homogeneous molecular nanoparticles with bioconjugation capabilities. For example it should be possible to modify the flexible Schiff base ligand (H_2L) to bear functional groups such as thiol (–SH), isothiocyanate (–NCS), amino ($-NH_2$), and carboxylate ($-CO_2H$). Adding functional groups would lead to homogeneous molecular nano-drums with bioconjugation capabilities, while maintaining their advantageous photophysical properties. This work is in progress.

Acknowledgements

This work was supported by the Welch Foundation (F-816) (RAJ), the Ministry of High Education (MOHE), Malaysia under High Impact Research (HIR) – MOHE project (E000013-20001) (to JV and KAB), NIH/NIAID 1U01AI078008-3 (to KAB), CPRIT R1003 (to LIRE), NIH-NCI CA68682 (to JLS), and National Institutes of Health, National Science Foundation, Cancer Prevention Research Institute of

Texas, and the Welch Foundation (F1515) (all to EMM). Single crystal X-ray data were collected using instrumentation purchased with funds provided by the National Science Foundation (CHE-0741973).

References

1 T. Kurkina and K. Balasubramanian, *Cell. Mol. Life Sci.*, 2012, **69**, 373.
2 C. L. Ventola, *P T*, 2012, **37**, 582.
3 M. R. Hartman, R. C. Ruiz, S. Hamada, C. Xu, K. G. Yancey, Y. Yu, W. Han and D. Luo, *Nanoscale*, 2013, **5**, 10141.
4 P. Tallury, A. Malhotra, L. M. Byrne and S. Santra, *Adv. Drug Delivery Rev.*, 2012, **62**, 424.
5 S. V. Eliseeva and J.-C. G. Bünzli, *Chem. Soc. Rev.*, 2010, **39**, 189.
6 S. U. Pandya, J. Yu and D. Parker, *Dalton Trans.*, 2006, 2757.
7 C. P. Montgomery, B. S. Murray, E. J. New, R. Pal and D. Parker, *Acc. Chem. Res.*, 2009, **42**, 925.
8 A. K. Hagan and T. Zuchner, *Anal. Bioanal. Chem.*, 2011, **400**, 2847.
9 X.-P. Yang, R. A. Jones and S. Huang, *Coord. Chem. Rev.*, 2014, **273–274**, 63–75.
10 X.-P. Yang, D. Schipper, R. A. Jones, B. J. Holliday and S. Huang, *J. Am. Chem. Soc.*, 2013, **135**, 8468.
11 F. Lam, J.-X. Xu and K.-S. Chan, *J. Org. Chem.*, 1996, **61**, 8414.
12 DENZO-SMN, Z. Otwinowski and W. Minor, in *Methods in Enzymology, 276: Macromolecular Crystallography*, ed. C. W. J. Carter, M. I. Simon and R. M. Sweet, Academic Press, 1997, p. 307, Part A.
13 G. H. Sheldrick, *SHELX 97, a software package for the solution and refinement of X-ray data*, University of Göttingen, Göttingen, Germany, 1997.
14 D. T. Cromer and J. T. Waber, *International Tables for X-Ray Crystallography*, Kynoch Press, Birmingham, 1974, vol. 4, table 2.2A.
15 (*a*) S. R. Meech and D. J. Philips, *J. Photochem.*, 1983, **23**, 193; (*b*) K. Nakamaru, *Bull. Chem. Soc. Jpn.*, 1982, **5**, 2697.
16 D. T. Thielemann, A. T. Wagner, E. Rösch, D. K. Kölmel, J. G. Heck, B. Rudat, M. Neumaier, C. Feldmann, U. Schepers, S. Bräse and P. W. Roesky, *J. Am. Chem. Soc.*, 2013, **135**, 7454.
17 (*a*) D. H. Metcalf, R. G. Ghirardelli and R. A. Palmer, *Inorg. Chem.*, 1985, **24**, 634; (*b*) M. Albin, R. R. Whittle and W. D. Horrocks, Jr, *Inorg. Chem.*, 1985, **24**, 4591.
18 (*a*) Q.-H. Xu, L.-S. Li, X.-S. Liu and R.-R. Xu, *Chem. Mater.*, 2002, **14**, 549; (*b*) Y. B. Katsimirski and N. K. Davidenko, *Coord. Chem. Rev.*, 1979, **27**, 223.
19 (*a*) X.-P. Yang, R. A. Jones and W.-K. Wong, *Chem. Commun.*, 2008, 3266; (*b*) X.-P. Yang, D. Lam, C. Chan, J. M. Stanley, R. A. Jones, B. J. Holliday and K.-W. Wong, *Dalton Trans.*, 2011, **40**, 9795.
20 (*a*) Y.-X. Chi, S.-Y. Niu, Z.-L. Wang and J. Jin, *Eur. J. Inorg. Chem.*, 2008, 2336; (*b*) J.-C. G. Bünzli and C. Piguet, *Chem. Soc. Rev.*, 2005, **34**, 1048; (*c*) S. Petoud, S. M. Cohen, J.-C. G. Bünzli and K. N. Raymond, *J. Am. Chem. Soc.*, 2003, **125**, 13324.
21 M. P. Coogan and V. Fernandez-Moreira, *Chem. Commun.*, 2014, **50**, 384.
22 J.-C. G. Bünzli, *Chem. Rev.*, 2010, **110**, 2729.

23 V. Biju, *Chem. Soc. Rev.*, 2014, **43**, 744.
24 T. V. Esipova, X. Yeb, J. E. Collins, S. Sakadzic, E. T. Mandeville, C. B. Murray and S. A. Vinogradov, *Proc. Natl. Acad. Sci. U. S. A.*, 2012, **109**, 20826.
25 I. Canton and G. Battaglia, *Chem. Soc. Rev.*, 2012, **41**, 2718.

Faraday Discussions

PAPER

Aptamer-conjugated theranostic hybrid graphene oxide with highly selective biosensing and combined therapy capability

Bhanu Priya Viraka Nellore, Avijit Pramanik, Suhash Reddy Chavva, Sudarson Sekhar Sinha, Christen Robinson, Zhen Fan, Rajashekhar Kanchanapally, Justin Grennell, Ivy Weaver, Ashton T. Hamme and Paresh Chandra Ray*

Received 19th April 2014, Accepted 9th July 2014
DOI: 10.1039/c4fd00074a

Cancer is a life-threatening disease, which is rapidly becoming a global pandemic. Driven by this need, here we report for the first time an aptamer-conjugated theranostic magnetic hybrid graphene oxide-based assay for highly sensitive tumor cell detection from blood samples with combined therapy capability. AGE-aptamer-conjugated theranostic magnetic nanoparticle-attached hybrid graphene oxide was developed for highly selective detection of tumor cells from infected blood samples. Experimental data indicate that hybrid graphene can be used as a multicolor luminescence platform for selective imaging of G361 human malignant melanoma cancer cells. The reported results have also shown that indocyanine green (ICG)-bound AGE-aptamer-attached hybrid graphene oxide is capable of combined synergistic photothermal and photodynamic treatment of cancer. Targeted combined therapeutic treatment using 785 nm near-infrared (NIR) light indicates that the multimodal therapeutic treatment is highly effective for malignant melanoma cancer therapy. The reported data show that this aptamer-conjugated theranostic graphene oxide-based assay has exciting potential for improving cancer diagnosis and treatment.

Introduction

Cancer has been familiar for more than three thousand years, however, in the 21st century, it is still one of the most life-threatening diseases known, causing 1 in 8 deaths worldwide.[1–5] With advances in our understanding of signalling pathways, we now realize that detection of circulating tumor cells (CTCs) could be an invaluable tool for early stage detection of cancer and for monitoring its progression.[4–10] Malignant melanoma is the most deadly human skin cancer and

Department of Chemistry and Biochemistry, Jackson State University, Jackson, MS, USA. E-mail: paresh.c.ray@jsums.edu; Fax: +16019793674

is responsible for 75.2% of skin cancer-related deaths. Malignant melanoma cell line G361 is an aggressive form of skin cancer that is highly resistant to conventional therapies and known to be resistant to apoptosis.[1-5] As a result, a simple blood test to detect circulating G361 melanoma cells can help to identify if melanoma has spread to other areas of the body.

Since there are very few CTCs, rare malignant cells, in blood containing several millions of red blood cells, extremely sensitive and specific methods are required to detect early stage CTCs.[8-17] In the last decade, due to its remarkable electronic and structural properties, graphene has revolutionized the scientific community.[19-38] However, the development of graphene-based material for selective CTC detection and therapy is still in its infancy.[39-41] This requires graphene with novel functionality that has selective bio-sensing and therapy capability. Driven by this need, here we report that an aptamer-conjugated theranostic graphene oxide-based assay has exciting potential for improving cancer diagnosis by detecting CTCs from infected blood. Recently, several articles have demonstrated that photoluminescence of graphene can be observed by modifying with various oxygen-containing groups or by reducing its size to the nanometer scale.[30-33] By using these wonderful tunable fluorescence properties of graphene oxide (GO), here we demonstrate that hybrid graphene can be used as a multi-color luminescence platform for selective imaging of G361 human malignant melanoma cancer cells, as shown in Scheme 1. Since AGE-aptamer is known to be selective for G361 human malignant melanoma cancer cells,[5] we have used AGE-aptamer-attached hybrid graphene oxide for selective imaging of malignant melanoma cancer cells. Our reported results also show that multimodal therapeutic action using ICG-bound hybrid graphene oxide is highly promising for malignant melanoma cancer therapy.

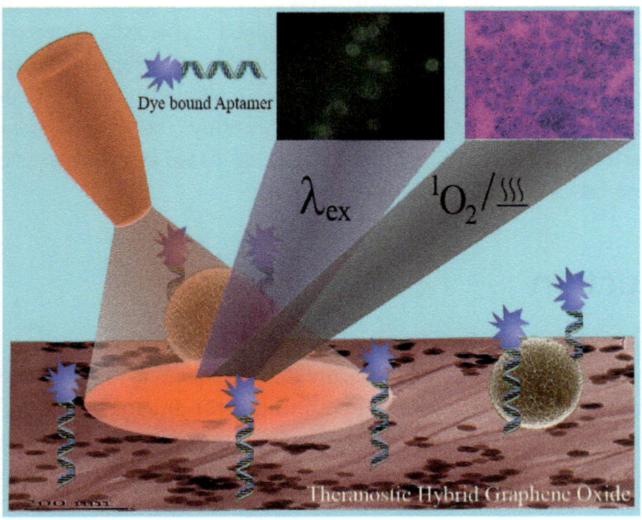

Scheme 1 Schematic representation showing aptamer-bound theranostic magnetic nanoparticle-attached hybrid graphene oxide with targeted separation, diagnostic and combined synergistic treatment capability for malignant melanoma cancer cells.

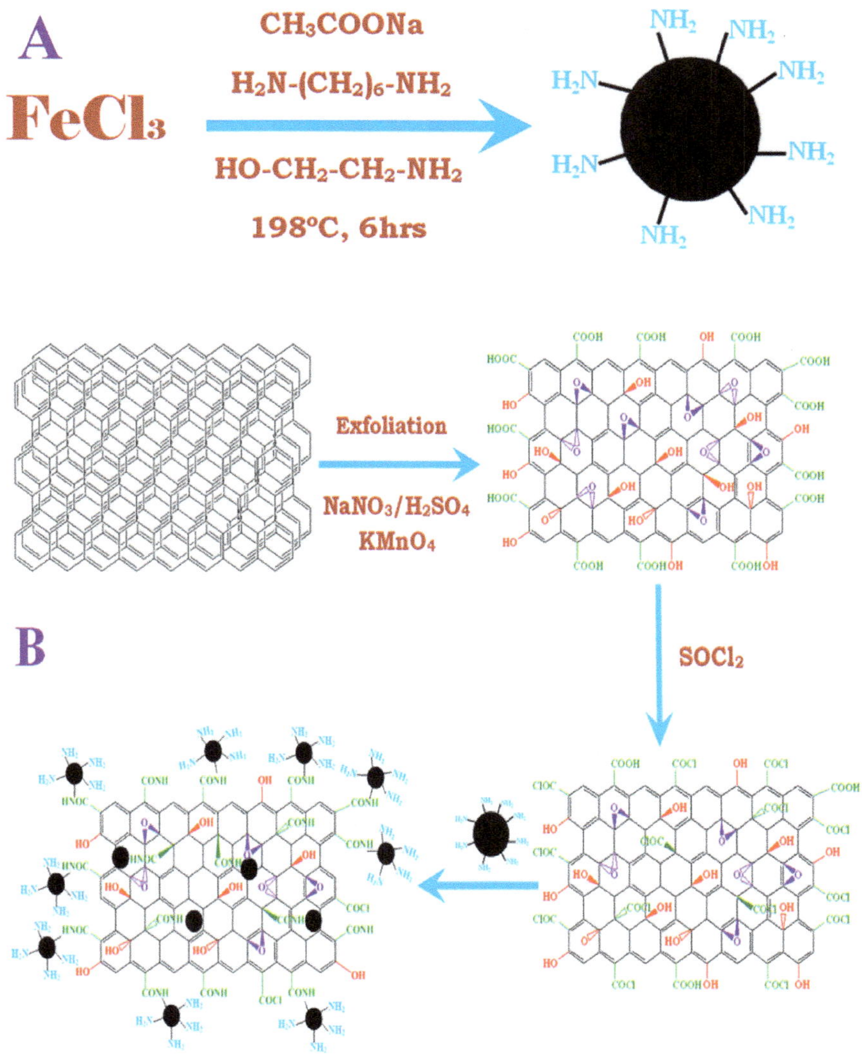

Scheme 2 (A) Schematic representation showing our synthesis procedure to develop magnetic nanoparticle. (B) Schematic representation showing our synthesis procedure to develop magnetic nanoparticle-attached hybrid graphene oxide.

Knowledge of the molecular underpinnings of cancer drug resistance has advanced significantly.[10–17] We now understand that a single therapeutic agent can have limited efficiency in a clinical environment due to drug-resistance profiles.[10–17] As a result, multimodal therapy is a highly promising strategy with minimal side effects.[10–17] Design of theranostic hybrid graphene oxide for selective diagnosis and therapeutic cancer therapy could have a revolutionary impact on the pharmaceutical industry.[24–30] In the last decade, several different types of nanoparticle-based therapeutic approaches have been developed for possible cancer treatment.[1–9] Amongst these, NIR light-driven photothermal therapy (PTT) and photodynamic therapy (PDT) have been shown to possess unique advantages,

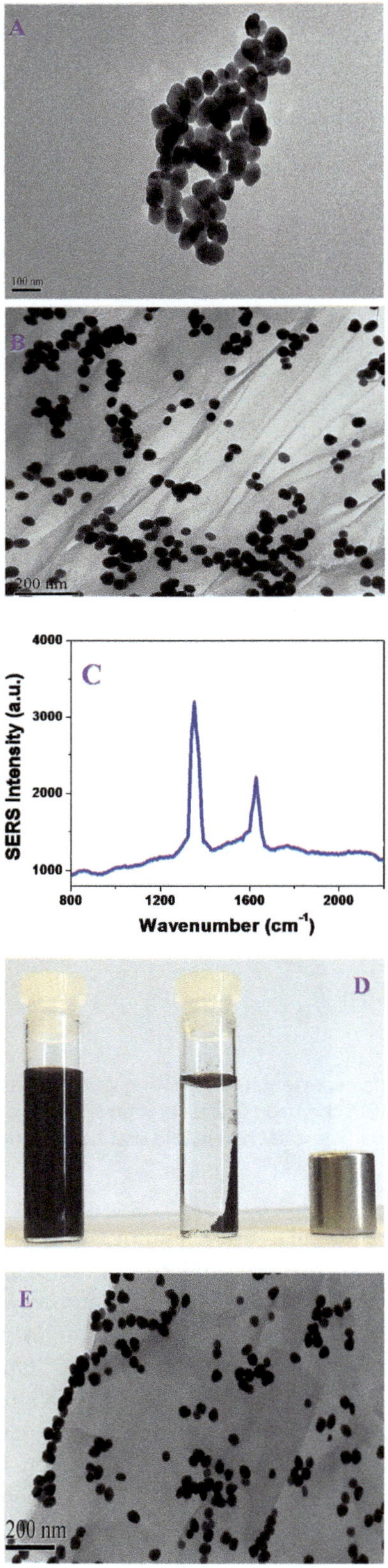
A
100 nm
B
200 nm
C
SERS Intensity (a.u.)
Wavenumber (cm^{-1})
4000
3000
2000
1000
800
1200
1600
2000
D
E
200 nm

including remote controllability, low systemic toxicity and side effects.[5–15,34–37] Since a combination of therapeutic approaches may cooperatively suppress cancer development,[10,34–37] here we report an ICG-bound magnetic nanoparticle-attached graphene oxide platform for combined photodynamic and photothermal synergistic targeted therapy of cancer, as shown in Scheme 1. ICG is a water-soluble photosensitizer approved by the U.S. Food and Drug Administration (FDA).[34–37] Experimental data demonstrated that, due to the synergistic effect, the therapeutic efficacy of the nano-platform was significantly enhanced compared to PDT or PTT alone.

Experimental

All chemicals, including graphite, $KMnO_4$, sodium borohydride, iron chloride, sodium citrate, were purchased from Sigma-Aldrich and Fisher Scientific. We have obtained G361 melanoma cells, growth media to grow cancer cells, buffered saline, trypsin, and fetal bovine serum from the American Type Culture Collection (ATCC, Rockville, MD).

Preparation of amine-functionalized magnetic nanoparticles

The magnetic nanoparticles were prepared using a two-step process, as shown in Scheme 2A. For this purpose, in the first step we dissolved $FeCl_3$ (1.0 g) in ethylene glycol (30 mL). In the next step, anhydrous sodium acetate (2.0 g) and 1,6-hexadiamine (6.5 g) were added and stirred vigorously to acquire a transparent solution. The mixture was sealed in a Teflon-lined stainless steel autoclave and was heated at 198 °C for 6 h. The product was washed with hot water and ethanol (3 times) under ultrasonic condition, using a magnet to remove the solvent and unbound 1,6-hexadiamine effectively, and then dried at 50 °C to obtain black powder. Magnetic nanoparticles were characterized using JEM-2100F transmission electron microscope (TEM) as shown in Fig. 1A, which shows that the average particle size is about 40 nm. Magnetic properties were determined using a superconducting quantum interference device (SQUID) magnetometer at room temperature, which indicated super-paramagnetic behavior with specific saturation magnetization of 43.5 emu g^{-1} for the amine-functionalized magnetite nanoparticles.

Preparation of magnetic nanoparticle-attached hybrid graphene oxide

Magnetic nanoparticle-attached graphene oxide was synthesized using a several step process, as shown in Scheme 2B. At first, graphene oxide was prepared from

Fig. 1 (A) TEM picture showing the morphology of freshly prepared magnetic nanoparticle. (B) TEM picture showing the morphology of freshly prepared magnetic nanoparticle-attached hybrid graphene oxide. (C) Raman spectra showing graphene oxide D-band at 1340 cm^{-1} and G-band at 1612 cm^{-1} in hybrid material. (D) Photograph showing that the magnetic nanoparticle-attached hybrid graphene oxide is highly magnetic, and as a result, we can separate it by using a bar magnet. (E) TEM picture showing the morphology of freshly prepared aptamer-bound magnetic nanoparticle-attached hybrid graphene oxide.

graphite. For this purpose, we have used a modified Hummers method[38] for graphite exfoliation.

We used strong oxidizing agents to yield graphene oxide as shown in Scheme 1B. For this purpose, 1 g of graphite powder was added to 1 g of $NaNO_3$ in 45 mL of H_2SO_4 under stirring in an ice bath. 3 g of $KMnO_4$ was then added very carefully, so as to not increase the temperature. Next, we transferred the reaction mixture container to a water bath at 35 °C and stirred it for 30 minutes. A thick paste was thus obtained. After that, 45 mL of water was very carefully added drop-wise. During the addition of water, the temperature rose to around 95 to 98 °C. We then added 140 mL of water and allowed the mixture to stand for 30 minutes. Next, we filtered the mixture and re-dispersed the obtained graphene oxide in 100 mL of water. After that, we have performed sonication for 1 hour for exfoliation. After that, we characterized the water soluble graphene oxide using IR and Raman spectroscopy, as we have reported recently.[18,23]

In the next step, to allow the attachment of amino-functionalized magnetic nanoparticle to the graphene oxide, acid chloride-functionalized graphene oxide was prepared. For this purpose, –COOH functionalized graphene oxide was treated with thionyl chloride in the presence of dimethylformamide catalyst. As shown in Scheme 1B, the acid chloride group was used as a chemical anchor for the connection with the amine-functionalized magnetic nanoparticles. TEM data (JEM-2100F instrument), as shown in Fig. 1B, clearly indicate that in the hybrid graphene oxide, magnetic nanoparticles are attached to the graphene oxide. The size of the magnetic nanoparticles on the hybrid graphene oxide is about 40 nm. From Fig. 1A and B, we can clearly see that the size and shape of the magnetic nanoparticles remain the same after conjugation, which indicates that the morphology of the magnetic nanoparticles remains unchanged during coupling between graphene oxide and the magnetic nanoparticles. Our experimental data also indicate that magnetic nanoparticle agglomeration is much lower when nanoparticles are attached to graphene oxide, which will be useful for biomedical applications. Fig. 1D indicates that our hybrid graphene oxide is highly magnetic and, as a result, we can separate it using a small bar magnet. SQUID measurements indicate that hybrid graphene oxide exhibits super-paramagnetic behavior with specific saturation magnetization 26.4 emu g^{-1}. Fig. 1C shows the Raman spectrum of hybrid graphene oxide, which displays a D-band at 1340 cm^{-1} and a G-band at 1612 cm^{-1}.[18–25] In the reported spectra, the strong D band clearly indicates that the degree of oxidation is high. To determine the amount of magnetic nanoparticles on the hybrid graphene oxide, we have carried out thermogravimetric analysis (TGA), using a Perkin-Elmer thermogravimetric analyzer. From the TGA data, we have estimated that the mass fraction of magnetic nanoparticles and graphene oxide are about 30 and 70 wt%, respectively.

Preparation of aptamer-conjugated hybrid graphene oxide

For targeted imaging of malignant melanoma G361 cells and synergistic therapy, we have modified the nanoplatform with MB-bound AGE-aptamer, which is known to be specific to malignant melanoma G361 cells. For this purpose, initially hybrid graphene oxide was coated with thiolated polyethylene glycol (HS-PEG), which helps to avoid nonspecific interactions with cells and cell media. After PEGylation, NH-modified AGE-aptamers were attached to acid chloride-

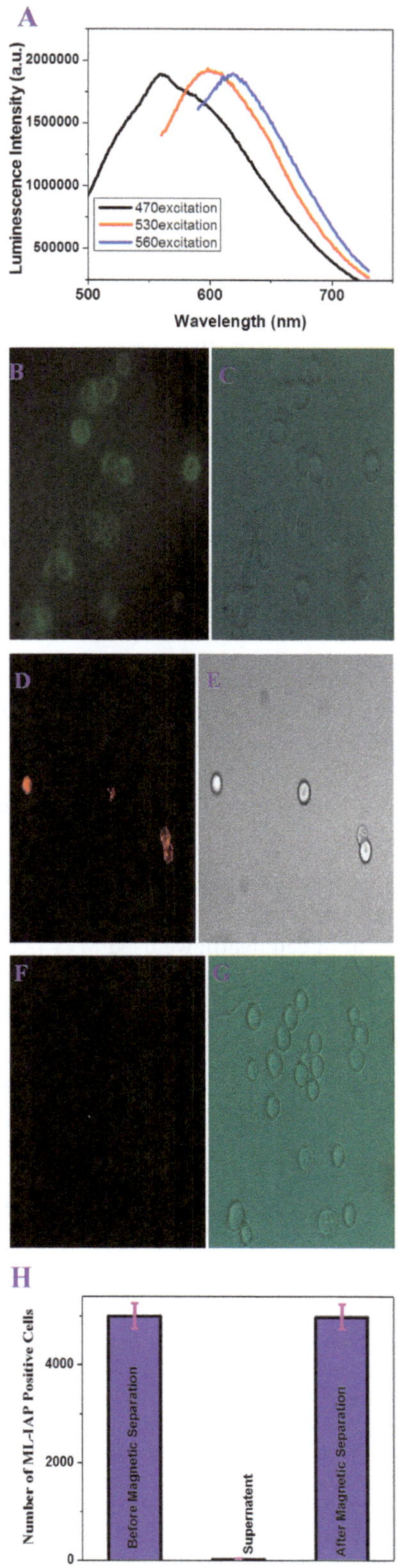
A
Luminescence Intensity (a.u.)
2000000
1500000
1000000
500000
470excitation
530excitation
560excitation
500
600
700
Wavelength (nm)
B
C
D
E
F
G
H
Number of ML-IAP Positive Cells
4000
2000
0
Before Magnetic Separation
Supernatent
After Magnetic Separation

functionalized graphene oxide. As shown in Fig. 1E, TEM data clearly indicate that, in aptamer-bound hybrid graphene oxide, magnetic nanoparticles are attached to the graphene oxide. From the TEM data shown in Fig. 1B and E, we clearly see that the morphology looks similar, which indicates that, during the aptamer-binding process, the morphology of hybrid graphene oxide remains the same.

Cell culture and incubation with hybrid graphene oxide

Malignant melanoma G361 cells were grown according to the ATCC procedure. We grew cells in a 5% CO_2 incubator at 37 °C using RPMI-1640 medium (ATCC, Rockville, MD) supplemented with 10% premium fetal bovine serum (FBS) (Lonza, Walkersville, MD) and antibiotics (10 IU mL^{-1} penicillin G and streptomycin) in 75 cm^2 tissue culture flasks. Once we had a culture of 10^5 cells mL^{-1}, different numbers of malignant melanoma G361 cells were spiked in citrated whole rabbit blood at various densities. After that, different concentrations of magnetic hybrid graphene oxide were mixed with infected blood for 30 minutes at room temperature, before performing the magnetic separation experiment. After magnetic separation, we performed TEM and fluorescence analyses.

Fluorescence analysis

After separation of malignant melanoma G361 cells using a magnet, we used an Olympus IX71 inverted confocal fluorescence microscope fitted with a SPOT Insight digital camera for fluorescence imaging.

Combined destruction of malignant melanoma G361 cells and determination of the percentage of live cells

For the PDT, PTT or combined destruction experiments, we used a continuous-wavelength OEM laser operating at 785 nm at 2 W cm^{-2} power for 10–25 minutes. To determine the amount of cell death, we used MTT (ATCC CA# 30-1010k) and trypan blue test, using our reported method.[9,10]

Reactive oxygen species (ROS) generation measurement

To understand how ROS formation varies with PTT, we measured the amount of ROS formed during PDT and combined therapy process. For this purpose, we

Fig. 2 (A) Excitation wavelength-dependent photoluminescence from freshly prepared magnetic nanoparticle-attached hybrid graphene oxide. Plot clearly shows that photoluminescence can be tuned just by varying the excitation energy, without changing the size. (B) Fluorescence images of aptamer-conjugated hybrid graphene oxide-attached G361 human malignant melanoma cancer cell after separation from blood sample using a magnet. We have used 470 nm light as the excitation source. (C) Bright-field image of the same cells after magnetic separation. (D) Fluorescence image of aptamer-conjugated hybrid graphene oxide-attached G361 human malignant melanoma cancer cell, after separation from blood sample using a magnet. We have used 560 nm light as an excitation source. (E) Bright-field image of the same cells after magnetic separation. (F) Fluorescence image of supernatant after magnet separation. Our result clearly shows that aptamer-attached hybrid graphene oxide does not bind with blood cells. (G) Bright-field image of same supernatant after magnetic separation. (H) Number of ML-IAP positive cells before and after magnetic separation, measured by immunosorbent assay.

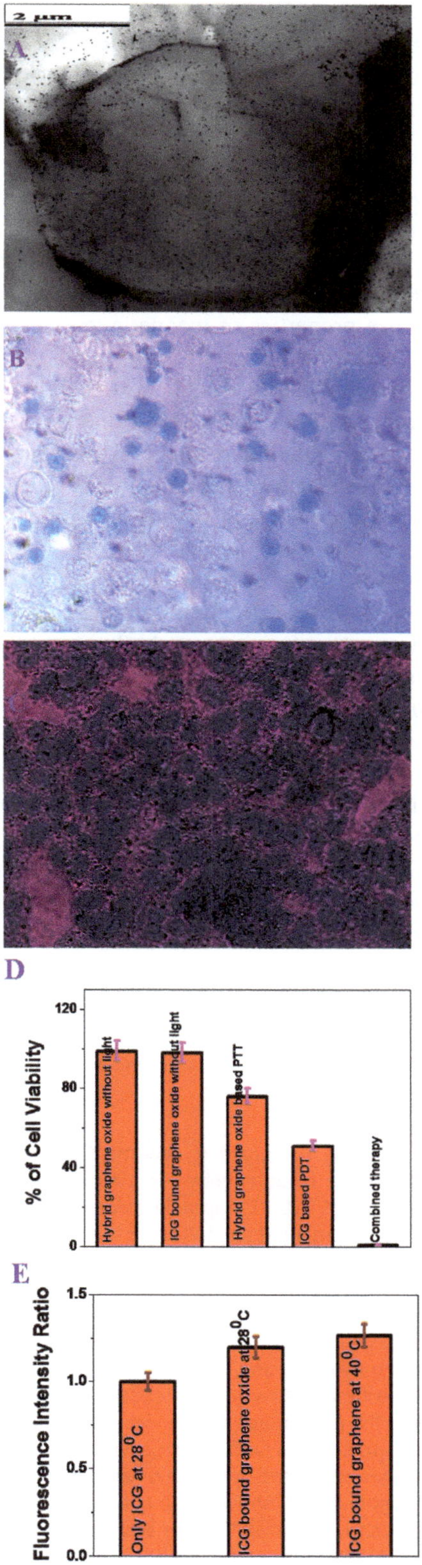
2 μm
A
B
D
% of Cell Viability
120
80
40
0
Hybrid graphene oxide without light
ICG bound graphene oxide without light
Hybrid graphene oxide based PTT
ICG based PDT
Combined therapy
E
Fluorescence Intensity Ratio
1.5
1.0
0.5
0.0
Only ICG at 28^{0}C
ICG bound graphene oxide at 28^{0}C
ICG bound graphene at 40^{0}C

used singlet oxygen sensor green reagent (SOSG, Sigma) and followed the manufacturer's protocol. For fluorescence measurement, we have used 485 nm light as the excitation source and the fluorescence intensity was measured at 528 nm using the microplate reader.

Results and discussion

Our label-free cancer cell imaging using aptamer-conjugated hybrid graphene oxide is based on the luminescence properties of graphene oxide. As shown in Fig. 2A, the fluorescence properties of magnetic graphene oxide can be tuned simply by varying the excitation energy without changing its chemical composition and size. Our observed excitation energy-dependent photoluminescence spectral shift for hybrid graphene oxide could be due to two factors,[30–33] which are as follows: (1) excitation wavelength dependent fluorescence from the –OH moiety in the graphene oxide sheets and (2) due to the local reorganization of photoexcited GO sheets in a polar solvent, the solvent relaxation time becomes comparable to the fluorescence lifetime.

To understand whether the aptamer-conjugated hybrid graphene oxide developed by us can be used for G361 human malignant melanoma cancer cell analysis from blood samples in settings similar to clinical diagnosis, G361 malignant melanoma cancer cells were spiked at various densities into suspensions of citrated whole rabbit blood. After 80 minutes of gentle shaking, we incubated 150 μL aptamer-conjugated hybrid magnetic graphene oxide with 2.5 mL of G361 malignant melanoma cells-infected blood sample. Next, we incubated the mixture for 40 minutes at room temperature under gentle shaking. After that, we used a bar magnet to separate and enrich malignant melanoma G361 cells attached to hybrid magnetic graphene oxide. Next, the supernatant blood sample was carefully removed with a pipette. After that, the suspensions of the magnetic graphene oxide-attached G361 malignant melanoma cells and supernatant were characterized using TEM, fluorescence imaging and enzyme-linked immunosorbent assay (ELISA) assay analysis kits, as shown in Fig. 2. The melanoma inhibitor of apoptosis protein (ML-IAP) is well known to over express in malignant melanoma cells, which is known to aid in disease progression and treatment

Fig. 3 (A) TEM image showing malignant melanoma G361 cells inside the aptamer-conjugated magnetic graphene oxide sheet after magnetic separation. (B) Bright-field inverted microscopic images of hybrid graphene oxide-attached malignant melanoma G361 cells, after separation by a magnet and following irradiation with 785 nm near IR light at 1 W cm^{-2} for 20 minutes. For imaging, we have used 485 nm excitation and the fluorescence has been collected between 520–535 nm. Bright field image clearly shows that some cancer cells are dead after PTT. (C) Bright-field inverted microscopic images of ICG-bound hybrid graphene oxide-attached malignant melanoma G361 cells after separation by a magnet and following by irradiation with 785 nm near IR light at 1 W cm^{-2} for 20 minutes. Bright field image clearly shows that most cancer cells are dead after combined therapy. (D) Plot showing the percentage of cell viability in ICG-bound hybrid graphene oxide-attached malignant melanoma G361 cells cell in the absence of laser light. It also shows how the cell viability varies in the presence of laser light. Plot shows no cytotoxicity in the absence of NIR light. It also shows dramatic synergistic action in the case of combined therapy. (E) Plot showing how ROS formation from ICG can vary in the presence of hybrid graphene oxide at different temperatures.

resistance. As shown in Fig. 2H, our ELISA results reveal no ML-IAP presence in the fractions of THE cell suspensions that did not bind to aptamer-conjugated magnetic graphene oxide, which indicates the absence of malignant melanoma G361 cells in cell suspensions that were not separated by the magnet. These are thus mainly blood cells. On the other hand, we found that ML-IAP was present in the magnetic graphene oxide-attached cell suspension. ELISA results, as shown in Fig. 2H, clearly indicate that malignant melanoma G361 cells are attached to the nanoplatform. From the ELISA experiments, we have estimated that the malignant melanoma G361 cell recovery by the bar magnet was about 97%.

Fig. 2A–G shows confocal fluorescence microscope imaging data, which clearly indicate that aptamer-conjugated graphene oxide can be used for label-free imaging of cancer cells. To understand whether the luminescence of graphene oxide can be tuned for multicolor cancer imaging, we have performed fluorescence imaging experiment using 470 nm and 560 nm excitation. Fig. 2B and D clearly show that the luminescence of graphene oxide can be used for multicolor cancer imaging, and this is due to the fact that single-photon photoluminescence from graphene oxide can be tuned simply by varying the excitation energy, without changing its chemical composition and size, as we have discussed before. As shown in Fig. 2E, no fluorescence was observed from the cell suspensions that were not separated by the magnet and this is due to the fact that malignant melanoma G361 cells are absent in cell suspensions that are not separated by the magnet. Our fluorescence imaging results also indicate that our hybrid magnetic graphene oxide can be used to separate malignant melanoma G361 cells from a blood sample in clinical environment. Similarly, TEM images, as shown in Fig. 3A, show magnetic graphene oxide-attached malignant melanoma G361 cells. All the experimental results described above clearly show that the aptamer-conjugated hybrid magnetic graphene oxide developed by us is highly selective for binding with the malignant melanoma G361 cell line and can be used for selective detection of malignant melanoma from blood sample.

After successful targeted malignant melanoma G361 cells separation, we have performed NIR light-driven therapy experiments using 785 nm excitation light. Before these therapy experiments, to understand the cytotoxicity of hybrid graphene oxide in the absence of NIR light, hybrid graphene oxide-attached G361 cells were incubated for 12 hours without any laser light. Our experimental data, as shown in Fig. 3D, clearly show that no cell death was observed even after 12 hours of incubation, which indicates that hybrid graphene oxide developed by us is not cytotoxic in the absence of external NIR light. Next, we performed 785 nm light-driven PTT experiments. To determine the amount of cell death due to the hybrid graphene oxide hyperthermia effect, we added trypan blue after NIR radiation exposure. Since living cells cannot bind with trypan blue, these were colorless. On the other hand, dead cells bind with the blue dye and became blue. Cell viability can therefore be qualitatively determined from the color of the cell monolayer. As shown in Fig. 3B, some of the cancer cells were dead after 20 min of PTT process. Using MTT test, we found that 23% cancer cells were dead due to PTT, as shown in Fig. 3D, when hybrid graphene oxide-attached G361 cells were irradiated with 785 nm light at 1 W cm^{-2} power for 20 minutes. Since graphene oxide absorbed 785 nm light and converted it into heat through nonradiative decay, G361 malignant melanoma cells were killed in the presence of 785 nm NIR light, mainly due to hyperthermia. Since graphene oxide absorption cross-section is very low at 785 nm, the observed hyperthermia effect is also low.

To improve the NIR light-driven cancer cell killing efficiency, we have synthesized ICG-bound AGE-aptamer-attached hybrid graphene oxide as a combined therapy material. To find out whether this combined therapy is much superior to single therapy, we performed several different experiments, which we will now discuss. At first, we measured the possible cytotoxicity of ICG-attached hybrid graphene oxide and for this purpose, ICG-bound hybrid graphene oxide-attached G361 malignant melanoma cells were incubated for 12 hours without any NIR light. As shown in Fig. 3D, our experimental data indicate no cell death even after 12 hours of incubation, which indicates that ICG-bound hybrid graphene oxide is not cytotoxic in the absence of external NIR light. On the other hand, as shown in Fig. 3D, NIR light induced experimental data clearly show that 99% of cells died when ICG-bound hybrid graphene oxide-attached G361 malignant melanoma cells were irradiated with 785 nm light at 1 W cm^{-2} power for 20 minutes. Bright-field inverted microscope images data as shown in Fig. 3C, clearly show G361 malignant melanoma cells were deformed during the combined therapy process. The cell death following nanoplatform exposure to 785 nm NIR light could be due to numerous factors, including ROS-induced cancer cell death due to the presence of ICG and thermal disintegration by graphene oxide. Fig. 3D also shows that about 50% cancer cells died when only ICG was used for PDT of G361 malignant melanoma cells using 785 nm light at 1 W cm^{-2} power for 20 minutes. Next, to understand how the temperature varies during the NIR radiation-based combined therapy process, we used a MikroShot Camera to measure thermal imaging at 1 minute intervals. Our thermal imaging experiments indicate that the temperature increased to about 40 °C when ICG-bound AGE-aptamer-attached hybrid graphene oxide with G361 malignant melanoma cells was exposed to 785 nm laser light at 1 W cm^{-2} power for 20 minutes. We also noted that the temperature increased to only 28 °C for G361 malignant melanoma cells in the absence of hybrid magnetic graphene oxide under the same light exposure conditions. All the therapy data reported above clearly indicate that malignant melanoma cell viability was much lower in the case of PDT and PTT combined therapy, than in the case of the individual therapies. Our experimental data show that when the two treatments were combined under a single 785 nm NIR light irradiation, synergistic therapeutic effects are expected in G361 malignant melanoma cells.

In ICG-bound AGE-aptamer-attached hybrid graphene oxide, ICG utilizes its ability to form ROS in the presence of light to kill G361 malignant melanoma cells. As a result, the efficiency of photodynamic killing by hybrid graphene oxide is highly dependent on the ROS forming capability of ICG. To understand the mechanism by which synergistic combined therapy destroys G361 malignant melanoma cells, we measured the cellular ROS formation during PDT and combined therapy processes using SOSG (Sigma). Fluorescence intensity was measured using a microplate reader with the excitation wavelength at 485 nm and the emission wavelength at 528 nm. As reported in Fig. 3E, our experimental data clearly show elevated ROS formation in the presence of AGE-aptamer-attached hybrid graphene oxide, which indicates that the photothermal effect of graphene oxide was able to increase the formation of singlet oxygen, which improved the photodynamic cancer cell killing efficiency. Due to this fact, we observed a synergistic therapeutic effect.

Conclusion

In conclusion, in this manuscript we have reported the development of aptamer-conjugated hybrid graphene oxide which can deliver targeted separation, diagnostic and combined synergistic photothermal and photodynamic treatment of malignant melanoma cancer cells. We found that the AGE-aptamer-attached magnetic graphene oxide can be used for the ultra-sensitive and label-free detection of G361 malignant melanoma cells from infected blood samples. Excitation wavelength-dependent photoluminescence imaging data indicate that aptamer-attached hybrid graphene can be used as a multicolor luminescence platform for selective imaging of malignant melanoma cells. We have shown that, in the absence of NIR light, no cytotoxicity has been observed from hybrid graphene oxide. In contrast, ICG-bound AGE-aptamer-attached hybrid graphene oxide can be used for 785 nm NIR light-activated combined PTT and PDT for G361 malignant melanoma cells. The reported experimental data indicate that a synergistic therapeutic effect can be obtained using ICG-bound AGE-aptamer-attached hybrid graphene oxide, which indicates that multimodal therapeutic treatment using ICG-bound hybrid graphene oxide may be a more effective system for malignant melanoma cancer therapy. Although we are at a relatively early stage in the development of aptamer-attached hybrid graphene oxide based CTC separation, imaging and synergistic combined therapy of malignant melanoma, we believe that the reported assay has enormous potential for real life applications, once it has been optimized properly in clinical settings.

Acknowledgements

Dr Ray thanks NSF-PREM (grant ##DMR-1205194) and NSF-REU grant # 1156111 for generous funding.

Notes and references

1 http://www.cancer.org/aboutus/globalhealth.
2 http://www.cancer.org/cancer/skincancer-melanoma.
3 K. Pantel, R. H. Brakenhoff and B. Brandt, *Nat. Rev. Cancer*, 2008, **8**, 329–340.
4 H. Lui, J. Zhao, D. McLean and H. Zeng, *Cancer Res.*, 2012, **72**, 2491–2500.
5 A. Ojima, M. Takanori, M. Sayaka, T. Masayoshi, I. Hiroyoshi Inoue, H. Yuichiro and Y. Sho-Ichi, *Lab. Invest.*, 2014, **94**, 422–429.
6 S. David, Q. Zhang, W. Xuexi, P. He, Y. J. Zhu, J. Zhao, O. M. Rennert and Y. A. Su, *Mol. Cancer Ther.*, 2009, **8**, 1292–1304.
7 E. C. Dreaden, A. M. Alkilany, X. Huang, C. J. Murphy and M. A. El-Sayed, *Chem. Soc. Rev.*, 2012, **41**, 2740–2779.
8 E. I. Galanzha, E. V. Shashkov, T. Kelly, J. Kim, L. Yang and P. V. Zharov, *Nat. Nanotechnol.*, 2009, **4**, 855–860.
9 Z. Fan, D. Senapati, A. K. Singh and P. C. Ray, *Mol. Pharmaceutics*, 2013, **10**, 857–866.

10 Z. Fan, X. Dai, Y. Lu, E. Yu, N. Brahmbatt, N. Carter, C. Tchouwou, A. K Singh, Y. Jones, H. Yu and P. C. Ray, *Mol. Pharmaceutics*, 2014, **11**, 1109–1116.
11 S. Lal, S. E. Clare and N. J. Halas, *Acc. Chem. Res.*, 2008, **41**, 1842–1851.
12 A. K. Singh, S. A. Khan, Z. Fan, T. Demeritte, D. Senapati, R. Kanchanapally and P. C. Ray, *J. Am. Chem. Soc.*, 2012, **134**, 8662–8669.
13 S. R. Panikkanvalappil, M. A. Mackey and M. A. El-Sayed, *J. Am. Chem. Soc.*, 2013, **135**, 4815–4821.
14 X. H. Huang, I. H. El-Sayed, W. Qian and M. A. El-Sayed, *J. Am. Chem. Soc.*, 2006, **128**, 2115–2120.
15 X. Dai, Z. Fan, Y. Lu and P. C. Ray, *ACS Applied Mater. Sci.*, 2013, **5**, 11348–11354.
16 Z. Fan, D. Senapati, S. A. Khan, A. K. Singh, A. Hamme, B. Yust, D. Sardar and P. C. Ray, *Chem.–Eur. J.*, 2013, **19**, 2839–2847.
17 K. Cheresh, D. A. Yang, H. M. Reisfeld and R. A. Disialoganglioside, *Cancer Res.*, 1987, **47**, 1098–1104.
18 R. Kanchanapally, Z. Fan, A. K. Singh, S. S. Sinha and P. C. Ray, *J. Mater. Chem. B*, 2014, **2**, 1934–1937.
19 B. Tian, C. Wang, S. Zhang, L. Feng and Z. Liu, *ACS Nano*, 2011, **5**, 7000–7009.
20 J. T. Robinson, S. M. Tabakman, Y. Liang, H. Wang, H. S. Casalongue, D. Vinh and H. Dai, *J. Am. Chem. Soc.*, 2011, **133**, 6825–6831.
21 M. Li, X. Yang, J. Ren, K. Qu and X. Qu, *Adv. Mater.*, 2012, **24**, 1722–1728.
22 M.-C. Wu, A. R. Deokar, J.-H. Liao, P.-Y. Shih and Y.-C. Ling, *ACS Nano*, 2013, **7**, 1281–1290.
23 Z. Fan, R. Kanchanapally and P. C. Ray, *J. Phys. Chem. Lett.*, 2013, **4**, 3813–3818.
24 A. K. Geim and K. S. Novoselov, *Nat. Mater.*, 2007, **6**, 183–191.
25 W. Gao, B. L. Alemany, L. Ci and P. M. Ajayan, *Nat. Chem.*, 2009, **1**, 403–408.
26 Y. Liang, Y. Li, H. Wang and H. Dai, *J. Am. Chem. Soc.*, 2013, **135**, 2013–2036.
27 A. Sinitskii and J. M. Tour, *J. Am. Chem. Soc.*, 2010, **132**, 14730–14732.
28 I. V. Lightcap and P. V. Kamat, *Acc. Chem. Res.*, 2013, **46**, 2235–2243.
29 M. Z. Hossain, J. E. Johns, K. H. Bevan, J. H. Karmel, Y. T. Liang, S. Yoshimoto, K. Mukai, T. Koitaya, J. Yoshinobu, M. Kawai, *et al.*, *Nat. Chem.*, 2012, **4**, 305–309.
30 K. P. Loh, Q. Bao, G. Eda and M. Chhowalla, *Nat. Chem.*, 2010, **2**, 1015–1024.
31 J. Kim, L. J. Cote, F. Kim and J. Huang, *J. Am. Chem. Soc.*, 2010, **132**, 260.
32 S. K. Cushing, M. Li, F. Huang and N. Wu, *ACS Nano*, 2014, **8**, 1002.
33 A. L. Exarhos, M. E. Turk and J. M. Kikkawa, *Nano Lett.*, 2013, **13**, 344–348.
34 X. Zheng, D. Xing, F. Zhou, B. Wu and W. R. Chen, *Mol. Pharm.*, 2011, **8**, 447–456.
35 M. A. Yaseen, J. Yu, B. Jung, M. S. Wong and B. Anvari, *Mol. Pharmaceutics*, 2009, **6**, 1321–1332.
36 J. Yu, D. Javier, M. A. Yaseen, N. Nitin, R. Richards-Kortum, B. Anvari and M. S. Wong, *J. Am. Chem. Soc.*, 2010, **132**, 1929–1938.
37 J. Lin, S. Wang, P. Huang, Z. Wang, S. Chen, G. Niu, W. Li, J. He, D. Cui, G. Lu, X. Chen and Z. Nie, *ACS Nano*, 2013, **7**, 5320–5329.
38 W. S. Hummers and R. E. Offeman, *J. Am. Chem. Soc.*, 1958, **80**, 1339.
39 H. J. Yoon, T. H. Kim, Z. Zhang, E. Azizi, T. M. Pham, C. Paoletti, J. Lin, N. Ramnath, M. S. Wicha, D. F. Hayes, *et al.*, *Nat. Nanotechnol.*, 2013, **8**, 735–741.

40 A. Pramanik, S. Reddy, R. Chavva, Z. Fan, S. S. Sinha, B. P. V. Nellore and P. C. Ray, *J. Phys. Chem. Lett.*, 2014, **5**, 2150–2154.
41 S. Yin, Y. L. Wu, B. Hu, Y. Wang, P. Cai, C. K. Tan, D. Qi, L. Zheng, W. R. Leow and N. S. Tan, *Adv. Mater. Interfaces*, 2014, **1**, 1300043.

Faraday Discussions

PAPER

Functionalised gold and titania nanoparticles and surfaces for use as antimicrobial coatings†

Sacha Noimark,[ab] Kristopher Page,[ab] Joseph C. Bear,[a] Carlos Sotelo-Vazquez,[a] Raul Quesada-Cabrera,[a] Yao Lu,[a] Elaine Allan,[b] Jawwad A. Darr[a] and Ivan P. Parkin*[a]

Received 16th May 2014, Accepted 3rd July 2014
DOI: 10.1039/c4fd00113c

We report the preparation, characterisation and antimicrobial functional testing of various titanium dioxide and gold modified titanium dioxide nanoparticles embedded into a polysiloxane polymer by a swell dip-coating procedure. We show that the surfaces are effective in killing both Gram-positive (*Staphylococcus aureus*) and Gram-negative (*Escherichia coli*) bacteria under different lighting conditions. The presence of the nanoparticles was of critical importance in improving the functional properties of the surface. These materials have the potential to reduce hospital-acquired infection, by killing bacteria on the polymer surface.

1 Introduction

The prevalence of hospital-acquired infections (HAIs) has been described as 'unacceptably high' by the National Institute for Health and Care Excellence.[1] Each year, approximately 300 000 patients treated at an NHS hospital in England acquire an HAI, including surgical site infections (15.7%), urinary tract infections (17.2%) and infections of the lower respiratory tract or pneumonia (22.8%).[1] An estimated one in 16 patients treated at a U.K. NHS hospital contracts an HAI,[1] with the highest infection rate incidence noted amongst patients in intensive care units.[2] Although interventions have achieved a decrease in the incidence of methicillin-resistant *Staphylococcus aureus* and *Clostridium difficile* infections, the rate of HAIs still poses a significant threat to both healthcare institutions and hospitalised patients.[1]

Hospital touch surfaces act as large reservoirs of bacteria, which can facilitate the spread of infection within healthcare institutions through a surface–

[a]*Materials Chemistry Research Centre, Department of Chemistry, University College London, 20 Gordon St, London, WC1H 0AJ, UK. E-mail: i.p.parkin@ucl.ac.uk; Tel: +44 (0)207 679 4669*

[b]*Division of Microbial Diseases, UCL Eastman Dental Institute, University College London, 256 Grays Inn Road, London, WC1X 8LD, UK*

† Electronic supplementary information (ESI) available. See DOI: 10.1039/c4fd00113c

healthcare worker–patient 'contact' cycle.[3,4] Rigorous hospital cleaning regimes may help reduce hospital surface contamination, however, it is difficult to completely prevent surface contamination or maintain low bacterial loads. One strategy to inhibit the cycle of the transmission of bacteria within a healthcare environment is the utilisation of self-sterilising surfaces. Examples of such surfaces include:[4] microbiocide-releasing surfaces,[5,6] silver and silver doped surfaces,[7,8] copper and copper alloy surfaces,[9–15] and photobactericidal surfaces.[16–32] Microbiocidal surfaces that utilise a light-activated mechanism in which a host of reactive oxygen species are generated are of particular interest, since the non-site specific mode of bacterial attack deems the emergence of bacterial resistance unlikely.[3]

Much research has been invested into the development of photobactericidal polymers activated by laser illumination for use in medical devices.[16–21] Similarly, this technology has also been harnessed for use in antimicrobial surfaces activated by standard hospital white light sources.[22–26] These surfaces were prepared using a simple dipping method, a "swell–encapsulation–shrink" strategy allowing either uniform encapsulation of the light-activated antimicrobial agents throughout the polymer bulk,[16–20,23–26] or both incorporation throughout the polymer bulk in addition to a high concentration at the polymer surface.[21,22] Although laser photo-activated bactericidal surfaces induce the lethal photosensitisation of bacteria in short time periods, laser technology is not practical for use in the sterilisation of touch surfaces commonly found in hospital environments. Conversely, white light-activated antimicrobial surfaces are ideal candidates for applications in general hospital touch surfaces and can help maintain low bacterial surface contamination levels. However, when microbial loads are high, the antimicrobial activity is effective over longer illumination time periods (up to 6 h for *E. coli*[22]). In cases where devices are frequently used and rapid surface sterilisation is preferable, polymers incorporated with UV-activated photocatalytic nanoparticles may prove more suitable.

Titania (TiO_2) is used in a wide range of products including white pigment, sunscreens and cosmetics. Inexpensive and robust, TiO_2 is an efficient photocatalyst under UV irradiation conditions[33,34] and this well established property has been utilised in applications such as drinking water sterilisation and in self-cleaning windows.[3,35–38] TiO_2 has a wide band gap (3.2 eV in the case of anatase TiO_2) and therefore requires <385 nm irradiation to activate its photocatalytic properties. The absorption of photons of energy greater than that of the semiconductor band gap results in the promotion of a valance band electron to the semiconductor conduction band. This produces electron–hole pairs, which can interact with adsorbed water and molecular oxygen at the surface of TiO_2, forming highly reactive radicals (hydroxyl groups and superoxide ions). These species participate in redox reactions for the degradation of organic pollutants and the killing of bacteria and viruses.[39–41] The mechanism followed in the inhibition of bacterial growth is still unclear, however, it is known that the species photogenerated on the TiO_2 surface have an active role in the destruction of the *Escherichia coli* (*E. coli*) cell outer membrane, causing the death of the bacteria.[41–43]

The system formed by TiO_2 modified with gold nanoparticles (henceforth Au/TiO_2) has been widely studied for environmental applications. These Au/TiO_2 systems are potentially efficient as visible-active photocatalysts due to their plasmon induced properties and the improved charge separation (Fermi level shift) which has been proposed previously for similar systems.[44–46] Nevertheless, Au/TiO_2 materials have received little attention in antimicrobial applications. Studies involving the sol–gel synthesis of gold-capped TiO_2 nanoparticles have shown a beneficial effect of Au-modified TiO_2 in the destruction of Gram-negative bacteria (*E. coli*).[47] On the contrary, different observations have been reported for Gram-positive bacteria. Fu *et al.*[47] observed similar inhibition rates in the growth of *Bacillus megaterium* for both TiO_2 and Au/TiO_2 systems, whereas the antimicrobial activity of the latter was reported to be retarded with respect to pure TiO_2 in the killing of *Bacillus subtilis*.[48]

In this paper, we report on the preparation of novel UV-activated photobactericidal polymers by incorporating TiO_2 and Au/TiO_2 nanoparticles into commercially available medical grade silicone, using a simple dipping technique. The nanoparticles were characterised by Raman, XRD, XPS and TEM and the photocatalytic activities of the TiO_2 and Au/TiO_2 nanoparticles were compared by investigating the photo-destruction of stearic acid upon UV illumination. The antimicrobial activity of the nanoparticle incorporated polymers was tested against both Gram-negative and Gram-positive bacteria commonly found in healthcare environments and a UVA source (365 nm) was used to activate the photobactericidal properties of these polymers. When tested against *S. aureus*, unprecedented kills were achieved with bacterial numbers reduced to below the detection limit within 15 minutes, without the need for any UV treatment (dark conditions). Although a smaller 'dark kill' effect was noted when tested against *E. coli*, exposure to UV illumination (365 nm) for 95 minutes reduced bacterial numbers to below the detection limit. To our knowledge, this is the first report of the 'dark kill' of bacteria using a titanium dioxide modified polymer. We attribute this kill to the surface functionalisation of the nanoparticles used.

2 Experimental

2.1 Materials synthesis

2.1.1 Synthesis of TiO_2 nanoparticles. TiO_2 nanoparticles were produced using continuous hydrothermal flow synthesis (CHFS). The pilot scale CHFS system has been described in detail elsewhere.[49,50] Four industrial diaphragm type chemical dosing pumps (Milton Roy, Primeroyal K) carried the precursors under controlled-flow conditions into a confined jet reactor. In the reactor, the precursors were mixed concurrently with supercritical water at 400 °C and 24.1 MPa. The temperature of the system was monitored at different points using type K thermocouples. The actual temperature at the reactor point, in the mixing of supercritical water and the precursor solutions, was 307 °C. Supercritical conditions for the water were reached using a custom built electrical heater arrangement (Watlow Cast X 2000, maximum thermal output 24 kW) and a back-pressure regulator in the system (BPR, Swagelok KHB series), was adjusted using a PID algorithm to maintain pressure. In the synthesis of the TiO_2 nanoparticles, each of the precursors, titanium(IV) oxysulfate ($TiOSO_4$ 0.2 M) with free hydrated sulphuric acid (Aldrich) and potassium hydroxide (KOH, 0.4 M) were pumped separately at 150 mL min^{-1} into a T-piece before meeting the flow of supercritical water. The flow rate of supercritical water was adjusted to 300 mL min^{-1}. The formed TiO_2 nanoparticles were rapidly quenched and cooled using an external cooling system (an outer jacket through which water flowed at 100 L min^{-1} and an inlet temperature of 15 °C). The nanoparticles were collected as a slurry and subsequently centrifuged and washed three times with deionised water (10 MΩ) until pH neutral and dried in a freeze drier (Virtis Genesis 35XL) by slowly heating from −60 to 25 °C over 24 h under a vacuum of 100 mTorr.

2.1.2 Deposition of gold nanoparticles on TiO_2. The deposition of gold nanoparticles was carried out following a procedure adapted from the literature.[51] 30 mg of chloroauric acid ($HAuCl_4$), 30 mL of 1% sodium citrate solution and 120 mL of DI water were refluxed for 4 h at 110 °C. 50 mL of the purple Au colloid was then stirred with 1 g of TiO_2 nanoparticles and 5.8 g of sodium chloride (NaCl) was suddenly added, resulting in the precipitation of Au metal on the TiO_2 particles. The purple precipitate (Au/TiO_2) was then filtered, repeatedly washed with deionised water (10 MΩ), centrifuged and lyophilized.

2.1.3 Functionalisation of TiO_2 and Au/TiO_2 nanoparticles. The functionalisation of TiO_2 and Au/TiO_2 nanoparticles was carried out following a procedure from the literature.[52] TiO_2 or Au/TiO_2 nanoparticles (1 g) were heated to 80 °C in excess oleic acid (120 mmol, 38.1 mL) over 24 hours, with a catalytic amount of triethylamine (8 mmol, 1.12 mL) added to encourage ester formation between the titanol groups on the particle surfaces and the oleic acid. The nanoparticles were precipitated with ethanol and centrifuged at 2500 *g* before being dispersed in toluene. The nanoparticles were precipitated re-dispersed in toluene three times in order to remove excess surfactant.

2.1.4 Preparation of antimicrobial polymers. A series of silicone samples were prepared for microbiological testing. Medical grade silicone polymer coupons (NuSil, Polymer Systems Technology Ltd, 1.21 cm^2) were immersed in toluene, oleic acid functionalised TiO_2 nanoparticles dispersed in toluene or Au/TiO_2 nanoparticles dispersed in toluene, and allowed to swell for 48 h. The samples were subsequently removed from the toluene swelling solutions, air-dried (48 h), washed and towel-dried. Where possible, the samples were maintained under dark conditions. An overview of the materials synthesis is presented in Fig. 1.

2.2 Materials characterisation

Transmission electron microscope images were recorded using a Jeol 2100 (high resolution) transmission electron microscope with a LaB_6 source operating at an acceleration voltage of 200 kV with an Oxford Instruments XMax EDX detector

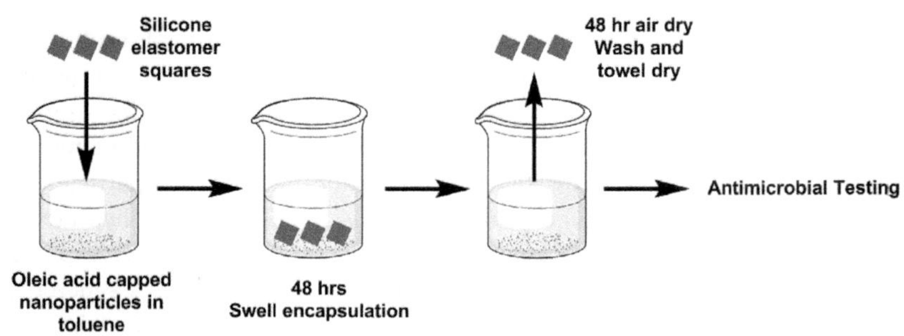

Fig. 1 Schematic to show the preparation of TiO_2 and Au/TiO_2 encapsulated silicone samples.

running AZTEC software. Micrographs were taken using a Gatan Orius Charge-coupled device (CCD). Samples for TEM were prepared by drop-casting a suspension of nanoparticles in hexane solvent onto a 400 Cu mesh holey carbon film TEM grid (Agar Scientific Ltd). UV/Vis spectroscopy was performed using a Perkin Elmer Lambda 950 UV/Vis/NIR spectrophotometer and a PerkinElmer Lambda 25 UV-Vis spectrometer. X-ray diffraction (XRD) data collection was performed using a Bruker-Axs D8 (GADDS) diffractometer. The instrument operates with a Cu X-ray source, monochromated ($K\alpha_1$ and $K\alpha_2$) and a 2D area X-ray detector with a resolution of 0.01°. Films were analysed with a glancing incident angle (θ) of 5°. The diffraction patterns obtained were compared with ICDD database standards. Raman spectra were obtained using a Renishaw Raman System 1000 calibrated using a silicon standard. X-Ray photoelectron spectroscopy (XPS) was performed using a Thermo Scientific K-alpha spectrometer with monochromated Al $K\alpha$ radiation, a dual beam charge compensation system and constant pass energy of 50 eV (spot size 400 μm). Survey scans were collected in the range 0–1200 eV. High-resolution scans were used for the principal peaks of Ti (2p), O (1s), Au (4f) and C (1s), and all spectra were calibrated against adventitious carbon (C1s, 284.8 eV).

2.3 Functional properties

2.3.1 Wetting properties. Equilibrium water contact angle measurements (~5 μL) were carried out on untreated silicone, solvent treated (control) silicone, unmodified TiO_2-encapsulated silicone (TiO_2) and gold modified titania-encapsulated silicone (Au/TiO_2). The contact angle measurement for each sample type was taken to be the average value over ≥5 measurements, using a droplet of deionised water dispensed by gravity from a gauge 27 needle and the samples were photographed side on. The data was analysed using SURFTENS software (V. 4.5).

2.3.2 Photocatalytic testing of the TiO_2 and Au/TiO_2 nanoparticles. The photo-activity of the nanoparticles was determined by the photo-oxidation of a model organic pollutant, octadecanoic (stearic) acid. In this test, films of TiO_2 and Au/TiO_2 nanoparticles (0.8 mg) were prepared by drop-casting a concentrated solution of nanoparticles onto borosilicate glass substrates and allowing to dry. A thin layer of stearic acid solution in chloroform (0.05 M) was subsequently dip-coated onto the samples. The samples were irradiated using a blacklight-blue (BLB) lamp (*Vilber Lourmat*, 365 nm) 2 × 8 W, 1.2 mW cm^{-2} and the photodegradation of the acid was monitored using a Perkin Elmer RX-I Fourier transform infrared (FTIR) spectrometer.

2.4 Microbiological testing

The following silicone elastomer samples (1.21 cm^2) were used in the microbiology experiments: solvent treated (control), oleic acid functionalised-TiO_2-encapsulated (TiO_2) and Au/TiO_2-encapsulated medical grade silicone. Immediately prior to microbiological testing, all samples were pre-irradiated for 18 hours using a 365 nm UV light source (Vilber Lourmat 365 nm BLB) emitting an average light intensity of 1.8 mW cm^{-2} at a distance of 10 cm from the samples, after which they were maintained under dark conditions (24 h). The samples were subsequently tested against *Staphylococcus aureus* 8325-4 and *Escherichia coli*

ATCC 25922. These organisms were stored at −70 °C in Brain-Heart-Infusion broth (BHI, Oxoid) containing 20% (v/v) glycerol and propagated on either mannitol salt agar (MSA, Oxoid) in the case of *S. aureus*, or MacConkey agar (MAC, Oxoid Ltd.) in the case of *E. coli*, for a maximum of two sub-cultures at intervals of two weeks.

BHI broth (10 mL) was inoculated with 1 bacterial colony and cultured in air at 37 °C for 18 h with shaking at 200 rpm. The bacterial pellet was recovered by centrifugation (21 °C, 1771 × *g*, 5 min), washed in phosphate buffered saline (PBS) and centrifuged again to recover the bacteria, which were finally re-suspended in PBS (10 mL). The washed suspension was diluted 1000-fold to obtain the inoculum (∼106 cfu mL^{-1}). The inoculum in each experiment was confirmed by plating ten-fold serial dilutions on agar for viable counts. Triplicates of each polymer sample type were inoculated with 20 μL of the inoculum and covered with a sterile cover slip (22 × 22 mm). The samples were then irradiated for the required time period using a 365 nm UV light source emitting an average light intensity of 1.8 mW cm^{-2} at a distance of 10 cm from the samples. A further set of samples (in triplicate) was maintained in the dark for the duration of the irradiation time.

Post irradiation, the inoculated samples and cover slips were added to PBS (180 μL) and vortexed. The neat suspension and ten-fold serial dilutions were plated on the appropriate agar for viable counts. The plates were incubated aerobically at 37 °C for 24 h (*E. coli*) or 48 h (*S. aureus*). Each experiment contained 3 technical replicates and the experiment was reproduced three times. The Mann–Whitney *U* test was used to determine the significance of the following comparisons: (i) the activity of each of the modified polymers compared to the control silicone sample when both were incubated in the dark and (ii) the activity of each of the irradiated modified polymers compared to the same material incubated in the dark.

3 Results and discussion

3.1 Materials synthesis and characterisation

Titania nanoparticles were synthesised by continuous hydrothermal flow synthesis (CHFS) and then gold addition was achieved through a modified Turkevich synthesis, which relies upon the gold forming upon the surface of the added titania nanoparticles after precipitation with salt. The titania particles provide a lower energy surface on which gold can grow, rather than undergo Ostwald ripening as they are precipitated. The citrate ligands on the Turkevich gold mean that the nanoparticles are electrostatically stabilised and as such are susceptible to salt flocculation due to a lowering of surface charge. Nanoparticle functionalisation was carried out according to Crick *et al.*,[52] which adds sterically stabilising oleic acid to the surface of the nanoparticles. A portion of the oleic acid is weakly bound (van der Waals interactions) and the rest is esterified with titanol groups on the nanoparticle surfaces.

Silicone samples were embedded with titania and gold–titania nanoparticles using a simple one-step dipping strategy. The modified surfaces were prepared using a "swell–encapsulation–shrink" technique, in which the samples were immersed in a toluene swelling solution made up to 20 mg mL^{-1} nanoparticles for 48 h. The toluene caused significant polymer swelling, enabling the diffusion of the nanoparticles through the polymer bulk. Removal from the swelling solution resulted in polymer shrinkage as residual solvent evaporated. The incorporation of neither the TiO_2 nanoparticles nor Au/TiO_2 nanoparticles impacted on polymer colouration. This is an attractive property in terms of future commercialisation of this technology for healthcare applications.

The toluene dispersed TiO_2 and Au/TiO_2 nanoparticle solutions and encapsulated silicone polymers were analysed using UV-Vis absorbance spectroscopy (Fig. 2). The UV-Vis data for the toluene dispersed nanoparticles were used to construct a Tauc plot—$(ahv)^{1/2}$ against energy (eV), where 'a' is the absorbance of the nanoparticles[29]—to approximate the TiO_2 and Au/TiO_2 nanoparticle band onset. The data indicate that the presence of gold shifts the band onset of TiO_2 towards the visible region of the spectrum (3.18 eV for Au/TiO_2 compared to 3.33 eV for TiO_2). The UV-Vis absorbance spectra of the TiO_2 and Au/TiO_2 encapsulated silicone polymers were measured and compared to that of a control silicone polymer. Fig. 2 shows that the band edge of the TiO_2 and Au/TiO_2-encapsulated silicone polymers is red shifted with respect to that of the control silicone polymer. This provides evidence of polymer nanoparticle encapsulation.

TEM micrographs showed general polydisperse populations of crystalline nanoparticles, dominated by anatase titania in both cases. Nanoparticles were measured to be an average size of 11.2 ± 4.7 nm and 10.9 ± 4.4 nm for the TiO_2 and Au–TiO_2 nanoparticle samples respectively, using Image J software. HR-TEM analysis showed highly crystalline samples, with d-spacings of *ca.* 0.34 nm and 0.33 nm (Fig. 3(b) and (d)). Using the Bragg equation, both spacings correspond to the ⟨101⟩ planes of anatase titania. From TEM analysis, nanoparticle size and shape were unaltered by both the gold and oleic acid functionalisation processes. This is possibly due to the small amount of Au (*ca.* 0.36%) being surface bound.

X-Ray diffraction supported the TEM analysis, with broad peaks due to small crystallite sizes compared with bulk anatase TiO_2. There was also no discernible change in peak positions on addition of Au. XRD analysis of the polymers

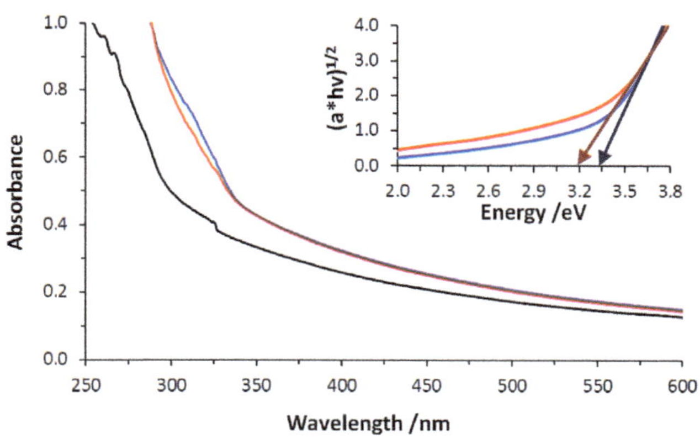

Fig. 2 UV-Vis absorbance spectra of a toluene treated silicone polymer (black line), a TiO_2-encapsulated silicone polymer (blue line) and a Au/TiO_2-encapsulated silicone polymer (red line). Inset: a Tauc plot to determine the band gap of the TiO_2 nanoparticles (blue line) and Au/TiO_2 nanoparticles (red line). The band onset of the TiO_2 nanoparticles was calculated to be 3.33 eV and the band onset of the Au/TiO_2 nanoparticles was calculated to be 3.18 eV.

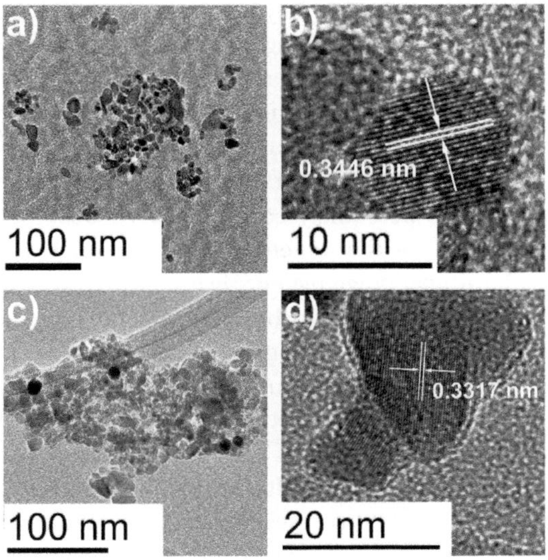

Fig. 3 (a) TEM image of oleic acid functionalised TiO_2 nanoparticles, (b) HR-TEM image of oleic acid functionalised TiO_2 nanoparticles, (c) TEM image of Au/TiO_2 nanoparticles and (d) HR-TEM image of Au/TiO_2 nanoparticles.

indicated no difference between the TiO_2-encapsulated silicone, Au/TiO_2-encapsulated silicone and control silicone. Raman spectra confirmed the anatase structure of the nanoparticles, with visible and characteristic E_g, B_{1g} and A_{1g} peaks at 146.7 and 636.7, 396 and 512.5 cm^{-1} for the TiO_2-based samples. There was no discernible shift in the Raman peaks between samples and there were no observable characteristic gold peaks (Fig. 4b). Raman spectra of the TiO_2- and Au/TiO_2-encapsulated silicone both before and after UV irradiation (365 nm, 18 h) were comparable to that of the control silicone sample. This indicates that neither the UV-irradiation, nor the photo-active nanoparticles photo-degrade the silicone polymer. XPS results showed the presence of Au in the Au/TiO_2 sample (Fig. 5), albeit in small quantities (*ca.* 0.3%), which is comparable to composition values obtained by EDX. Depth profile studies of the Au/TiO_2 powder showed that the gold *vs.* titanium composition did not change throughout the sample, up to a depth of 150 nm (ESI†), suggesting a uniform coating. It is worth noting that there

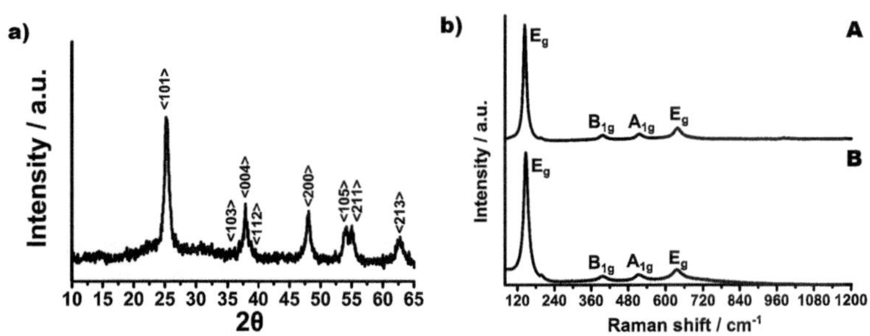

Fig. 4 (a) An X-Ray diffraction pattern of Au/TiO_2 showing distinctive peaks of anatase TiO_2 and (b) Raman spectra of the TiO_2 and Au/TiO_2 samples.[53]

is no discernible shift in the binding energies and chemical environments of the Ti 2p and O 1s in the two samples. XPS and EDX elemental composition information is available in the ESI.†

3.2 Functional properties

The wetting properties of a range of untreated and treated silicone samples were measured under the laboratory temperature and lighting conditions. The measurements indicated that the polymer itself is highly hydrophobic (120°) and the water contact angle is reduced upon encapsulation of the TiO_2 and Au/TiO_2 nanoparticles (90°). No photo-induced polymer hydrophilicity was observed upon incorporation of the nanoparticles. It can be speculated that this is because the nanoparticles do not form a thin film at the polymer surface, rather they are dispersed throughout the polymer matrix and therefore impact on the hydrophilicity of the polymer, rather than the UV-activated hydrophilicity.

The photoactivity of the films was evaluated during photodegradation of stearic acid, a model organic pollutant. Stearic acid is highly stable under UV irradiation in the absence of an underlying, effective photocatalyst film. The overall oxidation reaction is given by eqn (1) and it is monitored for transparent samples by infrared spectroscopy, following the disappearance of typical C–H bands at 2958, 2923 and 2853 cm^{-1} (Fig. 6a).

$$CH_3(CH_2)_{16}CO_2H + 26O_2 \xrightarrow{TiO_2(h\nu) \geq E_{bg}} 18CO_2 + 18H_2O \quad (1)$$

The number of acid molecules degraded is estimated using a conversion factor from the literature[54] (1 cm^{-1} = 9.7 × 10^{15} molecules per cm^2). The photoactivity rates are estimated from linear regression of the initial 20–30% degradation steps (zero-order kinetics regime) of the curves of integrated areas (Fig. 6b).

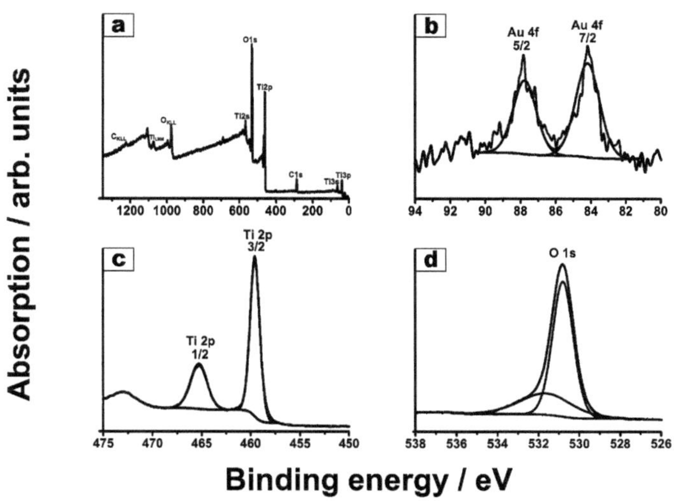

Fig. 5 Fitted XPS spectra of Au/TiO_2. (a) Labelled survey scan, (b) Au 4f scan, (c) Ti 2p scan and (d) O 1s scan.

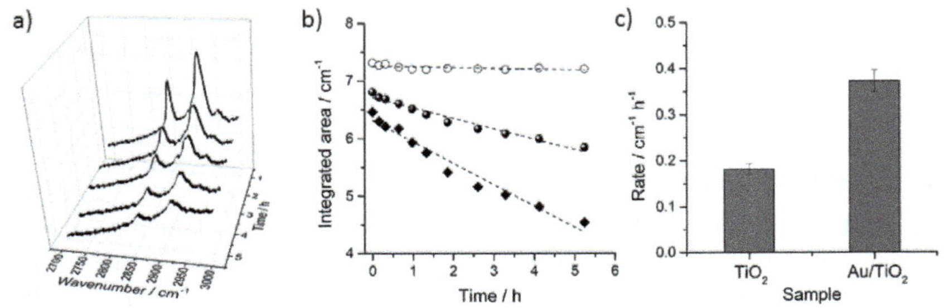

Fig. 6 (a) IR spectra of stearic acid upon UVA illumination (1.2 mW cm^{-2}) on a TiO_2 drop-cast film. (b) Integrated areas obtained during illumination of TiO_2 (full circles), Au/TiO_2 (full diamonds) and plain glass control (open circles) films. (c) Photo-activity rates (given as rate over irradiance) of TiO_2 and Au/TiO_2 films obtained during UVA irradiation (BLB 365 nm).

Preliminary photocatalytic tests of drop-cast films of TiO_2 and Au/TiO_2 powders showed an enhanced activity in the case of the latter material (Fig. 6c), in agreement with the literature.[54] The enhanced activity of TiO_2-based systems containing noble metal nanoparticles is typically explained in terms of an efficient charge separation as shown in Fig. 7.

3.3 Photobactericidal activity

We report the light-activated antimicrobial activity of titania nanoparticle encapsulated silicone sample systems when tested against representative Gram-positive and Gram-negative bacteria, *S. aureus* and *E. coli* respectively, both of which are commonly found in healthcare environments. The samples were

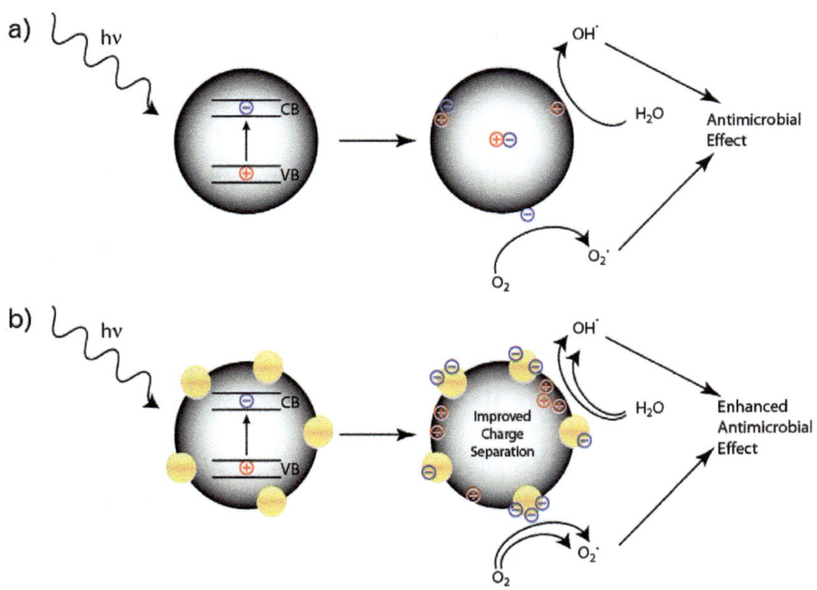

Fig. 7 Proposed mechanism of photocatalytic antimicrobial activity of (a) the TiO_2 encapsulated silicone sample and (b) the Au/TiO_2 encapsulated silicone sample.

pre-irradiated (18 h) with a 365 nm UV lamp prior to testing to photodegrade residual organic precursor material, after which they were stored under dark conditions (24 h). In each experiment the samples were irradiated with a 365 nm UV lamp emitting an average of 1.8 mW cm^{-2} at a distance of 10 cm, whilst a control sample set was maintained under dark conditions for the same duration, to determine whether bacterial kills are attributed to a photocatalysed effect or intrinsic bactericidal properties of the surfaces.

The antimicrobial activity of medical grade silicone encapsulated with TiO_2 or Au/TiO_2 nanoparticles was compared to that of a control, solvent treated silicone sample. Interestingly, when tested against *S. aureus*, both titania incorporated samples demonstrated potent bactericidal activity under dark conditions, with bacterial numbers reduced to below the detection limit ($P < 0.001$, Fig. 8a). As a result, no additional photosensitisation effects could be determined. TiO_2 is a photo-activated antimicrobial agent and thus, it was surprising to observe bactericidal activity under dark conditions. Hence, it can be speculated that the exhibited antimicrobial properties of the TiO_2 and Au/TiO_2 encapsulated polymers can be attributed to the oleic acid capping agent on the nanoparticles, which has been noted to inhibit growth of Gram-positive bacteria including *S. aureus*.[55] Further experiments with different oleic acid capped nanoparticles would be of benefit in elucidating the mechanism of 'dark kill' observed in this study.

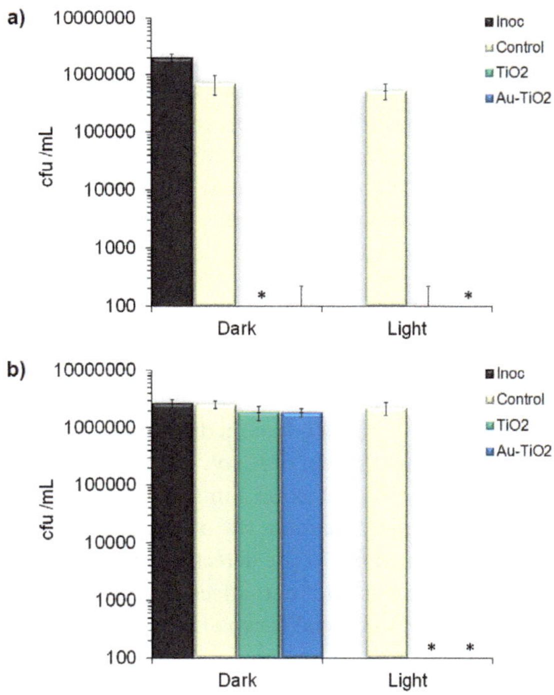

Fig. 8 Viable counts of bacteria (colony forming units (cfu) per mL) after incubation on modified silicone polymers exposed to UV irradiation (365 nm): (a) *S. aureus* (15 min illumination) and (b) *E. coli* (95 min illumination). The UV source emitted an average light intensity of 1.8 ± 0.1 mW cm^{-2} at a distance of 10 cm from the samples. * symbols indicate that the bacterial numbers were reduced to below the detection limit of 500 cfu mL^{-1}.

Fig. 8b shows that under dark conditions, both the TiO_2-encapsulated silicone and the Au/TiO_2-encapsulated silicone exhibited limited antimicrobial activity against *E. coli*, although these effects were not highly significant ($P < 0.01$). This observation indicates that the bactericidal properties of the capping agent are not as effective against this bacterium. However, the data show that when exposed to a 365 nm UV lamp, both the TiO_2 and Au/TiO_2 samples exhibited efficacious photo-activated properties, reducing viable bacterial numbers to below the detection limit within 95 minutes ($P < 0.001$).

TiO_2, a UV-activated photocatalyst, generates electron–hole pairs upon absorption of photons with energy greater than that of its band gap energy. The electrons can interact with molecular oxygen in the vicinity generating superoxide anions, whereas the positive holes at the nanoparticle surface can interact with adsorbed water, generating hydroxyl radicals. The photo-generated radical species can subsequently react further, resulting in the production of a wide range of radicals. These radicals can initiate a multi-site attack against bacteria in the environs, oxidising organic matter and subsequently photo-destroying bacterial cellular membranes and components. Due to the non-site specific attack mechanism, the emergence of bacterial resistance is highly unlikely. This is a highly desirable property in antimicrobial surfaces developed for applications in healthcare settings.

In this study, it was not possible to determine whether the presence of Au enhanced the photo-activity of the TiO_2 nanoparticles. However, rapid kills were achieved on both the TiO_2 and Au/TiO_2 samples when tested against *S. aureus* and *E. coli*, with viable bacterial numbers reduced to below the detection limit. Further microbial investigations involving white light activation or shorter illumination times may shed light onto whether the Au shifts the nanosized TiO_2 towards visible light photocatalysis or improves the antimicrobial efficacy over that of TiO_2 alone. Moreover, microbial testing on surfaces incorporated with a range of oleic acid-capped nanoparticles may be of interest. It would also be advantageous in further investigations to study organisms with known antibiotic resistance or recent clinical isolates.

4 Conclusion

We report a new class of potent UV-activated antimicrobial polymers by incorporating nano-TiO_2 and nano-Au/TiO_2 into medical grade silicone. These surfaces induced the lethal photosensitisation of *E. coli* within 95 minutes, with bacterial numbers reduced to below the detection limit. They also exhibited efficacious antimicrobial activity against *S. aureus via* a non-light-activated mechanism, reducing bacterial numbers to below the detection limit within 15 minutes. The incorporation of nanoparticles did not affect polymer colouration, a highly desirable property from a commercial viewpoint.

Acknowledgements

The authors would like to thank the EPSRC for their financial support. IPP and SN would also like to thank Ondine Biomedical for their financial support. The authors would like to thank Mr Martin Vickers for assistance in XRD, Dr Robert

Palgrave for assistance in XPS and Dr Steven Firth for assistance in TEM and Raman.

References

1 Doctors and nurses must redouble hygiene efforts to bring down 'unacceptable and avoidable' infection rates, http://http://www.nice.org.uk/newsroom/pressreleases/DoctorsAndNursesMustRedouble\\HygieneEffortsToBringDownInfectionRates.jsp, accessed: 27-04-2014.

2 HCAI and Antimicrobial Point Prevalence Survey England, http://www.hpa.org.uk/Topics/InfectiousDiseases/InfectionsAZ/HCAI/HCAIPointPrevalenceSurvey/, accessed: 27-04-2014.

3 S. Noimark, C. W. Dunnill and I. P. Parkin, *Adv. Drug Delivery Rev.*, 2013, **65**, 570–580.

4 K. Page, M. Wilson and I. P. Parkin, *J. Mater. Chem.*, 2009, **19**, 3819–3831.

5 C. E. Edmiston, F. C. Daoud and D. Leaper, *Surgery*, 2013, **154**, 89–100.

6 D. L. Williams, K. D. Sinclair, S. Jeyapalina and R. D. Bloebaum, *J. Biomed. Mater. Res., Part B*, 2013, **101**, 1078–1089.

7 AcryMed Inc., SilvaGardTM Technology Summary., http://www.acrymed.com/pdf/SilvaGard%20Technical%20Summary.pdf, accessed: 2013-07-22.

8 AgION Technologies Inc., AgIONTM Technology, http://www.agion-tech.com, accessed: 2013-07-22.

9 F. Marais, S. Mehtar and L. Chalkley, *J. Hosp. Infect.*, 2010, **74**, 80–82.

10 S. A. Wilks, H. Michels and C. W. Keevil, *Int. J. Food Microbiol.*, 2005, **105**, 445–454.

11 T. J. Karpanen, A. L. Casey, P. A. Lambert, B. D. Cookson, P. Nightingale, L. Miruszenko and T. S. J. Elliott, *Infect. Control Hosp. Epidemiol.*, 2012, **33**, 3–9.

12 C. D. Salgado, K. A. Sepkowitz, J. F. John, J. R. Cantey, H. H. Attaway, K. D. Freeman, P. A. Sharpe, H. T. Michels and M. G. Schmidt, *Infect. Control Hosp. Epidemiol.*, 2013, **34**, 479–486.

13 J. O. Noyce, H. Michels and C. W. Keevil, *J. Hosp. Infect.*, 2006, **63**, 289–297.

14 L. Weaver, H. T. Michels and C. W. Keevil, *J. Hosp. Infect.*, 2008, **68**, 145–151.

15 M. G. Schmidt, H. H. Attaway, P. A. Sharpe, J. John, K. A. Sepkowitz, A. Morgan, S. E. Fairey, S. Singh, L. L. Steed, J. R. Cantey, K. D. Freeman, H. T. Michels and C. D. Salgado, *J. Clin. Microbiol.*, 2012, **50**, 2217–2223.

16 S. Perni, C. Piccirillo, J. Pratten, P. Prokopovich, W. Chrzanowski, I. P. Parkin and M. Wilson, *Biomaterials*, 2009, **30**, 89–93.

17 S. Perni, C. Piccirillo, A. Kafizas, M. Uppal, J. Pratten, M. Wilson and I. P. Parkin, *J. Cluster Sci.*, 2010, **21**, 427–438.

18 S. Perni, P. Prokopovich, C. Piccirillo, J. Pratten, I. P. Parkin and M. Wilson, *J. Mater. Chem.*, 2009, **19**, 2715–2723.

19 S. Perni, J. Pratten, M. Wilson, C. Piccirillo, I. P. Parkin and P. Prokopovich, *J. Biomater. Appl.*, 2011, **25**, 387–400.

20 S. Noimark, C. W. Dunnill, C. W. M. Kay, S. Perni, P. Prokopovich, S. Ismail, M. Wilson and I. P. Parkin, *J. Mater. Chem.*, 2012, **22**, 15388–15396.

21 S. Noimark, M. Bovis, A. J. MacRobert, A. Correia, E. Allan, M. Wilson and I. P. Parkin, *RSC Adv.*, 2013, **3**, 18383–18394.

22 S. Noimark, E. Allan and I. P. Parkin, *Chem. Sci.*, 2014, **5**, 2216–2223.

23 A. J. T. Naik, S. Ismail, C. Kay, M. Wilson and I. P. Parkin, *Mater. Chem. Phys.*, 2011, **129**, 446–450.

24 V. Decraene, J. Pratten and M. Wilson, *Appl. Environ. Microbiol.*, 2006, **72**, 4436–4439.

25 V. Decraene, J. Pratten and M. Wilson, *Curr. Microbiol.*, 2008, **57**, 269–273.

26 M. Wilson, *Infect. Control Hosp. Epidemiol.*, 2003, **24**, 782–784.

27 J. Bozja, J. Sherrill, S. Michielsen and I. Stojiljkovic, *J. Polym. Sci., Part A: Polym. Chem.*, 2003, **41**, 2297–2303.

28 M. Wainwright, M. N. Byrne and M. A. Gattrell, *J. Photochem. Photobiol., B*, 2006, **84**, 227–230.

29 C. W. Dunnill, K. Page, Z. A. Aiken, S. Noimark, G. Hyett, A. Kafizas, J. Pratten, M. Wilson and I. P. Parkin, *J. Photochem. Photobiol., A*, 2011, **220**, 113–123.

30 Z. A. Aiken, G. Hyett, C. W. Dunnill, M. Wilson, J. Pratten and I. P. Parkin, *Chem. Vap. Deposition*, 2010, **16**, 19–22.

31 C. W. Dunnill, Z. Ansari, A. Kafizas, S. Perni, D. J. Morgan, M. Wilson and I. P. Parkin, *J. Mater. Chem.*, 2011, **21**, 11854–11861.

32 C. W. Dunnill, Z. A. Aiken, J. Pratten, M. Wilson and I. P. Parkin, *Chem. Vap. Deposition*, 2010, **16**, 50–54.

33 A. Mills and S. LeHunte, *J. Photochem. Photobiol., A*, 1997, **108**, 1–35.

34 M. R. Hoffmann, S. T. Martin, W. Y. Choi and D. W. Bahnemann, *Chem. Rev.*, 1995, **95**, 69–96.

35 A. Mills, R. H. Davies and D. Worsley, *Chem. Soc. Rev.*, 1993, **22**, 417–425.

36 A. Mills, N. Elliott, I. P. Parkin, S. A. O'Neill and R. J. H. Clark, *J. Photochem. Photobiol., A*, 2002, **151**, 171–179.

37 S. A. O'Neill, I. P. Parkin, R. J. H. Clark, A. Mills and N. Elliott, *J. Mater. Chem.*, 2003, **13**, 56–60.

38 A. Mills, G. Hill, M. Crow and S. Hodgen, *J. Appl. Electrochem.*, 2005, **35**, 641–653.

39 J. C. Yu, W. K. Ho, J. Lin, K. Y. Yip and P. K. Wong, *Environ. Sci. Technol.*, 2003, **37**, 2296–2301.

40 L. K. Adams, D. Y. Lyon and P. J. J. Alvarez, *Water Res.*, 2006, **40**, 3527–3532.

41 K. Sunada, Y. Kikuchi, K. Hashimoto and A. Fujishima, *Environ. Sci. Technol.*, 1998, **32**, 726–728.

42 D. M. Blake, P. C. Maness, Z. Huang, E. J. Wolfrum, J. Huang and W. A. Jacoby, *Sep. Purif. Methods*, 1999, **28**, 1–50.

43 Z. X. Lu, L. Zhou, Z. L. Zhang, W. L. Shi, Z. X. Xie, H. Y. Xie, D. W. Pang and P. Shen, *Langmuir*, 2003, **19**, 8765–8768.

44 Y. Tian and T. Tatsuma, *J. Am. Chem. Soc.*, 2005, **127**, 7632–7637.

45 K. Kimura, S.-i. Naya, Y. Jin-nouchi and H. Tada, *J. Phys. Chem. C*, 2012, **116**, 7111–7117.

46 M. J. Uddin, F. Cesano, D. Scarano, F. Bonino, G. Agostini, G. Spoto, S. Bordiga and A. Zecchina, *J. Photochem. Photobiol., A*, 2008, **199**, 64–72.

47 G. F. Fu, P. S. Vary and C. T. Lin, *J. Phys. Chem. B*, 2005, **109**, 8889–8898.

48 L. Armelao, D. Barreca, G. Bottaro, A. Gasparotto, C. Maccato, C. Maragno, E. Tondello, U. L. Stangar, M. Bergant and D. Mahne, *Nanotechnology*, 2007, **18**, 375709.

49 C. J. Tighe, R. I. Gruar, C. Y. Ma, T. Mahmud, X. Z. Wang and J. A. Darr, *J. Supercrit. Fluids*, 2012, **62**, 165–172.

50 R. I. Gruar, C. J. Tighe and J. A. Darr, *Ind. Eng. Chem. Res.*, 2013, **52**, 5270–5281.

51 A. Mills, *J. Chem. Soc., Chem. Commun.*, 1982, 367–368.

52 C. R. Crick, J. C. Bear, A. Kafizas and I. P. Parkin, *Adv. Mater.*, 2012, **24**, 3505–3508.

53 W. F. Zhang, Y. L. He, M. S. Zhang, Z. Yin and Q. Chen, *J. Phys. D: Appl. Phys.*, 2000, **33**, 912–916.

54 A. Mills and J. Wang, *J. Photochem. Photobiol., A*, 2006, **182**, 181–186.

55 F. Dilika, P. D. Bremner and J. J. Meyer, *Fitoterapia*, 2000, **71**, 450–452.

Faraday Discussions

DISCUSSIONS

Other Nanoparticles: general discussion

Zoe Pikramenou, Catriona McCallion, Sara Carreira, Peter Dobson, Katherine Brown, Yuri Antonio Diaz Fernandez, Maha Abdollah, Dejian Zhou, Dan Sun, Sandhya Moise, Lucio Litti, Lanry L. Yung, Stefan Borsley, Nadiya Dragneva, Annette Barchanski, Mostafa El-Sayed, Amelie Heuer-Jungemann, Roger M. Pallares, Edman Tsang, Nicolas Barry, Scott Mitchell, Nguyen T. K. Thanh, Maya Thanou, Ivan Parkin, Paresh Ray and Richard Jones

DOI: 10.1039/C4FD90077D

Sara Carreira opened the discussion of the paper by Nicolas Barry: Why are your **RuMs** and **OsMs** so specific to cancer cells without using any targeting moiety?

Nicolas Barry replied: We believe that the specificity comes from the size of the particles – passive targeting. Clinically-validated therapeutic and imaging NPs usually target cancer cells in a passive way. This is achieved by taking advantage of the enhanced permeability and retention (EPR) effect in tumor tissues. Tumor vasculature is highly disorganised, compared to the vasculature in normal tissues, and the vascular endothelium in tumors proliferates rapidly and discontinuously. This results in a higher number of fenestrations and open junctions (from 200 nm to 1.2 μm) than in normal vessels. Particles with a typical size of a few nanometres can therefore passively cross the tumor endothelial barrier through fenestrations, and accumulate at particular sites through blood hemodynamic forces and diffusion mechanisms.

One of our objectives is to increase the size of our particles from 15 nm to a few hundred nanometres, in order to maximize this passive targeting. We also wish to introduce an active targeting moiety (*e.g.* specific peptides, antibodies) on the corona of the particles to increase this selectivity.

Peter Dobson asked: Have you looked to see if any of your compounds are luminescent?

Nicolas Barry responded: There are numerous examples of ruthenium compounds that are luminescent. Usually, arene Ru(II) complexes are not luminescent owing to the arene–metal interactions. Nonetheless, it is possible to introduce a luminescent ligand (such as a pyrene derivative) by functionalizing the 16-electron complexes (to make an 18-electron complex).

Peter Dobson added: The reason for my previous question is that there is growing interest in making screens for visual detection of neutrons, and I wondered if your compounds might be used for this purpose?

Nicolas Barry answered: This is an interesting remark, especially since our compounds are also boron-rich (boron–neutron interactions being of potential interest).

Ivan Parkin commented: Do you have any idea where the anticancer properties of these materials orginates from?

Nicolas Barry responded: The polymer used for the assembly of the metallated micelles belongs to the Pluronic® family, a class of block co-polymers widely used for the design of drug delivery systems, or used as pharmaceutical ingredients and biological response modifiers. In the absence of an encapsulated complex, the micelles made of Pluronic® P123 do not exhibit any toxicity towards the two cell lines studied in this paper. On the other hand, the carborane-containing organometallic complexes are highly cytotoxic. Although entrapment of the 16-electron complexes in Pluronic® micelles leads to a reduction in their anticancer potency, the micelles exhibit enhanced selectivity towards cancer cells compared to normal cells. These results therefore suggest that the anticancer properties of these materials originate from the organometallic complexes, but that the micellar formulation has a direct impact on their biological mode(s) of action.

Ivan Parkin asked: Why did you initially choose to study these materials?

Nicolas Barry replied: It is clear that dicarba-*closo*-dodecarborane derivatives can give rise to interesting and unusual properties as ligands in organometallic complexes. These clusters are remarkably stable in biological media and can be recognised by various bio-targets. They can be easily functionalised *via* organic reactions on the CH vertices and they have useful probe properties. Due to the high number of boron atoms in their globular structures, they have been extensively studied as potential boron neutron capture theory (BNCT) agents. Different clinical trials have been carried out in Japan, Europe and the United States, and sodium borocaptate ($Na_2B_{12}H_{11}SH$) has been used clinically. Nevertheless, selective and effective delivery of boron agents is still a critical issue. For this reason, it is of central importance to explore new concepts that are able to take advantage of these unique pharmacophores. The combination of dicarba-*closo*-dodecarboranes with half-sandwich complexes of ruthenium, osmium, rhodium and iridium could provide the expected breakthrough in dicarba-*closo*-dodecarborane biochemistry. However, these complexes being highly hydrophobic, their biological applications are impaired by their lack of solubility in water. Here we have formulated these complexes in micelles made of polymer, and reported their encapsulation in Pluronic® polymer micelles dispersed in water.

Zoe Pikramenou enquired: Can you identify the interactions between your metal complex and the polymer by NMR?

Nicolas Barry responded: We cannot identify the interactions between the metal complexes and the polymer by proton or carbon NMR since the resonances of the complexes are too broad and unresolved to be visible. The encapsulation does not lead to noticeable shifts in the polymer signals. We are, however, in the process of studying the micelles by diffusion NMR (DOSY).

Edman Tsang said: Do you know how these complexes locate in the core?

Nicolas Barry answered: The complexes are encapsulated in the PPO core of the micelles, through hydrophobic interactions between the hydrophobic blocks of the Pluronic® polymers and the organometallic complexes. There are *ca.* 50 complexes per micelles (and each micelle is made of *ca.* 50 unimers).

Dejian Zhou queried: When your micelles are taken up by cells, where are they located intracellularly?

Nicolas Barry answered: This is a question we urgently wish to address. At the moment, we do not know where the micelles are located, nor do we know about their speciation in cells. Are the organometallic complexes released from the micellar cores, where are they released, and at what rate? We are currently trying to image the micelles in cells (which is possible due to the high contrast of Os and Ru in microscopy) to try and address this question.

Dejian Zhou added: How stable are the micelles? After they are taken up by cells, are they going to gradually degrade? This is important for biosafety.

Nicolas Barry responded: We showed by SAXS studies that the micelles are stable in aqueous solution and in RPMI media at various concentrations (1 to 10 mg mL^{-1}) and at various temperature (10 to 50 degrees C). However, we do not know how stable the micelles are in cells. Are they going to gradually degrade? Are they going to release the complexes and, if so, where? These are important questions we need to answer.

Sara Carreira asked: Do you know why the **RuMs** are more potent than the **OsMs**? Is there possibly a difference in electronic structure between the two elements that could affect toxicity?

Nicolas Barry answered: No, we don't know. However, osmium, the heavier congener of ruthenium and a third row transition metal, commonly exhibits slower kinetics than ruthenium, and is often considered to be relatively inert.

Ivan Parkin opened the discussion of the paper by Richard Jones: Can you say something about the generic properties of these materials? Perhaps in "Chem 101" terms in the form of hard acids and bases? Cadmium (2+) would be expected to be a soft acid, whilst lanthanides would be hard acids (3+).

Richard Jones responded: I'm not sure I understand the question about generic properties. Yes, Cd^{2+} is generally considered a soft acid and Ln^{3+} a hard acid, but they both coordinate to the available N and O donor atoms of the ligands. The overall structure is effectively a giant coordination compound (complex) with chemical and physical properties that you would expect for these ions bound by organic based ligands – soluble in polar organic solvents (MeOH, MeCN *etc.*), crystalline, fairly high melting points and stable in air.

Ivan Parkin said: Is there a particular architecture that results in the lanthanides and cadmium occupying specific sites in the lattice?

Richard Jones answered: Careful examination of the crystal structures reveals a 1 : 3 ratio of $Ln^{3+}$to Cd^{2+}. In compound **1**, for example, the 1 : 3 motif is repeated three times as you progress around the top and bottom rings of metal ions. The coordination environments are quite complex and involve donor atoms from both the ligands, as well as acetates. The lanthanides (in **1**) are 8-coordinate, while the Cd ions have lower coordination numbers (7 and 6). Other than that, there does not seem to be any particularly well-recognized architecture that results in metals occupying specific sites.

Ivan Parkin remarked: How sure are you of the absolute configurations of the structures? The lanthanides and the cadmium have similar electron densities and might be imagined to have a variety of site occupancies within the crystal.

Richard Jones responded: Yes, you are absolutely correct when you say that the Ln and Cd ions have similar electron densities, and so site occupancies could be a concern. When we first discovered these structures we obtained good reproducible ICPMS data on the Ln : Cd ratios and have confirmed that the Ln : Cd ratio is 1 : 3 for the bulk materials. While it is possible that some Ln and Cd ions could partially occupy the same sites in a structure, the best refinements (lowest R values) of single crystal X-ray data have the ions in the locations published.

Zoe Pikramenou asked: Do you get different structures for the different lanthanides, such as Eu and Yb?

Richard Jones replied: Generally speaking, we get very similar structures when we go from one lanthanide to another. However, we do get drums of different dimensions and compositions when we vary the nature of the lanthanide salt starting material (*e.g.* acetate *vs.* chloride) or when we vary the length of the flexible linker of the ligand. See, for example, Yang *et al.*[1] and Yang *et al.*[2]

1 X. Yang, D. Schipper, R. A. Jones, L. A. Lytwak, B. J. Holliday and S. Huang, *J. Am. Chem. Soc.*, 2013, **135**, 8468–8471.
2 X. Yang, D. Schipper, L. Zhang, K. Yang, S. Huang, J. Jiang, C. Su and R. A. Jones, *Nanoscale*, 2014, **6**, 10569–10573.

Sara Carreira enquired: Could you tune the size of the cavity in your nanodrum and therefore make it possible to load drugs?

Richard Jones answered: This would be a very interesting and potentially useful approach to drug delivery. However, at present the size of the cavity is fairly small and probably not large enough to load drugs. We are exploring ways to tune the size of the cavity.

Scott Mitchell said: Out of curiosity, have you performed any theoretical studies on the stability of the clusters in solution? The presence of multiple lanthanide centres obviously makes calculations more demanding, but perhaps you might be able to obtain at least some indication of the solution stability of your nano-drums using density functional theory, if you can find someone who is undaunted by modelling molecules with so many atoms.

Richard Jones replied: Thank you, Scott, we appreciate the question. The drums have 32 metals and their molecular weights are well over 12,000 amu. Our DFT programs are currently not capable of handling this size of molecule with any level of accuracy. We would need far more computer power, more sophisticated programs and, as you say, someone with expertise in modelling such large structures.

Edman Tsang remarked: These lanthanides are very sensitive to chemical environment – is there any design which could maintain their photophysical properties?

Richard Jones answered: Thanks for your question. We are working on multidentate ligand systems designed to encapsulate them firmly and to shield the lanthanides from external solvents, such as water, which can quench the luminescence.

Sandhya Moise asked: How do the nano-drums interact with cells?

Richard Jones replied: We do not know at this time, but given that the dimensions of the nano-drums described here are approximately 30 Å, uptake is likely to involve one or more endocytotic pathways.

Sandhya Moise noted: It may be possible to use total internal reflection fluorescence (TIRF) to study the interactions of the nano-drums with cell membranes.

Nicolas Barry enquired: Have you tried to collect your crystals from the TEM?

Richard Jones replied: Thank you. We have not tried this and to be completely honest I did not know that this might be possible. I will ask our TEM experts how we might go about attempting this experiment.

Sara Carreira asked: Is the "rarity" of these lanthanides reflected in their price?

Richard Jones replied: Despite their name, rare earth elements (with the exception of radioactive promethium) are relatively plentiful in the Earth's crust. Although they are expensive to isolate by modern techniques, and future geopolitical issues may affect the price of the bulk ores, the amount of material used in a typical bioassay is very small. The lanthanides we use in these studies may offer a more affordable alternative to precious metals, such as gold.

Ivan Parkin added: The rare earths are in general not that "rare". The issue is that they are often widely distributed, making extraction costs high.

Peter Dobson also noted: Rare earth materials are generally not "rare" in the sense that they are scarce. They are very plentiful, and the fluorescent tubes in this lecture room probably contain enough rare earths to make a large batch of the nano-drum-type materials studied here.

The situation is made clear in ref. 1.

1 M. Humphries, *Rare Earth Elements: The Global Supply Chain*, http://fas.org/sgp/crs/natsec/R41347.pdf

Nguyen T. K. Thanh addressed Peter Dobson: Thank you Peter, it is a very useful and interesting document. I also later received ref. 1 from a colleague, which is quite interesting as well.

Looks like, for now, we can safely assume we have enough resources to make nanomaterials. The final decision would be based on the cost of the materials and the economic benefit from each application. Thank you for the discussion.

1 F. W. Clarke and H. S. Washington, The composition of the Earth's crust, 1924, *U. S. G. S. Prof. Paper.*, 127.

Ivan Parkin added : According to Wikipedia, 95% of the world's rare earth supply comes from China.

Dejian Zhou continued the discussion of Richard Jones' paper: When you want to biofunctionalise the nanodrums, do you make the nano-drums first and then biofunctionalise them, or do you conjugate the ligands first and let them self-assembly?

Richard Jones responded: That's a really good question. We intend to explore and evaluate both routes to bio-functionalized nanoparticles.

Dejian Zhou commented: Biomolecules are typically much bigger than your nano-drums. If you attach a biomolecule to the nanodrum, will that affect the assembly properties?

Richard Jones answered: We plan to carry out conjugation of assembled nano-drums if possible. However, the self-assembly properties could change, although at this point we do not know whether alternative assemblies would necessarily affect the utility of our compounds.

Dejian Zhou asked: What are the coordination numbers for the lanthanides in the complexes? If they are coordination saturated then it would be difficult to bind other species.

Richard Jones responded: The coordination numbers for the lanthanides are high. However, we are proposing to bind other species (*i.e.* biomolecules) *via* functionalized ligand systems, not directly to the Ln ions.

Stefan Borsley said: Have you tried reducing the imines? If so, does this lead to increased stability of the nanodrums?

Richard Jones replied: Yes, the imine portion of the flexible Schiff base ligand can be reduced easily with $NaBH_4$. We are currently exploring the use of saturated ligands for the synthesis of self-assembled nano-drums.

Peter Dobson asked: How stable are these nanodrum compounds to heating?

Richard Jones answered: The nano-drums we have isolated and characterized so far appear to be stable to moderate heating and we do not anticipate that thermal degradation would be a problem for these compounds in biological applications. Some indication of their thermal stability can be gleaned from melting point data. The nano-drums do not melt when heated, but decompose. Their decomposition points range from 176 to 212 °C. (See also supplementary data in ref. 1 and 2.)

1 X. Yang, D. Schipper, R. A. Jones, L. A. Lytwak, B. J. Holliday and S. Huang, *J. Am. Chem. Soc.*, 2013, **135**, 8468–8471.
2 X. Yang, D. Schipper, L. Zhang, K. Yang, S. Huang, J. Jiang, C. Su and R. A. Jones, *Nanoscale*, 2014, **6**, 10569–10573.

Lanry L. Yung enquired: Since it was shown in the talk that the nano-drum assembly forms spontaneously and is very stable, has the dissociation constant of the assembly been measured? If not, is there any difficulty in measuring it?

Richard Jones responded: As the nano-drums form during crystallization, we have no information on the dissociation constant of the complex, although we predict that it will be very small as the nano-drums are very stable. We agree that understanding the assembly mechanism of these compounds would be very valuable for developing additional strategies for controlling the self-assembled properties of these compounds. This is an area of research we are interested in exploring further.

Amelie Heuer-Jungemann opened the discussion of the paper by Paresh Ray: Do you see a future for graphene-based materials for *in vivo* applications, despite the fact that their size and surface chemistry still cannot be controlled very well?

Paresh Ray answered: All the reported results are from experiments within the past year, whereas it will take one to two decades to discover the future of graphene-based materials. The synthesis procedures need to evolve to allow control of the size and surface chemistry of the material. One needs to understand their interactions with the environment and with biomolecules, before the materials can be used for real life applications.

Peter Dobson remarked: I don't understand what you are tuning with graphene oxide. It is not like a semiconductor quantum dot and these are flakes with variable lateral dimensions.

Paresh Ray responded: Thank you for your comment. These are carbon dots, not quantum dots. The quantum effects that we have observed in semiconductor nanoparticles are not significant in carbon dots. In graphene oxide, the emission wavelength can be tuned by changing the excitation wavelength, which is mainly due to the huge red-edge effect and excited state proton transfer, as reported by several groups.[1–3]

1 C. T.Chien, S. S. Li, W. J. Lai, Y. C. Yeh, H. A. Chen, I. S. Chen, L. C. Chen, K. H. Chen, T. Nemoto and S. Isoda, Tunable Photoluminescence from Graphene Oxide, *Angew. Chem., Int. Ed.*, 2012, **51**, 6662–6666.
2 J. Shang, L. Ma, J. Li, W. Ai, T. Yu and G. G. Gurzadyan, The Origin of Fluorescence from Graphene Oxide, *Sci. Rep.*, 2012, **2**, 792.
3 S. K. Cushing, M. Li, F. Huang and N. Wu, Origin of Strong Excitation Wavelength Dependent Fluorescence of Graphene Oxide, *ACS Nano*, 2014, **8**, 1002–1013.

Nadiya Dragneva asked: It seems like the modified Hummers method is the most popular method for graphene oxide production. Are there any defects present in your structure as a result of using this method?

Paresh Ray replied: Yes, Hummers method produced defects in the structure and, as a result, we have reported a strong defect band in the Raman spectra from graphene oxide. The observed wavelength dependent photo-luminescence in graphene oxide is associated with defects in single- and/or multiple-layers. Scientists believe that defects and their effects on electronic transitions are the main cause for the bright photoluminescence emissions that we have used for sensing.

Nadiya Dragneva queried: What kind of defects (vacancies, impurities *etc.*) can be found in your graphene oxide structure?

Paresh Ray responded: Graphene oxide has oxygenated functional groups, combined with non-oxidized regions, where most of the carbon atoms preserve sp^2 hybridization. On the other hand, in the oxidized region, the oxygen atoms attach to graphene sites randomly and convert sp^2 carbon bonds in graphene to sp^3 bonds. Several reports also indicate vacancies or line defects, which are due to reconstruction of the graphenic lattice.

Nadiya Dragneva remarked: Have you tried reduced graphene oxide? How may this affect the cytotoxic or luminesence properties of the resulting hybrid structure?

Paresh Ray replied: We have not tried to reduce graphene oxide and, as a result, we do not know how the cytotoxic and luminescence properties change with reduction.

Nadiya Dragneva commented: Could you please explain more about 3D graphene oxide? Is it more like graphite oxide (multilayered graphene)? Are there any unique properties of 3D graphene oxide?

Paresh Ray answered: The 3D graphene oxide has been prepared by cross-linking of 2D graphene oxide. Due to its highly porous structure, 3D graphene oxide will be valuable for membrane applications, such as separating bacteria and cancer cells from a blood sample.

Lucio Litti remarked: What is the efficiency of your nano-systems and your protocols for targeting circulating tumor cells (CTCs), in terms of selectivity and number of CTCs recognized to total number of cells in the sample?

Paresh Ray answered: We have demonstrated that AGE-aptamer attached magnetic graphene oxide can be used for the selective separation of G361 malignant melanoma cells from infected blood sample. We have tested with different cancer cells and found that our developed nano-system is highly selective in separating cancer cells from infected blood. Our result shows 97% malignant melanoma G361 cell separation from infected blood, when the number of cells was 100 cells mL^{-1}.

Lucio Litti asked: Have you tried to compare your method to other FDA-approved methods, such as Veridex, for example? In your opinion, what are the principal limitations of these methods and what are the advantages of your approach compared to them?

Paresh Ray answered: We have not compare our assay results with FDA-approved methods. In future, we plan to optimise our system in a clinical environment and we will then compare our assay result with FDA-approved technology.

Nguyen T. K. Thanh queried: What is the mechanism of fluorescence of graphene oxide? How high a quantum yield do you get from your system?

Paresh Ray responded: The origin of the fluorescence can be attributed to electron–hole recombination from the lowest conduction band and the wide-range of the valence band. The emission band between 400 and 800 nm for graphene oxide is mainly due to the presence of functionalized groups C–O, C=O and O=C–OH on the graphene sheet. Our experimental data indicate excitation wavelength dependent fluorescence and multiexponential fluorescence decay kinetics. Water-soluble graphene oxide exhibits a quantum yield of 0.28 with respect to fluorescein as a standard.

Catriona McCallion commented: It has been shown that the structure of graphene oxide, when prepared using the Hummers method or a related protocol, actually consists of a poorly oxidised graphene sheet covered with oxidative debris, which can be removed by washing with a base, such as NaOH.[1] It has also been shown that it is the oxidative debris that imparts the fluorescence properties and that the graphene oxide is no longer fluorescent after the oxidative debris has been removed.[2] Have you tried base washing your graphene oxide to see if it affects its fluorescence properties at all?

1 J. P. Rourke, P. A. Pandey, J. J. Moore, M. Bates, I. A. Kinloch, R. J. Young and N. R. Wilson, The real graphene oxide revealed: stripping the oxidative debris from the graphene-like sheets, *Angew. Chem., Int. Ed*, 2011, **50**, 3173–3177.
2 H. R. Thomas, C. Vallés, R. J. Young, I. A. Kinloch, N. R. Wilson and J. P. Rourke, Identifying the fluorescence of graphene oxide, *J. Mater. Chem C*, 2013, **1**, 338–342.

Paresh Ray answered: We have not washed our graphene oxide with NaOH and, as a result, I can not comment about the reported results.

However, we have measured the fluorescence below pH 10 in the presence of NaOH solution. We have observed huge fluorescence even at pH 10 to 12. As we have discussed, the fluorescence intensity and position can both be tuned by changing the pH or by changing the excitation wavelength. As a result, one may need to excite at different wavelength to obtain good fluorescence intensity at a higher pH, or after washing by NaOH.

Dejian Zhou queried: The use of different excitation wavelength results in different emission – do you see any lifetime-dependency on the excitation wavelength?

Paresh Ray replied: Several reports using time resolved experiments clearly show multiexponential decay features for graphene oxide fluorescence. The reported data indicate that the graphene oxide emission peaks shifted due to the decrease in lifetime. Experimental results shows that with the change in excitation, the emission lifetime decreases from the lower to higher excited states.

1 C. T.Chien, S. S. Li, W. J. Lai, Y. C. Yeh, H. A. Chen, I. S. Chen, L. C. Chen, K. H. Chen, T. Nemoto and S. Isoda, Tunable Photoluminescence from Graphene Oxide, *Angew. Chem., Int. Ed.*, 2012, **51**, 6662–6666.
2 Z. Luo, P. M. Vora, E. J. Mele, A. T. C. Johnson and J. M. Kikkawa, Photoluminescence and band gap modulation in graphene oxide, *Appl. Phys. Lett.*, 2009, **94**, 111909.
3 S. K. Cushing, M. Li, F. Huang and N. Wu, Origin of Strong Excitation Wavelength Dependent Fluorescence of Graphene Oxide, *ACS Nano*, 2014, **8**, 1002–1013.
4 J. Shang, L. Ma, J. Li, W. Ai, T. Yu and G. G. Gurzadyan, The Origin of Fluorescence from Graphene Oxide, *Sci. Rep.*, 2012, **2**, 792.

Dejian Zhou asked: What are the typical quantum yields of your graphene dots?

Paresh Ray responded: Reported typical quantum yields of the graphene dots are from 0.1 to 0.15.

Nguyen T. K. Thanh said: What is the excitation and emission range for graphene oxide – can you extend it by doping?

Paresh Ray responded: The excitation range for graphene oxide lies between 280 and 550 nm and the emission band varies from 360 to 660 nm. Several reports indicate that the emission range can be extended by doping with nitrogen, boron or other atoms.

Annette Barchanski opened the discussion of the paper by Ivan Parkin: Do you have any information about the particle distribution in your polymer?

Ivan Parkin answered: We have not investigated this in detail for this system. However, for related zinc oxide nanoparticles we have found that, although there are particles throughout the polymer, most of them reside close to the surface, in the first few microns.

Maya Thanou said: What is the reaction between the surface and the bacteria? How are the bacteria killed? Is the surface altered afterwards?

Ivan Parkin answered: The surface seems unaltered afterwards. The killing method is a mixture of oleic acid disrupting the bacteria and some free radical-based killing, engendered by the titanium dioxide when it is illuminated with UV light.

Maha Abdollah commented: How long will the surfaces covered with these polymers last and do they need to be stimulated with light every time before use?

Ivan Parkin answered: We have not studied this for the current system. In previous work using nanoparticles and dyes, we have shown that they are effective for many years when exposed to room lighting. However, in direct sunlight the dyes can degrade in two weeks.

Sacha Noimark, with whom I discussed the matter, stated that the antimicrobial properties are activated by the generation of reactive oxygen species at the sample surface, instead of through a "leaching" mechanism by which the antimicrobial agent is released from the polymer. It is therefore expected that these surfaces should maintain antimicrobial efficacy for many years. Illumination is required to maintain low surface bacterial contamination levels, however, these polymers have also demonstrated potent antimicrobial activity under dark conditions against *Staphylococcus aureus*.

Yuri Antonio Diaz Fernandez commented: Thank you very much for presenting these interesting results. I was particularly impressed by the antimicrobial effect of this material under dark conditions. Could you please comment on the fate of oleic acid? Is oleic acid released from the material? Have you characterized the extent and the kinetics of this release? Do you have any idea of the mechanism for antimicrobial activity in absence of light and on the long-term duration of this effect?

Ivan Parkin responded: From this study, we have found that oleic acid has a fairly potent antimicrobial effect on its own. Indeed, we have now made films with embedded oleic acid and these do show marked kill in the dark. However, the kill rates we have obtained are not as high as for titanium dioxide particles coated with oleic acid. We presume that the oleic acid is used up in killing the bacteria, but we have not done any kinetic studies.

Yuri Antonio Diaz Fernandez asked: Considering the application of these materials in hospital environments, could you please comment on the efficiency of the photo-activated reaction under normal indoor light?

Ivan Parkin answered: The titanium dioxide–gold nanoparticles made here work in the dark and hence would also work (but not be enhanced by) in visible light. We think this is due in part to the oleic acid component. To get a photocatalytic effect, we would need to use UV-light. We have, however, published previous work on making visible light-activated titanium dioxide based coatings.[1–2] In those cases, we use N or S dopants in the TiO_2. They give the material a colour and allow it to function with indoor lighting.

1 C. W. Dunhill and I. P. Parkin, *Dalton Trans.*, 2011, **40**, 1635–1640.

2 C. W. Dunnill, Z. A. Aiken, J. Pratten, M. Wilson, D. J. Morgan, I. P. Parkin, *J. Photochem. Photobiol. A*, 2009, **207**, 244–253.

Yuri Antonio Diaz Fernandez asked: While combining noble metal nanoparticles with TiO_2, have you observed any antenna effect or catalytic enhancement due to the plasmonic structures? Are these effects correlated to the wavelength of the incident light?

Ivan Parkin replied: We have not found any specific plasmon effects for the nanoparticles under UV irradiation conditions and have not observed an interaction with the titanium dioxide. Normally, the smallest size of Au nanoparticles have the greatest effect in stimulating bacterial kill. Notably, from previous work we have shown that gold nanoparticles of mean size 2 nm have the greatest effect in enhancing bacteria kill, this size of particle is below the surface plasmon resonance limit for gold – that is the particles are colourless. We hence think the enhancements we see are due to catalytic effects, and also the fact that the gold nanoparticles can help stabilise the lifetime of the e^-–h^+ exciton pair, as the metal can act as a reservoir for electron limiting recombination reactions.

Raul Quesada-Cabrera, with whom I discussed the matter, added that the investigation of plasmonic effects in gold-modified TiO_2 materials is planned as future work. The current work explored these materials in antimicrobial applications, aiming not only to increase the efficiency of these systems in the UV range, but to evaluate their performance against materials containing silver nanoparticles, which we have investigated extensively in our group. The examination of plasmonic effects will require action spectra studies and the optimisation of gold particle size. However, the latter experiments are very time-consuming, particularly when involving antimicrobial testing.

Katherine Brown enquired: Are your antimicrobial surfaces effective against the spore form of *Clostridium difficile*?

Ivan Parkin responded: Experiments are in progress at present against spore-forming organisms. However, these surfaces seem to be effective against a wide range of organisms, including the eight types of bacteria and virus mimics we have tested against.

Katherine Brown asked: Have you done a systematic comparison of how your antimicrobial coatings perform at concentrations of microbial pathogens found in actual hospital settings?

Ivan Parkin answered: Hospital contaminated surfaces tend to have much lower inoculum values than the ones we test. We have previously done clinical trials at the Eastman Dental Hospital and also Univeristy College London Hospitals using coated keyboards and settle plates. Both show a marked effect in these environments. We have also done studies at lower inoculum values and the coatings perform just as well. The reason we use such high bacterial counts in our studies is that it makes the data more statistically reliable.

Mostafa El-Sayed said: Have you determined the action spectrum, *i.e.* the effect of using light of different wavelengths, but of the same intensity?

Ivan Parkin replied: Good question. We have not done many studies relating to action spectra herem but we have done so for photovoltaic devices – it looks like a good experiment to try. We have, however, done studies using filters to cut out the UV portion and to prove that we have created a visible light photocatalyst in the case of N-doped TiO_2.

Nguyen T. K. Thanh queried: In Fig 7 of your paper (*Faraday Discuss.*, 2014, DOI: 10.1039/C4FD00113C), the proposed mechanism of photocatalytic antimicrobial activities involved both water and oxygen – which route is more dominant, please?

Ivan Parkin replied: It is hard to say, as we have not done these experiments for this system. Previous work has shown that both singlet oxygen and reactive oxygen species are important – hence both O_2 and H_2O are key.

Peter Dobson commented: Due to the limited penetration of ultraviolet light, in order to stimulate TiO_2 inserted into the body to create reactive oxygen species, we have been trying to use X-rays to activate TiO_2 to kill tumours. We use rare earth-doped titania and the work has been published and patented as a possible way of enhancing radiotherapy.[1–2]

1 H. E. Townley, J. Kim and P. J. Dobson, *Nanoscale*, 2012, **4** 5043–5050.
2 H. E. Townley, E. Rapa, G. Wakefield and P. J. Dobson, *Nanomedicine*, 2012, **8**, 526–536.

Ivan Parkin responded: This is an interesting development that could provide a new method for reducing tumours. However, it is probably a long way from clinical application.

Roger M. Pallares asked: Do you expect any harmful effects to people exposed to the reactive oxygen species produced on the treated surfaces?

Ivan Parkin answered: Not really – mammalian cells are much more resistant to reactive oxygen species than bacteria. Skin also has a dead layer that provides protection against these radicals.

Dan Sun opened the discussion of the concluding remarks by Mostafa El-Sayed: What sort of temperature rise do you get from the gold nanoparticles in cancer cells? How did you measure the temperature?

Mostafa El-Sayed responded: In the cells we did not measure it, however, in the mice we did, as high temperature burns the skin. We tried to keep the temperature below 50 °C. We used a red light laser thermometer.

Dan Sun asked: How do you control the temperature of the gold nanoparticles inside the cell nuclei?

Mostafa El-Sayed responded: By the intensity of the near-IR light and the amount of gold nanorods we inject.

Dan Sun added: How do you determine when the cell has been killed?

Mostafa El-Sayed responded: When we succeed in stopping the growth of the cancer spot.

Dan Sun asked: Have you tried to extend the application of this plasmonic effect to fields other than biomedical applications? What is the highest temperature the gold nanoparticle can produce?

Mostafa El-Sayed replied: Yes, we have used it in studying the activation energy of chemical reactions. Reagrding the highest temperature that can be achieved, this depends on the size of the rods and the light intensity.

Peter Dobson commented: Can you use the optical tweezers effect on your nanoparticles - you could manipulate the gold nanoparticles inside the cell?

Mostafa El-Sayed responded: This is a very good question. I am sure in the hands of a good physicist it could be done. To observe biochemical effects, one would have to move many particles.

Lucio Litti remarked: You showed your gold nanoparticles inside living cells. You were also able to recover SERS spectra from these particles. As you know, in order to penetrate the cell membrane and not damage the cell, the nanoparticles have to be small. At the same time, smaller nanoparticles are characterized by smaller scattering and so lower the SERS efficiency. What are the ideal dimensions, where the nanoparticles have good penetration in cells while still remaining good SERS substrates?

Mostafa El-Sayed responded: You are quite correct, the nanoparticles have to be small to penetrate into the cell, but as we shown very recently in a physical chemistry letter, the nanoparticles plasmonically (not chemically) aggregate when they localize around the nucleus and this aggregation gives them a high scattering probability.

Amelie Heuer-Jungemann queried: Your particles, which are presumably taken up by endocytosis, are targeted to the nucleus – how do they escape the endosomes in order to reach their target? If they do not escape the endosomes, but are transported to the perinuclear region simply by being in late endosomes or lysosomes, then do you even need to have the peptide coating?

Mostafa El-Sayed responded: Good question. What we think happens is that the particles do not make it to the nucleus surface, but get stuck around the nuclear membrane and aggregate, giving us strong Raman enhancement.

Sandhya Moise asked: Is the phenylalanine signal from the dying cells typical of all dying cells, or does it probably only occur due to the denaturation of proteins during heat generation from the gold nanoparticles?

Mostafa El-Sayed replied: This is an excellent question. As you know phenylalanine is embedded in a very hydrophobic medium. Once death occurs, this opens up and it gets exposed to the plasmonic field which enhances its Raman signal. This could be due to the heat ,but we also saw it happening in apoptosis caused by addition of hydrogen peroxide.

Maha Abdollah remarked: Do the gold nanoparticles enter the nucleus or stay in the perinuclear region? If they stay around the nucleus, how do they cause DNA damage?

Mostafa El-Sayed responded: We really do not know where the gold nanoparticles are distributed. We are trying to see if we can determine the distribution now. I expect that most of them are outside, but some monomers could pass through the membrane and these are the ones that affect the DNA structure.

Faraday Discussions

PAPER

Concluding remarks: Summary of some of our recent studies in the field of conjugating plasmonic gold nanoparticles to single cancer cells and their molecular and cellular dynamics

Mostafa A. El-Sayed*

Received 10th November 2014, Accepted 9th December 2014
DOI: 10.1039/c4fd00213j

This Faraday meeting was indeed very educational and important. We should all thank the organizing committee, and especially our Chair, Professor Nguyen TK Thanh, for a job well done. The Faraday Discussion format is certainly very creative and effective. Getting each presenter of one of the excellent posters to summarize the topic of their poster in five minutes was very important in encouraging the delegates to spend their time in the poster sessions with the speakers working on research that overlapped with his or her own.

The Opening Lecture did a very good job in focusing the participants on the important issues in the field of conjugating nanoparticles and was very relevant to the work of those attended the meeting. The content of the presentations was excellent, especially in the field of magnetic nanoparticles, but was slightly anaemic in the field of plasmonic nanoparticles. For this reason, and after talking with the chair and a member of the organizing committee, I suggested that I spent most of my concluding remarks discussing our work in this field. This was accepted by the organizers and I hope the participants found it useful.

Over the past 20 years, our group worked in the field of metallic nanoparticles with applications in the fields of nano-catalysis, nano-sensors and nano-medicine. Due to the current interests of the group, I will focus in the short time I have on some of our recent cancer research using plasmonic gold nano-particles in cancer cell nano-medicine and single cell molecular dynamics. Due to time and space limitations, I will summarize the work, but give references so the reader can go to the actual publications for more detailed information.

Our extensive early studies on the plasmonic properties of gold and silver nanoparticles[1–4] enabled our group to use their findings to develop methods for

Laser Dynamics Laboratory, School of Chemistry and Biochemistry, Georgia Institute of Technology, Atlanta, Georgia, 30305, USA. E-mail: melsayed@gatech.edu

cancer diagnostics and therapy *in vitro*[5] and *in vivo*.[6] We have used the properties of gold nanoparticles (AuNPs) in photo-thermal therapy, in which moderately strong light is absorbed, converted into intense electromagnetic fields and decays either by producing strong scattered light (used in imaging applications) or is converted into high temperature heat used in photo-thermal destruction of the cancer cells, or any diseased cells to which the gold nanoparticles are conjugated.

Only near-infrared light can penetrate long distances into the human body. Nanorods are known to absorb in the near-IR at different wavelengths depending on the aspect ratios of the rods. We have developed[7] the best synthetic method to make nanorods of an excellent homogeneous aspect ratio or a distribution of different aspect ratios. This method is now the most frequently used method for making nanorods for use in photo-thermal and many other photonic applications.

Recently, our group has made use of enhanced light scattering, not only to image cellular environments, but also to continually observe in real time intracellular molecular changes during various cellular processes. In 2010, AuNP scattering in conjunction with dark field imaging was used to determine particle localization near the nucleus and to monitor cellular response after a cytotoxic level of AuNPs were delivered.[8] This work was expanded upon in 2012, as non-cytotoxic concentrations of nuclear-targeted AuNPs were used to probe the real-time molecular changes near the nucleus during cell cycle progression.[9] In particular, AuNP scattering was exploited to simultaneously acquire structural and molecular information from a single cell through Rayleigh and Raman imaging. This revealed a strong correlation between DNA and protein Raman vibrations and cell cycle phase progression. Intracellular AuNP-coupled Raman spectroscopy was also used to determine drug efficacy[10] and drug delivery dynamics[11] in cancer cells after treatment with chemotherapeutic agents. This work demonstrated the possibility of using AuNP light scattering to continually assess the efficiency of drug delivery and drug efficacy in real-time, as well as reducing the time required to obtain cytotoxicity drug profiles.

More recently, nuclear AuNP scattering and Raman spectroscopy were used to reveal real-time molecular changes in human cancer cells during apoptosis.[12] The ability to directly compare morphological changes (*i.e.* Rayleigh scattering images) with molecular changes (*i.e.* Raman spectroscopy analysis) of the DNA/protein composition around the cell nucleus revealed the occurrence and dynamics of three apoptotic molecular events: protein denaturation, proteolysis, and DNA fragmentation. A temporal profile of these molecular events was also created. Together, these results elucidated apoptotic cell death mechanism and demonstrated the potential of AuNP scattering to advance molecular and cellular biology methods to accurately assess therapeutic imaging and response.

In conclusion, from the presented material and discussions at this meeting, I believe that the progress of research in the synthesis and applications of magnetic and plasmonic nanoparticles is progressing at a very impressive and encouraging rate indeed. The progress of developing new conjugating organic or biochemical molecules is progressing at a slower rate, or is not as strongly coupled as it could be with the community most concerned with nano applications. Of course, the two communities come from different fields and go to different meetings under different titles. The organizers of the present meeting have made a greater effort than usual to attract a number of good speakers in this area.

I wish to thank the Chair, Prof. Nguyen, and the rest of the organizing committee, the speakers and the participants, as well as the staff, for the efforts that they have all put into this important and timely meeting, which indeed made it worthy of a Faraday Discussion meeting.

Acknowledgements

I wish to thank all my graduate students, whose hard work and creativity were essential to develop and finish the exciting work I have described above. The contributions of the different granting agents, such as NSR-DMR, in supporting the material development of this field and NIH–NCI for supporting the cellular and animal studies are most appreciated. Finally, I would like to thank the organizers and the Faraday society for inviting me to this scientifically rich meeting.

References

1 S. Link and M. A. El-Sayed, Spectral properties and relaxation dynamics of surface plasmon electronic oscillations in gold and silver nanodots and nanorods, *J. Phys. Chem. B*, 1999, **103**(40), 8410–8426.
2 M. A. El-Sayed, Some interesting properties of metals confined in time and nanometer space of different shapes, *Acc. Chem. Res.*, 2001, **34**(4), 257–264.
3 S. Link and M. A. El-Sayed, Shape and size dependence of radiative, non-radiative and photo thermal properties of gold nanocrystals, *Int. Rev. Phys. Chem.*, 2000, **19**(3), 409–453.
4 P. K. Jain, K. S. Lee, I. H. El-Sayed and M. A. El-Sayed, Calculated absorption and scattering properties of gold nanoparticles of different size, shape, and composition: applications in biological imaging and biomedicine, *J. Phys. Chem. B*, 2006, **110**(14), 7238–7248.
5 X. Huang, I. H. El-Sayed, W. Qian and M. A. El-Sayed, Cancer cell imaging and photothermal therapy in the near-infrared region by using gold nanorods, *J. Am. Chem. Soc.*, 2006, **128**, 2115–2120.
6 E. B. Dickerson, E. C. Dreaden, X. Huang, I. H. El-Sayed, H. Chu, S. Pushpanketh, *et al.* Gold nanorod assisted near-infrared plasmonic photothermal therapy (PPTT) of squamous cell carcinoma in mice, *Cancer Lett.*, 2008, **269**, 57–66.
7 B. Nikoobakht and M. A. El-Sayed, Preparation and growth mechanism of gold nanorods (NRs) using seed-mediated growth method, *Chem. Mater.*, 2003, **15**(10), 1957–1962.
8 B. Kang, M. A. Mackey and M. A. El-Sayed, Nuclear Targeting of Gold Nanoparticles in Cancer Cells Induces DNA Damage, Causing Cytokinesis Arrest and Apoptosis, *J. Am. Chem. Soc.*, 2010, **132**, 1517–1519.
9 B. Kang, L. A. Austin and M. A. El-Sayed, Real-Time Molecular Imaging Throughout the Entire Cell Cycle by Targeted Plasmonic-Enhanced Rayleigh/Raman Spectroscopy, *Nano Lett.*, 2012, **12**, 5369–5375.
10 L. A. Austin, B. Kang and M. A. El-Sayed, A New Nanotechnology Technique for Determining Drug Efficacy Using Targeted Plasmonically Enhanced Single Cell Imaging Spectroscopy, *J. Am. Chem. Soc.*, 2013, **135**, 4688–4691.

11 B. Kang, M. M. Afifi, L. A. Austin and M. A. El-Sayed, Exploiting the Nanoparticle Plasmon Effect: Observing Drug Delivery Dynamics in Single Cells *via* Raman/Fluorescence Imaging Spectroscopy, *ACS Nano*, 2013, **7**, 7420–7427.
12 B. Kang, L. A. Austin and M. A. El-Sayed, Observing Real-Time Molecular Event Dynamics of Apoptosis in Living Cancer Cells Using Nuclear-Targeted Plasmonically Enhanced Raman Nanoprobes, *ACS Nano*, 2014, **8**, 4883–4892.

Poster titles

Quantifying DNA on the surface of gold nanoparticles – evidence for interference of the dye when using dye-labelled DNA, **Kristian Göeken and Ron Gill**, *University of Twente, The Netherlands*

Modeling of biomolecule-graphene interactions in water, **Nadiya Dragneva, Dennis Stauffer, Wely B. Floriano, Robert C. Mawhinney, Stephen French, Giovanni Fanchini and Oleg Rubel**, *Lakehead University, Canada*

Comparison of the synthesis and heating ability of biofunctionalized gold nanoprisms and nanorods for bioapplications, **Scott Mitchell, Gabriel Alfranca Ramon, Maria Moros and Jesus Martinez de la Fuente**, *University of Zaragoza, Spain*

Novel superparamagnetic iron oxide nanoparticles for targeted PET/SPECT-MR imaging, **Lydia Sandiford, Mark Green and Rafael Torres Martin de Rosales**, *King's College London, UK*

The effect of varying graphene surface chemistries on the growth of primary lymphocytes, **Catriona McCallion, Karen Rees-Unwin, Julie Adams, Alexander Golovanov, Alain Pluen and John Burthem**, *University of Manchester, UK*

Magnetic labelling of stem cells with a surface functionalised magnetic protein, **Sara Carreira, James Armstrong, Adam Perriman, Annela Seddon and Walther Schwarzacher**, *University of Bristol, UK*

Direct surface functionalization of superparamagnetic cobalt ferrite nanoparticles by oligonucleotides, **Alexandra Pershina, Anna A. Magaeva, Volya I. Itin and Alexei E. Sazonov**, *Siberian State Medical University, Russia*

The influence of silver nanoparticle ageing process on the utilization in surface enhanced Raman spectroscopy, **Ariana Fargasova, Robert Prucek, Anna Balzerova, Michaela Sopuchova, Panacek Ales, Libor Kvitek and Radek Zboril**, *Palacky University, Czech Republic*

Development of novel functional magnetite nanoparticles for evaluation of human mesenchymal stem cell (hMSC) labelling in a tissue-engineered airway, **Roxanne Hachani, Mark Lowdell, Martin Birchall and Nguyen Thanh**, *University College London, UK*

Simple peptide-mediated synthesis of cell-targeted platinum nanoparticles for imaging and therapy, **Stefanie Papst, Margaret A. Brimble, Clive W. Evans, Richard D. Tilley and David E. Williams**, *The University of Auckland, New Zealand*

Nanoparticle agglomeration and cytotoxicity, **Sebastian Moerz, Henrike Peuschel, Annette Kraegeloh and Tobias Kraus**, *Leibniz Institute for New Materials, Germany*

Development of novel organic thermo-responsive nanogels for skin drug delivery applications, **Yolanda Salinas-Soler and Marina Resmini,** *Queen Mary University of London, UK*

Methods to transfer iron nanoparticles into water: what is the best way to reach good relaxometric properties?, **Catherine Amiens, Marlene Branca, Marzia Marciello, Diana Ciuculescu-Pradines, Marc Respaud, Maria del Puerto Morales, Raphael Serra and Marie-José Casanove,** *CNRS, France*

Functionalized luminescent nanoparticles for cellular delivery, **Chris Stepanek, Amy Davies, David J. Lewis, Iain B. Styles, Nikolas J. Hodges, Steven G. Thomas, Stephen P. Watson and Zoe Pikramenou,** *University of Birmingham, UK*

Nanoproteomics. Activity-based protein profiling with gold nanoparticles, **Álvaro Martínez, Fabrizio Mancin and Paolo Scrimin,** *University of Padova, Italy*

Near-infrared fluorescent ribonuclease-A-encapsulated gold nanoclusters: preparation, characterization, cancer targeting and imaging, **Yifei Kong, Dejian Zhou, Jun Chen, Feng Gao, Rik Brydson, Benjamin Johnson, Yue Zhang and Lin Wu,** *University of Leeds, UK*

Quantum dot-bioconjugates for bioimaging and robust, specific, biosening, **Weili Wang, Yuan Guoa and Darren Tomlinson,** *University of Leeds, UK*

The influence of 1,2-alkanediol on the synthesis and growth of magnetite nanoparticles by thermal decomposition of iron(III) acetylacetonate, **Liane Rossi, Fernando Effenberger and Pedro Kiyohara,** *University of São Paulo, Brazil*

A universal magnetic carrier for use in cancer prodrug therapy, **Robert Hobbs, Vanessa Gwenin, Ellen Freeborn, Jennifer Halliwell, Thomas Caspari and Christopher Gwenin,** *Bangor University, UK*

Study of the immobilization of silver nanoparticles onto the textile fabric, **Martina Karlikova, Jana Soukupova, Anna Balzerova, Olga Balzerova, Klara Cepe and Libor Kvitek,** *Palacky University Olomouc, Czech Republic*

Synthesis and properties of gadolinium oxide nanoparticles as contrast agents for magnetic resonance imaging, **Alicja Kownacka, Benjamin Burke, Juozas Domarkas, Mark Lorch, Chris Cawthorne and Stephen Archibald,** *University of Hull, UK*

In-situ synthesis of magnetic nanoparticles in electrospun nanofibres for applications in regenerative medicine, **Luke Burke, Chris Mortimer and Chris Wright,** *Swansea University, UK*

CuInS2-based quantum dots for biomedical applications, **Riccardo Marin and Patrizia Canton,** *Università Ca' Foscari Venezia, Italy*

Targeted T1- and T2- dual-modal contrast agent for *in-situ* study of neural stem cell harvesting, **Yung-Kang Peng, Nga-Ping Lui, Kin-Lam Yung and Shik-Chi Tsang,** *University of Oxford, UK*

Monitoring the behaviour of DNA-modified nanoparticles using tunable resistive-pulse sensing, **Emma Blundell and Mark Platt,** *Loughborough University, UK*

Peptide-capped conjugated polymer nanoparticles for biosensing and bioimaging **Carina dos Santos Almeida, Philip Howes, Inge Herrmann and Molly Stevens,** *Imperial College London, UK*

Interactions of cross – linked and uncross – linked chitosan hydrogels with surfactants for biomedical applications, **Mojisola Nkiko, Janet T. Bamgbose, Akeem A. Bamigbade and Enoch O. Dare,** *Elizade University, Nigeria*

Functionalization of selected Raman reporters and their application in SERRS-based sensing devices, **Lucio Litti,** *University of Padova, Italy*

Single nucleotide polymorphism sensing on gold nanoparticles, **David Kershaw,** *University of Birmingham, UK*

Gallium-68 labelled silica coated iron-oxide nanorods as PET/MR multimodal imaging agents, **Neazar Baghdadi, Benjamin Burke, Gonçalo Clementebc, Alicja Kownackaa, Juozas Domarkasab, Stephen Archibald, Mark Lorcha, Shubhanchi Nigam, Amaury Guilloud, Nathalie Camus, Zakaria Halimed, Raphaël Tripierd and Chris Cawthorne,** *University of Hull, UK*

Understanding the mutual energy transfer in water soluble QD-dye conjugates witha fluorescent pH-probe, **Sandra Ast, Matthew H. Todd and Peter J. Rutledge,** *University of Sydney, Australia*

Luminescent nanoparticles for single particle imaging in cells, **Sunil Claire, Nicola J. Rogers, Robert M. Harris, Shiva Farabi, Gerald Zikeli, Iain B. Styles, Nikolas J. Hodges and Zoe Pikramenou,** *University of Birmingham, UK*

Functionalisation and characterisation of magnetic nanoparticles for application in thermo-chemotherapy of cancer, **Aziliz Hervault and Thanh Nguyen,** *UCL-Healthcare Biomagnetics Laboratories, UK*

Dextran nanodroplets for oxygen release to hypoxic cutaneous tissues, **Mauro Prato, Chiara Magnetto, Jithin Jose, Amina Khadjavi, Federica Cavallo, Elena Quaglino, Alice Panariti, Ilaria Rivolta, Emilio Benintende, Gianfranco Varetto, Monica Argenziano, Adriano Troia, Roberta Cavalli and Caterina Guiot,** *University of Torino, Italy*

Two-steps preparation approach for nano-calcium hydroxyapatite towards medical applications, **Aivaras Kareiva, Aleksandra Prichodko, Audrone Jankeviciute, J. C. Yang, Thomas C. K. Yang and Aldona Beganskiene,** *Vilnius University, Lithuania*

Reactive oxygen generation by silicon nanoparticles: a size-dependent study, **Daniel Wolverson, Jamaree Amonkosolpan, Gazi Aliev, John Davies and Paul Snow,** *University of Bath, UK*

Electrostatic self-assembly of oppositely charged metal nanoparticles: the marriage of Capulets and Montagues, **Yuri Antonio Diaz Fernandez, Tina A. Gschneidtner, Svetlana Syrenova, Carl Wadell, Louise H. Fornander, Fredrik Westerlund, Christoph Langhammer, Kasper Moth-Poulsen and Samuel Lara Avila,** *University of Liverpool, UK*

Hofmeister effects on the synthesis and fine-tuning of gold nanorods: controlling the surfactant template, **Roger M. Pallares, Xiaodi Su and Nguyễn T. K. Thanh**, *University College London, UK*

Preparation of 2-Deoxy-D-glucose coated SPION nanoparticles and characterizationof their physical, chemical and biological properties, **Ulviyya Hasanova, Aysel Ahadova, Matthias Kloor, Juergen Kopitz, Alimamed Shabanov, Mahammadali Ramazanov, Abel Maharramov and Magnus von Knebel Doeberitz**, *Baku State University, Azerbaijan*

Triphenylarsonium-conjugated gold nanoparticles: potential nanocarriers formitochondria-targeted therapeutics, **Nikhil Lalwani, Neil Bricklebank and Neil Cross**, *Sheffield Hallam University, UK*

Multifunctional and theranostic nanoparticles by laser-based, one-step fabricationand their physicochemical functionalization aspects, **Annette Barchanski, Csaba L. Sajti, Boris N. Chichkov**, *Laser Zentrum Hannover e.V., Germany*

Nano-immunoassay for diagnosis of active schistosomal infection, **Manal M. Kamel, Hanan G. El-Baz, Zeinab A. Demerdash, Engy M. Abd El-Moneem, Mohamed A. Hendawy and Ibrahim R. Bayoumi**, *Research Institute, Egypt*

Boronic esters for dynamically reconfigurable nanoparticle functionalization and assembly, **Stefan H. Borsley and Euan R. Kay**, *The University of St Andrews, UK*

Functional two-dimensional nanoarchitectures based on chemically converted graphene oxide and hematoporphyrin under sulphuric acid treatment, **Nora Grincevičiūtė, Boris Snopok and Valentinas Snitka**, *Kaunas University of Technology, Lithuania*

Folate conjugated magnetic nanocarriers for targeted drug delivery and hyperthermia treatmentfor cancer, **Suman Rana, K. C. Banch and P. A. Hussan**, *Bhabha Atomic Research Centre, India*

A study of the cellular–nanoparticle interactions and assessment of the magnetic response of novel bacterially synthesized substituted ferrites, **Sandhya Moise, Alicia El Haj and Neil Telling**, *Keele University, UK*

Click chemistry – atool for the programmed ligation of DNA-coated nanoparticles, **Amelie Heuer-Jungemann, Afaf H. El-Sagheer, Andreas Riedinger, Yi Xie, Teresa Pellegrino, Tom Brown and Antonios G. Kanaras**, *University of Southampton, UK*

The use of AC-susceptometry to investigate the magnetic nanoparticle stability in various biologically relevant solutions, **D. Soukup and N. D. Telling**, *Keele University, UK*

Synthesis, characterisation and photo-catalytic studies of cobalt ferrite–silica–titania nanocomposites, **Raquel Serrano-Garcia, David Greene, Joseph Govan and Yurii Gun'ko**, *Trinity College Dublin, Ireland*

The Skinner Prize for the best poster was awarded to Mr Yung-Kang Peng of University of Oxford, UK, for his poster on a targeted T1- and T2- dual-modal contrast agent for *in-situ* study of neural stem cell harvesting.

List of participants

Miss Maha Abdollah, *UCL Cancer Institute*, *United Kingdom*
Mrs Siti Fatimah Abdul Ghani, *King's College London*, *United Kingdom*
Professor A. Paul Alivisatos, *University of California, Berkeley*, *U.S.A*
Professor Catherine Amiens, *The National Center for Scientific Research (CNRS)*, *France*
Dr Steve Archibald, *University of Hull*, *United Kingdom*
Dr Sandra Ast, *University of Sydney*, *Australia*
Mr Neazar Baghdadi, *University of Hull*, *United Kingdom*
Mrs Anna Balzerova, *Palacky University Olomouc*, *Czech Republic*
Ms Annette Barchanski, *Laser Zentrum Hannover e.V.*, *Germany*
Dr Nicolas Barry, *University of Warwick*, *United Kingdom*
Miss Mehrnaz Behray, *University of East Anglia*, *United Kingdom*
Dr Tom Berkleman, *Bio-Rad Laboratories*, *U.S.A*
Miss Emma Blundell, *Loughborough University*, *United Kingdom*
Mr Stefan Borsley, *School of Chemistry, The University of St Andrews*, *United Kingdom*
Dr Katherine Brown, *Cavendish Laboratory*, *United Kingdom*
Dr Benjamin Burke, *University of Hull*, *United Kingdom*
Mr Luke Burke, *Swansea University*, *United Kingdom*
Ms Sara Carreira, *University of Bristol*, *United Kingdom*
Mr Thomas Carter, *UCL Cancer Institute*, *United Kingdom*
Professor Kerry Chester, *UCL*, *United Kingdom*
Mr Sunil Claire, *University of Birmingham*, *United Kingdom*
Dr Yuri Antonio Diaz Fernandez, *University of Liverpool*, *United Kingdom*
Professor Peter Dobson, *Queen's College Oxford*, *United Kingdom*
Miss Carina Dos Santos Almeida, *Imperial College London*, *United Kingdom*
Miss Nadiya Dragneva, *Lakehead University*, *Canada*
Prof Ifeanyi Elekwa, *Abia State University*, *Nigeria*
Dr Mostafa El-Sayed, *Georgia Institute of Technology*, *USA*
Ms Ariana Fargasova, *Department of Physical Chemistry, Faculty of Science, Palacky University*, *Czech Republic*
Miss Soraia Fernandes, *University of Regensburg*, *Germany*
Professor Asterios Gavriilidis, *University College London*, *United Kingdom*
Miss Giulia Glorani, *University Ca Foscari of Venice, Italy*, *Italy*
Mr Kristian Göeken, *University of Twente*, *Netherlands*
Professor Mark Green, *King's College London*, *United Kingdom*
Miss Nora Grincevičiūtė, *Research Centre for Microsystems and Nanotechnology, Kaunas University of Technology*, *Lithuania*
Miss Roxanne Hachani, *University College London*, *United Kingdom*
Dr Peter Harvey, *University of Manchester*, *United Kingdom*
Professor Dr Ulviyya Hasanova, *Baku State University*, *Azerbaijan*
Ms Aziliz Hervault, *UCL Healthcare Biomagnetics Laboratories*, *United Kingdom*
Miss Amelie Heuer-Jungemann, *University of Southampton*, *United Kingdom*

Mr Robert Hobbs, *Bangor University*, *United Kingdom*
Dr Hanna Jankevics, *Malvern Instruments*, *United Kingdom*
Professor Richard Jones, *University of Texas at Austin*, *U.S.A*
Professor Dr Manal Kamel, *Department of Immunology, Research Institute*, *Egypt*
Professor Dr Aivaras Kareiva, *Vilnius University*, *Lithuania*
Mrs Martina Karlikova, *Palacky University Olomouc*, *Czech Republic*
Mr David Kershaw, *University of Birmingham*, *United Kingdom*
Dr Khalid Khan, *Islamia College University Peshawar*, *Pakistan*
Mr Yifei Kong, *School of Chemistry, University of Leeds*, *United Kingdom*
Miss Alicja Kownacka, *University of Hull*, *United Kingdom*
Mr Nikhil Lalwani, *Sheffield Hallam University*, *United Kingdom*
Dr Alain Laurent, *bioMérieux*, *France*
Dr Anna Lesniak, *University College Dublin*, *Ireland*
Dr Lucio Litti, *University of Padova*, *Italy*
Miss Kirsty Lunnon, *Royal Society of Chemistry*, *United Kingdom*
Mr Riccardo Marin, *Università Ca' Foscari Venezia*, *Italy*
Mr Álvaro Martínez, *University of Padova*, *Italy*
Professor Hedi Mattoussi, *Florida State University*, *U.S.A*
Ms Catriona McCallion, *University of Manchester*, *United Kingdom*
Dr L J Michie, *Royal Society of Chemistry*, *United Kingdom*
Dr Scott Mitchell, *University of Zaragoza*, *Spain*
Dr Sebastian Moerz, *Leibniz Institute for New Materials*, *Germany*
Ms Sandhya Moise, *Keele University*, *United Kingdom*
Professor Dr Thanh T. K. Nguyen, *UCL Healthcare Biomagnetic and Nanomaterials Labs*, *United Kingdom*
Dr Mojisola Nkiko, *Elizade University*, *Nigeria*
Ms Chloe Oakland, *University of Manchester*, *United Kingdom*
Mr Roger M. Pallares, *University College London (UCL)*, *United Kingdom*
Dr Stefanie Papst, *University of Auckland*, *New Zealand*
Professor Ivan Parkin, *UCL Chemistry*, *United Kingdom*
Mr Yung-Kang Peng, *Department of Chemistry, University of Oxford*, *United Kingdom*
Mrs Alexandra Pershina, *Siberian State Medical University*, *Russia*
Professor Zoe Pikramenou, *University of Birmingham*, *United Kingdom*
Dr Mauro Prato, *University of Torino*, *Italy*
Miss Chloe Pritchard, *Royal Society of Chemistry*, *United Kingdom*
Mrs Suman Rana, *Bhabha Atomic Research Centre*, *India*
Professor Paresh Ray, *Jackson State University, U.S.A*
Professor Dr Oliver Reiser, *Universität Regensburg*, *Germany*
Professor Liane Rossi, *University of São Paulo*, *Brazil*
Dr Yolanda Salinas-Soler, *School of Biological and Chemical Sciences, Queen Mary University of London*, *United Kingdom*
Miss Lydia Sandiford, *King's College London*, *United Kingdom*
Miss Neha Sen, *Jiwaji University*, *India*
Ms Raquel Serrano Garcia, *Trinity College Dublin, Ireland*
Professor Dr Sushil Singh, *Indian Institute of Technology (BHU)*, *India*
Mr Dalibor Soukup, *Keele University*, *United Kingdom*
Mr Chris Stepanek, *University of Birmingham*, *United Kingdom*
Dr Dan Sun, *Queen's University Belfast*, *United Kingdom*
Dr Maya Thanou, *King's College London*, *United Kingdom*
Dr Matthew Todd, *University of Sydney*, *Australia*
Professor Edman Tsang, *University of Oxford*, *United Kingdom*
Professor Donus Tuncel, *Bilkent University*, *Turkey*
Mr Weili Wang, *School of Chemistry, University of Leeds*, *United Kingdom*

Miss Dannielle Whittaker, *Royal Society of Chemistry*, *United Kingdom*
Dr Daniel Wolverson, *University of Bath*, *United Kingdom*
Dr Chris Wright, *Swansea University*, *United Kingdom*
Professor Lanry L. Yung, *National University of Singapore*, *Singapore*
Mrs Jerina Zain, *Bhabha Atomic Research Centre, Mumbai*, *India*
Dr Dejian Zhou, *University of Leeds*, *United Kingdom*